D1162085

The Census Data System

The Census Data System

Edited by

Philip Rees
University of Leeds

David Martin
University of Southampton

Paul Williamson
University of Liverpool

JOHN WILEY & SONS, LTD

National 01243 779777
International (+44) 1243 779777

e-mail (for orders and customer service enquiries): cs-books@wiley.co.uk

Visit our Home Page on http://www.wileyeurope.com or http://www.wiley.com

Other Wiley Editorial Offices

John Wiley & Sons, Inc., 605 Third Avenue,
New York, NY 10158-0012, USA

Jossey-Bass, 989 Market Street
San Francisco, CA 94103-1741, USA

Wiley-VCH Verlag GmbH
Pappelallee 3, D-69469 Weinheim, Germany

John Wiley & Sons Australia, Ltd
33 Park Road, Milton, Queensland 4064, Australia

John Wiley & Sons (Asia) Pte Ltd, 2 Clementi Loop #02-01,
Jin Xing Distripark, Singapore 129809

John Wiley & Sons (Canada) Ltd, 22 Worcester Road,
Rexdale, Ontario, M9W 1L1, Canada

Library of Congress Cataloging-in-Publication Data

Philip Rees, David Martin, and Paul Williamson.
The Census Data System.
p. cm.
Includes bibliographical references and index.
ISBN 0 470 84688 7 (alk.paper)
1. Asynchronous transfer mode. I. Title.

TK5105.35.P48 2001
004.6′6—dc21 2001026646

British Library Cataloguing in Publication Data

A catalogue record for this book is available from the British Library

ISBN 0 470 84688 7

Typeset in 10/12pt Times Roman by Laserwords Private Limited, Chennai, India.
Printed and bound in Great Britain by TJ International Ltd, Padstow, Cornwall
This book is printed on acid-free paper responsibly manufactured from sustainable forestry in which at least two trees are planted for each one used for paper production.

For Stan Openshaw

Contents

List of Contributors

Seraphim Alvanides
Department of Geography, University of Newcastle, Newcastle NE1 7RU
tel 0191 222 6436, fax 0191 222 5421, email s.alvanides@newcastle.ac.uk

Marcus Blake
The National Key Centre for Social Applications in GIS, University of Adelaide, South Australia, 5005 and 3 Lange Place, Bruce, ACT 2617, Australia
tel +61 2 6251 8587, email mblake@gisca.adelaide.edu.au

Reg Carr
Bodleian Library, Broad Street, Oxford, OX1 3BG
tel 01865 277166, fax 01865 277187, email reg.carr@ulib.ox.ac.uk

Jackie Carter
Manchester Computing, University of Manchester, Oxford Road, Manchester, M13 9PL
tel 0161 275 6725, fax 0161 275 6040, email j.carter@mcc.ac.uk

John Charlton
Office for National Statistics, 1 Drummond Gate, London, SW1V 2QQ
tel 020 8533 6239, fax 020 8533 6238, email john.charlton@ons.gov.uk

Keith Cole
Manchester Computing, University of Manchester, Oxford Road, Manchester, M13 9PL
tel 0161 275 6066, fax 0161 275 6040, email k.cole@mcc.ac.uk

Mike Coombes
CURDS, University of Newcastle, Claremont Bridge, Newcastle upon Tyne, NE1 7RU
tel 0191 222 8014, fax0191 232 9259, email mike.coombes@ncl.ac.uk

Rosemary Creeser
Centre for Longitudinal Studies, Institute of Education, 20 Bedford Way, London WC1 OAL
tel 020 7612 6875, fax 020 7612 6880, email ls@cls.ioe.ac.uk

Marie Cruddas
Office for National Statistics, Segensworth Road, Titchfield, Fareham, PO15 5RR
tel 01329 813 512, fax 01329 813 638, email marie.cruddas@ons.gov.uk

Angela Dale
CCSR, Economic & Social Studies, University of Manchester, Dover Street, Manchester, M13 9PL
tel 0161 275 4876, fax 0161 275 4722, email angela.dale@man.ac.uk

Chris Denham
Census Services Branch, Office for National Statistics, Segensworth Road, Titchfield, Fareham, PO15 5RR
tel 01329 813 720, fax 01329 813 587, email chris.denham@ons.gov.uk

Ian Diamond
Department of Social Statistics, University of Southampton, Southampton, S017 1BJ
tel 023 8059 2518, fax 023 8059 3846, email: idd@socsci.soton.ac.uk

John Dixie
DSEE, Mowden Hall, Darlington, Co. Durham, DL3 9BG
tel 01325 392420, fax 01325 392989, email j.dixie@dsee.gsi.gov.uk

Brian Dodgeon
Centre for Longitudinal Studies, Institute of Education, 20 Bedford Way, London WC1 OAL
tel 020 7612 6877, fax 020 7612 6880, email bd@cls.ioe.ac.uk

Danny Dorling
School of Geography, University of Leeds, Leeds, LS2 9JT
tel 0113 2333347, fax 0113 233 3308, email d.dorling@geog.leeds.ac.uk

Oliver Duke-Williams
School of Geography, The University of Leeds, Leeds, LS2 9JT
tel 0113 233 3392, fax 0113 233 3308, email o.duke-williams@geog.leeds.ac.uk

Jason Dykes
School of Informatics, City University, Northampton Square, London EC1V 0HB
tel 020 7040 8906, fax 020 7040 8587, email jad@is.city.ac.uk

Martin Frost
School of Geography, Birkbeck College, University of London, 7-15 Gresse Street, London, W1P 2LL
tel 0207 631 6470, fax 0207 631 6498, email m.frost@bbk.ac.uk

James Harris
Netcentricity Ltd., 40 Bowling Green Lane, Clerkenwell, London, EC1R 0NE
tel 020 7415 7041, email james@netcentricity.co.uk

Justin Hayes
Manchester Computing, University of Manchester, Oxford Road, Manchester, M13 9PL
tel 0161 275 6060, fax 0161 275 6040, email justin.hayes@man.ac.uk

Heather Joshi
Centre for Longitudinal Studies, Institute of Education, 20 Bedford Way,
London WC1 OAL
tel 020 7612 6874, fax 020 7612 6880, email hj@cls.ioe.ac.uk

Neil Lander-Brinkley
Office for National Statistics, Segensworth Road, Titchfield, Fareham, PO15 5RR
tel 01329 813522, fax 01329 813532, email neil.lander-brinkley@ons.gov.uk

William Mackaness
Geography & RRL, University of Edinburgh, Drummond Street, Edinburgh, EH8 9XP
tel 0131 650 8163, fax 0131 650 2524, email Wam@geo.ed.ac.uk

David Martin
Department of Geography, University of Southampton, Southampton, SO17 1BJ
tel 023 8059 3808, fax 023 8059 3295, email d.j.martin@soton.ac.uk

Stan Openshaw
Maureen Rosindale, School of Geography, University of Leeds, Leeds, LS2 9JT
tel 0113 233 6834, fax 0113 233 3308, email m.rosindale@geog.leeds.ac.uk

David Owen
Centre for Research in Ethnic Relations, University of Warwick, Coventry, CV4 7AL
tel 024 7652 4259, fax 024 7652 4324, email d.w.owen@warwick.ac.uk

John Pullinger
Office for National Statistics, 1 Drummond Gate, London, SW1V 2QQ
tel 020 7533 5750, fax 020 7533 5808, email john.pullinger@ons.gov.uk

Philip Rees
School of Geography, University of Leeds, Leeds, LS2 9JT
tel 0113 233 3341, fax 0113 233 3308, email p.rees@geog.leeds.ac.uk

Linda See
School of Geography, University of Leeds, Leeds, LS2 9JT
tel 0113 34 33300, fax 0113 34 33308, email 1.see@geog.leeds.ac.uk

Martyn Senior
Department of City and Regional Planning, University of Wales, Cardiff,
PO Box 906, Colum Drive, Cardiff, CF1 3YN
tel 029 2087 4000 ext. 6088, fax 029 2087 4845, email seniorml@cf.ac.uk

Stephen Simpson
41 Park Crescent, Bradford, BD3 OJZ and CCSR, Economic & Social Studies, University of Manchester, Dover Street, Manchester, M13 9PL
tel 01274 642 838/0161 275 4721, fax 01274 754 252, email ludi.simpson@man.ac.uk

Jillian Smith
Office for National Statistics, Room B7/11, 1 Drummond Gate, London, SW1V 2QQ
tel 020 7533 5184, fax 020 7533 5103, email jillian.smith@ons.gov.uk

Andy Teague
Office for National Statistics, Segensworth Road, Titchfield, Fareham, PO15 5RR
tel 01329 81 3404, fax 01329 81 3407, email andy.teague@ons.gov.uk

Frank Thomas
General Register Office Scotland, Census, Ladywell House, Ladywell Road, Edinburgh, EH12 7TF
tel 0131 314 4217, fax 0131 314 4344, email frank.thomas@gro-scotland.gov.uk

Alistair Towers
Data Library, Computing Service, University of Edinburgh, George Square, Edinburgh, EH8 9LJ
tel 0131 650 1383, fax 0131 650 330, email a.towers@ed.ac.uk

Ian Turton
School of Geography, University of Leeds, Leeds, LS2 9JT
tel 0113 233 3392, fax 0113 233 3308, email i.turton@geog.leeds.ac.uk

Paul Williamson
Department of Geography, University of Liverpool, Roxby Building, Liverpool, L69 3BX
tel 0151 794 2854, fax 0151 794 2866, email william@liv.ac.uk

Tom Wilson
Department of Social Statistics, University of Southampton, Southampton, SO17 1BJ
tel 023 8059 4382, fax 023 8059 3846, email tgw@socsci.soton.ac.uk

Jennet Woolford
Office for National Statistics, Segensworth Road, Titchfield, Fareham, PO15 5RR
tel 01329 813 947, fax 01329 813 407, email jennet.woolford@ons.gov.uk

Acknowledgements

This book is the collective product of researchers who use census data in the UK academic community, working in collaboration with staff from the UK Census Organizations. All of the *census data* quoted in the book are Crown Copyright and are reproduced with permission of Her Majesty's Stationery Office and the Census Organizations. We are very grateful for the support and encouragement of the Office for National Statistics (Chris Denham and colleagues), the General Register Office Scotland (Frank Thomas and colleagues) and the Northern Ireland Statistics and Research Agency (Robert Beatty and colleagues) throughout the book's gestation. The EWCPOP mid-1991 small area estimates are copyright of the Estimating with Confidence project. Census data and EWCPOP estimates are made available by the Census Dissemination Unit through the Manchester Information and Associated Services (MIMAS) of Manchester Computing, University of Manchester. The 1991 Census data have been purchased for academic research purposes by ESRC and JISC.

Digital boundaries of Census maps in the book are based on data provided by the United Kingdom Boundary Outline and Reference Database for Education and Research Study (UKBORDERS) via Edinburgh University Data Library (EDINA) with the support of ESRC and the Joint Information Systems Committee of Higher Education Funding Councils (JISC) and boundary material, which is Copyright of the Crown, the Post Office and the ED-LINE consortium (including Ordnance Survey).

Lookup tables are made available by the Census Dissemination Unit through the MIMAS service of Manchester Computing, University of Manchester. The Postcode-Enumeration District Directory (PCED) was originally created by OPCS and has been updated by ONS. The Central Postcode Directory (CPD) (Postzon) is based on a file originally created by the Department of Transport and has been enhanced by the Post Office, OPCS, GRO(S), Ordnance Survey, the Welsh Office and various local authorities; permission to use the CPD has been given by the UK Data Archive. The All-Fields Postcode Directory (AFPD) is produced by ONS with information from ONS, GRO(S), NISRA, the Post Office and Department of Health. The Updated UK Area Masterfiles ESRC funded project (H507255164) has re-engineered the AFPD to link census geographies to other administrative geographies. Data in these lookup tables are Crown Copyright, ESRC purchase.

The Samples of Anonymised Records from the 1991 Census are Crown Copyright and supplied by the Census Microdata Unit, Cathie Marsh Centre for Census and Survey Research, University of Manchester. The data were commissioned by ESRC and JISC for academic use and are licensed for non-academic use by the CMU.

A large part of the research reported in the book was supported by the Economic and Social Research Council (ESRC) and the Joint Information System Committee (JISC) of the Higher and Further Education Councils. ESRC and JISC have jointly supported three major programmes of dissemination and support for UK census data (1991–1996, 1996–2001 and 2001–2006); ESRC has funded three smaller programmes of research and development (1994–1996, 1996–1998 and 1999–2001), while JISC has funded a major learning and teaching project (2000–2003), the Collection of Historical and Contemporary Census (CHCC) materials, to bring census data and knowledge to HE/FE students via the Internet. Many of the book's chapters have their origins in these programmes. Most of the census data quoted in book chapters derives from the ESRC/JISC purchase or licensing from the Census Organisations, as provided by the ESRC/JISC Data Units: the Census Dissemination Unit, the Census Microdata Unit, the UKBORDERS Service and the Longitudinal Study Support Programme (see Table 1.1 for contact details). We acknowledge the permission of Malcolm Campbell to reproduce Figure 12.1, taken from *www.mimas.ac.uk/focus/99dec/*. All other figures are produced by the Chapter authors from documentary, census or boundary material whose copyright is acknowledged above.

Finally, we recognize the seminal contribution of our colleague Stan Openshaw, who suffered a disabling stroke in 1999 when book chapters were being drafted. Although Stan was unable to contribute to the book beyond May of that year, we have included him as author of Chapters 4 and 11, reflecting his innovative (and sometimes iconoclastic) work on the tools for analysis of census data. We sincerely hope that others will take forward his pioneering of new methods of deriving knowledge from large geographical data sets (zone design, neural net classification, cluster searching and data mining) in the use and application of 2001 Census information.

This book provides the reader with a platform for the analysis of the 2001 Census as the data are released during 2002, 2003 and 2004 and for the comparison with information from the previous 1991 Census. We hope that researchers, practitioners and students will find it helpful over the next decade.

Philip Rees, David Martin and Paul Williamson
February 2002

Forewords

John Pullinger, Executive Director, Office for National Statistics

The national census is an enormous undertaking, involving everyone in the country in one way or another. It generates an enormous quantity of information designed to help us understand the state of the nation and inform decisions about the allocation of billion pounds to services provided to communities across the country.

For the cost and effort involved in collecting census information to be worthwhile, it is crucial that the quality and validity of that information is understood and that it is used to its fullest potential. That potential is huge, but it must also be recognized that census taking in the United Kingdom is getting harder—there are more households than ever before, with more diverse and complex lifestyles and often with a great reluctance to share personal information about themselves, even within the strict confidentiality restrictions maintained by the Registrars General. Learning lessons from the experience of 1991, recognizing changes in society and harnessing new technologies, the data systems for 2001 have been designed to deliver results well targeted to the expressed needs of users in government, academic and business sectors.

This book provides a wide-ranging and thoughtful exploration of the census data system in the United Kingdom that, I am sure, will prove invaluable to those seeking to make good use of census information. Like all aspects of the census, it has been put together as a partnership between a large number of experts both within and outside government. The editors have done a great job in bringing so many distinguished contributors together in this volume.

Ian Diamond, Chairman, Research Resources Board, Economic and Social Research Council

The census remains the only dataset that covers the entire nation and provides a basis for a wealth of social scientific research. The improvements in computing power in the past decade mean that these fascinating and valuable data can now be utilized by an increasing number of researchers for an ever-wider range of uses. For the 1991 Census, the Economic and Social Research Council was very pleased to fund a large number of projects that have helped make the data more available to the user community, to improve its quality and to provide model illustrative and innovative analyses. Very many ESRC-funded researchers have made use of census data for a wide variety of uses. In addition, ESRC was delighted to be associated with the development of the Sample of Anonymised Records.

For the 2001 Census, ESRC will again be pleased to support the delivery and use of the data from what is expected to be one of the most exciting censuses ever. New questions, such as those on carers and household structure, will provide researchers with new and exciting challenges, and it is essential that everything is done to ensure that as many potential users as possible access the data. This book provides a valuable tool to aid researchers to maximize use of these data and, as Chair of the ESRC Research Resources Board, I am delighted to welcome its publication.

Reg Carr, Chairman, JISC Committee on Electronic Information

The Joint Information Systems Committee has consistently supported the acquisition and dissemination of population census data for the past decade, and continues to see it as a key national resource for teachers, learners and researchers in the HE (and now the FE) communities. This book represents an impressive return on the JISC's investment over the years, and illustrates how important the population census data really are. The contributors—all closely involved with various ESRC/JISC programmes and projects—show what value can be added to the data, what new products can be developed. Working in close collaboration with the Census Office, they have contributed effectively to the design and dissemination of the 2001 Census.

This book, together with its associated web links and CD materials, will undoubtedly be a key reference source for the development of the Census-related learning and teaching resources, which JISC is also currently funding. It provides an essential guide to a wealth of on-line information of great value to study and research in this important subject field. On behalf of JISC, I congratulate the editors and authors on a very fine piece of work.

How to Use the Book

The Text

Like all books this can be read from the first to the last chapter. However, readers will probably be interested in selected chapters. All the chapters stand by themselves to provide an account of the census data resource being described. These are enhanced (usually) with pointers to the rich collection of software resources, datasets and information hosted on a wide variety of Web sites. Where appropriate there is a box at the end of each chapter providing the Web links.

The CD-ROM

In addition, the authors and editors have assembled, with permission of the Census Organizations, a large collection of files containing software, datasets and information based on the 1991 and earlier censuses, or describing the outputs proposed for the 2001 Census. Once again, a box at the end of each chapter lists the additional resources to be found on the CD accompanying the book as appropriate.

The Website

The Web links will inevitably become out-dated as Web sites are usually under continuous revision. To address this issue, the editors have compiled a book Web site containing an up-to-date version of the Web links cited in the book. This Web site is to be found at *http://www.mimas.ac.uk/cdu/censusdatasystem/* which will be maintained for 3 years.

Phil Rees, Dave Martin, Paul Williamson
May 2002

1

Census Data Resources in the United Kingdom

Philip Rees, David Martin, and Paul Williamson

SYNOPSIS

This chapter provides the reader with an overview of the census data system, which ensures that the census is successfully administered, processed, corrected, turned into useful statistics, disseminated, and used. Census statistics—comprehensive in their coverage of a national population, the only data set to yield a rich picture of the population at small area scale—are described in terms of key features, dissemination funding and organization, the different types of output, how the Census Offices fulfil their guarantee of maintenance of individual confidentiality, the formats in which outputs are produced and the software that can be used to extract subsets and carry out analysis. The chapter then provides an outline of the actual census data resources associated with the different chapters and placed on the accompanying compact disk (CD) and concludes with an account of the logic of the structure of the whole book.

1.1 INTRODUCTION

1.1.1 Census Data Resources

The census is a device for counting people and recording their characteristics. Censuses are normally carried out once every decade for the whole population resident in a national territory. Although the census employs a fairly simple questionnaire, social science researchers are making increasingly sophisticated use of the comprehensive data on national, regional, local, and small area populations provided by the UK Censuses.

Census data resources are the raw census counts, the added value statistics, and the tools for accessing, manipulating, and visualizing the wealth of information provided. To produce the resources, a sequence of components is needed. The first consists of the people who complete census returns. The second component consists of the Census Offices that administer the data collection exercise, an extensive survey that delivers census forms to 24 to 25 million households in the United Kingdom. The Census Offices organize data entry, check and edit the input data, place the data into suitable databases, and produce many gigabytes of census output statistics and associated materials. The third component is made up of the user consortia that license use of the data from the

Census Offices and disseminate the data and added value products to user organizations. These organizations in turn distribute the data to individual users, who use the statistics for research, teaching, administrative, or commercial purposes. Individual users of the printed volumes and computer files may be researchers, teachers, students or pupils, local or central government officers, market analysts, or software developers. The academic community covering higher education institutions and research institutes makes up one customer sector that licenses most census outputs and distributes them to several thousand individual users.

Arrangements for disseminating census data from the 2001 Census will differ from procedures used with earlier censuses (1971, 1981, and 1991). In 2001, ONS announced the intention of providing free access via the Worldwide Web (referred to hereafter as the 'Web') to Key Statistics, Census Area Statistics (CAS), and Standard Tables (area statistics from the 2001 Census), in association with Neighbourhood Statistics, statistics derived from administrative sources that fill the gap between censuses in knowledge about small areas. The Cabinet Office announced in April 2001 the provision of a Web-based licensing facility for value-added use of official data, to be administered by Her Majesty's Stationery Office (HMSO) (see *http://www.hmso.gov.uk/*).

1.1.2 The Aims of the Book

Comprehensive reviews of the 1991 Census in the United Kingdom have been given in two previous books. The *1991 Census User's Guide* (Dale and Marsh, 1993) gave full details of census topics—taking, processing, and products. The *Census User's Handbook* (Openshaw, 1995a) described methods for extracting, processing, visualizing, and analysing census data, and contained a critique of what had been achieved (Openshaw, 1995b). The dissemination arrangements for UK 1991 Census data were described in detail in Chapter 2 of the Handbook (Rees, 1995a). This book builds on these accounts by detailing the additional data products and software that have been developed since original release of 1991 Census data. Although census data in the United Kingdom are produced only once in a decade, data sets evolve over time. For example, census data need to be reaggregated to new administrative policy or analysis regions; better ways of accessing the data for a wider set of users have been developed and new classifications of the data have been produced.

The principal aim of this book is to provide an account of these value-added products, created with the support of the Economic and Social Research Council (ESRC) and the Joint Information Systems Committee (JISC), that have been subsequently added to the basic census data to create a *Census Data System.* The book also provides information on plans for producing outputs from the 2001 Census, carried out on Sunday 29, April 2001. All of these new value-added products are to be found on computers connected to the Joint Academic Network (JANET), a subset of the Internet, established by JISC. They can be tracked down and used on-line via the Internet (see *http://census.ac.uk/*). However, users may be, at first, overwhelmed by the volume and variety of data provided via the ESRC/JISC Census Programme. This book provides a guide to the new derived data and value-added products associated with the census, held on different servers. It acts, therefore, as a guide to UK census data available in computer readable form. By providing links to the relevant Web pages and data locations, the intention is to future-proof the guide. Inevitably, data products and software will be enhanced over time and the Web references (Unique Resource Locators or URLs) will change. The book editors will maintain a set of Web pages for the three years 2002, 2003, and 2004, in which the links referred to in the text will be kept up to date. The URL for this resource will be: *http://census.ac.uk/censusdatasystem/*.

Users outside the academic community may wonder what there is in this book for them. Many of the on-line resources consist of metadata (information about data sets) and basic statistics and they are made available in the public domain and are accessible without the need to register. These public domain resources have been gathered together and placed on a Compact Disk that accompanies the book, which users can use on their own personal computers. There is, of course, a great deal of detailed data for which potential users will need to obtain licences from the Census Offices. As indicated earlier, more and more census data will become available, free at the point of use, via the Web under new dissemination arrangements. There is likely to be a transition period during which 1991, 1981, and 1971 Census data sets are transferred to new licensing arrangements. New arrangements should be fully in place by the time 2001 Census outputs are released.

1.1.3 Chapter Organization

Section 1.2 describes the essential attributes of the census, and assesses its strengths and weaknesses as an information resource. Section 1.3 outlines how the licensing of data sets is funded and how dissemination is organized. Section 1.4 defines the fundamental data types used in census outputs and their internal structure. Section 1.5 briefly discusses the measures used by Census Offices to protect the confidentiality of the individual records embedded in the census outputs, measures that users need to be aware of when employing the data. Section 1.6 of the chapter gives details of output formats and the main software packages used to extract subsets of data from larger files. Section 1.7 gives guidance on the URLs that link to census data resources and on the contents of the CD. The final Section 1.8, outlines the organization of the book.

1.2 CENSUS DATA: ESSENTIAL ATTRIBUTES

The census is a simple questionnaire *survey* of the whole of the UK population that has been held every ten years since 1801, except in 1941, and with the addition of 1966 when the brave experiment of a mid-decade ‘sample census’ was attempted. The most recent census was held on 29 April, 2001. The census is administered separately in England and Wales, Scotland, and Northern Ireland, but the majority of the statistics published are common to all countries. The census covers a wide range of *topics* describing the characteristics of the population of the UK. Subjects covered include demography, households, families, housing, ethnicity, birthplace, migration, illness, economic status, occupation, industry, workplace, transport mode to work, cars, and language. Census data are available in computer format for a variety of geographies and spatial scales: (1) for *small areas* used in the collection of the census or in its most detailed output, (2) for *administrative areas* such as districts, counties, regions, and countries, (3) for *postal areas* such as unit postcodes, postal sectors, postal districts and postal areas, and (4) for *electoral areas* such as wards within local government areas, and Parliamentary and European constituencies.

Why are census data important? Census data describe the state of the *whole* nation, area by area. No other data set provides such comprehensive *spatial* coverage. The data are extremely relevant for *policy analysis*. They are used by government in the allocation of billions of pounds of *public expenditure*. The data are very valuable *commercially*. They are essential ingredients in marketing analysis and retail modelling. Finally, census data are important teaching resources, as students need to be trained in census use and analysis to meet the demands in public and private organizations.

It is helpful to provide a few simple definitions. A census is a complete counting of a population at a point in time. National states have long regarded such counts as providing essential information for such purposes as congressional apportionment (USA), resource allocation from central to local government (UK), planning of welfare schemes, pensions, school provision by public bodies, and the planning of delivery systems by private retail, finance, and information organizations (most developed countries).

Users of census data need to be aware of their strengths and weaknesses. Population censuses held by national statistical offices have the following *strengths*: (1) they are geographically comprehensive; (2) they represent the 'gold standard' of data collection; (3) they provide data for many geographical scales; (4) they provide objective attributes for the population; and (5) they have the confidence of the people. However, they have the following *weaknesses*: (1) they increasingly suffer from underenumeration (see Chapter 12); (2) there is debate about how to estimate and locate the missing population; (3) the data are only collected at periodic intervals; (4) the range of characteristics gathered is limited; (5) respondents make many mistakes when they fill in the census questionnaire and (6) output is modified to protect respondent confidentiality, with 'raw' microdata never being released. Work has been carried out and is ongoing, both in the Census Offices and in user communities, to overcome these deficiencies.

The UK Censuses use rather simple household questionnaires. The 1991 Census of Great Britain asked only 19 individual and 5 household questions. However, the questions were asked of 22 million households and an additional 0.8 million individuals living in communal establishments (Central Statistical Office, 1995). The 2001 Census in England, for example, asked 36 individual and 10 household questions. Because a census questionnaire is completed by the householder (and not by an interviewer), it must be simple, clear, and easy to complete.

From these simple beginnings, an enormous volume and staggering variety of information can be produced. How many cross-tabulations might be produced from the 24 questions of the 1991 Census? The number of possible cross-tabulations is the sum of all combinations of 24 variables taken 2 at a time, 3 at a time, and so on up to 24 at a time, which is 16 777 215. However, these tables can be designed in a variety of ways as each variable can be classified in numerous ways. There are, therefore, millions of possible tables that can be generated containing millions of cell counts. Most of these tables will never sensibly be requested but there are still a very large number that make sense. There is no problem in providing the software necessary to construct such tables. The Census Offices have licensed from Space-Time Pty, a suite of fast cross-tabulation programs called SUPERSTAR, SUPERCROSS, and SUPERTABLE.

So, for any census there is enormous scope for analysis and for the production of outputs, which we examine in Sections 1.4, 1.5, and 1.6. First, we discuss how census statistics are provided to the individual user.

1.3 CENSUS DATA: FUNDING AND DISSEMINATION

We now address the issue of how all the census outputs from the 1991 Census (and earlier censuses) can be accessed by the researcher in the United Kingdom with information on arrangements for the 2001 Census outputs. Two aspects are discussed: funding and agreements, and then access and support. It is perfectly possible for individual researchers or organizations to arrange the purchase or licensing of census output data directly with the appropriate UK Census Office. However, there is considerable value in making collective arrangements. The UK Census Offices have agreements covering data purchase and use with Central Government departments, with Local Governments, the Health Service, and marketing analysis firms acting as Census Agencies (value-added re-sellers of Census data with their own products). Arrangements made for supply to and distribution within the Higher Education community of census information in place for the 1991 to 2001 period are described in the following section. However, for the 2001 to 2011 period, the arrangements are likely to be much simplified. We try to indicate how they might change, although at the time of writing there is much to be negotiated.

1.3.1 Funding and Agreements for Outputs from the 1991 and Earlier Censuses

In connection with the 1981 and 1991 Censuses, two academic bodies, ESRC and JISC, have collaborated to make agreements with the UK Census Offices and fund collective purchase, for academic users, research, and teaching purposes. ESRC funds research in the social sciences in UK universities and prefers to avoid the expense of each researcher bidding separately for support to buy census data. JISC funds central computer provision for individual universities, larger servers for collective use, and JANET. JISC aims to provide the information to populate the IT systems funded for research and teaching use. In total, about £1.5 million was spent by ESRC and JISC in acquiring the 1991 Census of Population outputs for academic use.

The census in the United Kingdom is administered by three separate offices, which have recently undergone reorganization. Figure 1.1 gives full contact details for the Census Offices. The 1991 Census in England and Wales was carried out by the Office for Population Censuses and Surveys (OPCS). This merged in April 1996 with the Central Statistical Office to form the Office for National Statistics (ONS). The 1991 Census in Scotland was carried out by the General Register Office Scotland (GROS) based in Edinburgh. The Census Office Northern Ireland, which administered the 1991 Census in Northern Ireland, has since been included in 1996 within the Northern Ireland Statistics and Research Agency (NISRA). Further bodies involved in the negotiations were the ED (enumeration districts)-Line consortium and Ordnance Survey Northern Ireland (OSNI), which provide access to digital boundary data for census areas.

The key advantages for the Census Offices in negotiating collective agreements with the academic community are reduced administration and guaranteed purchase of all data sets for the whole country. The ESRC/JISC constitute the Census Offices' largest individual customer. Other advantages include the high degree of expertise available in the academic community to help check census outputs.

Licensing agreements were agreed separately for each of the separate data sets (discussed in Section 1.4). The general model for the data agreements is shown in Figure 1.2. This general model has been applied to some data sets (e.g., SARs, DBD), but others are

England and Wales Census Data

People:	Census Customer Services
Address:	Census Marketing, Office for National Statistics, Segensworth Road, Titchfield, Fareham, Hampshire PO15 5RR
URL:	*http://www.statistics.gov.uk/*
Email:	*census.customerservices@ons.gov.uk*
Tel:	01329 813800
Fax:	01329 813532

England and Wales Census Area Boundaries (1991)

People:	The ED-line Consortium
Address:	Data Management and Analysis Group, Greater London Authority, 81 Black Prince Road, London SE1 7SZ
URL:	*http://www.london-research.gov.uk/dshome.htm*
Email:	*dsinfo@london.gov.uk*
Tel:	020 7 787 5500
Fax:	020 7 787 5606

Scotland Census Data and Boundaries

People:	Census Customer Services
Address:	General Register Office for Scotland, Ladywell House, Ladywell Road, Edinburgh EH12 7TF
URL:	*http://www.open.gov.uk/gros/groshome.htm* or *http://wood.ccta.gov.uk/grosweb/grosweb.nsf*
Email:	*customer@gro-scotland.gov.uk*
Tel:	0131 314 4254, 0131 334 0380
Fax:	0131 314 4696

Northern Ireland Census Data

People:	Customer Services
Address:	Northern Ireland Statistics and Research Agency, McAuley House, 2-14 Castle Street, Belfast BT1 1SA
URL:	*http://www.nisra.gov.uk/census/*
Email:	*Censusnisra@dfpni.gov.uk*
Tel:	028 90 348160
Fax:	028 90 348161

Northern Ireland Census Area Boundaries:

People:	Customer Services
Address:	Ordnance Survey Northern Ireland, Colby House, Stranmillis Court, Belfast BT9 5BJ
URL:	*http://www.osni.gov.uk/*
Email:	*osni@nics.gov.uk*
Tel:	028 90 255755
Fax:	028 90 255700

(Note: Contact details may change over time but up-to-date references can be easily found using a search engine such as Google and Web sites *www.open.gov.uk* or *www.ukonline.gov.uk*.)

Figure 1.1 UK Census Offices and mapping agencies: contact information.

Level 1: data supplier
Office for National Statistics
General Register Office Scotland
Northern Ireland Statistics and Research Agency
⇓
Agreement
⇑
Level 2: the academic community
Economic and Social Research Council
Joint Information Systems Committee
⇓
Agreement
⇑
Level 3: the data distributor
[Data Unit, UK University]
⇓
Agreement
⇑
Level 4: the academic community organization
[University, Research Institute]
⇓
Agreement
⇑
Level 5: The User
[Individual Researcher or Student]

Figure 1.2 The general model of data agreements, 1991 Census.

simpler (e.g., SMS, SAS/LBS) or have different arrangements (LS). In total, ten agreements cover the census data sets from the 1981 and 1991 Census. Users had to fill in ten registration forms and institutions had to agree to four contracts in order to gain access to all the information. The process has been streamlined to some extent by making the individual registration forms accessible via the Web and a Web Site linking to all registration systems has been set up (see *http://census.ac.uk/*).

1.3.2 Funding and Agreements for Outputs from the 2001 Census

ESRC and JISC have set aside resources for a 2001 to 2006 Census Programme that will fund the purchase of census data sets and their support. The 2001 CAS, which the Census Offices intend to make freely available on the Web, still needs to be commissioned under the Census Acts and funded. This will be done through the *Census Access Project*, set up by ONS with a Treasury 'Invest and Save' grant in partnership with organizations representing the major census user communities. These organizations are ESRC/JISC representing the Higher and Further Education and Research Community, the Department of the Environment, Transport, and the Regions (until reorganization in 2001) representing central government departments, the Department of Health representing the National

Health Service and the Local Government Association representing member authorities. A partner representing the private sector may be added. Partners contribute a share of project costs and receive free access to the area statistics data, along with any other user. The major data sets to be produced from the 2001 Census, such as the Origin-Destination Statistics (ODS) (see Chapter 18) and the Samples of Anonymized Records (see Chapter 14), will need to be purchased by user organizations and licences negotiated.

Where census data are placed on Web sites for universal access, as ONS intend for the area statistics, then a user will need to obtain a click-use licence from HMSO, for the use of data in publications and other products. The click-use licence may follow the general model set up by HMSO or may be a tailored licence, the contents of which will be negotiated.

1.3.3 Access and Support

The Census Offices have, to date, supplied data to and interacted with a handful of people, appointed and funded by ESRC/JISC in order to disseminate census data to all academic users (the Data Units). In the 1981 Census round, the Regional Computing Centres (funded by JISC) were the data distributors along with the ESRC supported Data Archive. In the 1991 and 2001 Census rounds, national data distributors were appointed for the data sets as set out in Table 1.1, which provides full contact details for the Data Units. The *Census Dissemination Unit* (CDU), part of the Manchester Information and Associated Services (MIMAS) service at the University of Manchester, has principal responsibility for the area statistics, and has also looked after the interaction statistics and some DBD. Digital boundary data for census areas are supported by *the UKBORDERS service* within the University of Edinburgh's EDINA Data Centre. The *Census Microdata Unit* has responsibility for the Samples of Anonymized Records for the 1991 and 2001 Censuses. The *Centre for Longitudinal Studies* looks after project users who access the complex

Table 1.1 Arrangements for the dissemination of UK Census data in higher and further education.

Data unit	Data sets and functions supported
Census Dissemination Unit (CDU) (1992–2006)	
Keith Cole (Director), Justin Hayes MIMAS, Manchester Computing, University of Manchester, Oxford Road, Manchester M13 9PL web: *http://Census.ac.uk/cdu/* email: *info@mimas.ac.uk* tel: 0161 275 6109, fax 0161 275 6040	Key Statistics UK, 2001 Small Area Statistics GB, 1981, 1991, NI 1991 Census Area Statistics UK, 2001 Local Base Statistics GB, 1991 Standard Tables UK, 2001 Special Migration Statistics GB, 1981, 1991 Special Workplace Statistics GB, 1981, 1991 Lookup Tables 1991, 2001

Table 1.1 *(continued)*

Data unit	Data sets and functions supported
Census Interaction Data Service (CIDS) (2001–2006)	
John Stillwell (Director), Paul Boyle, Oliver Duke-Williams, Zhiqiang Feng School of Geography, University of Leeds, Leeds LS2 9JT web: *http://www.geog.leeds.ac.uk/* email: *j.stillwell@geog.leeds.ac.uk, o.duke-williams@geog.leeds.ac.uk* tel 0113 233 3315, fax 0113 233 3308	Special Migration Statistics GB, 1981, 1991 Special Workplace Statistics GB, 1981, 1991 Origin-Destination Statistics UK, 2001
UKBORDERS Service (1996–2006)	
David Medyckyj-Scott (Director), Alistair Towers EDINA, Edinburgh University Data Library, George Square, Edinburgh EH8 9LJ web: *http://edina.ac.uk/ukborders/* email: *ukborders@ed.ac.uk, edina@ed.ac.uk* tel: 0131 650 3302, fax: 0131 650 3308	Digital Boundary Data (England & Wales) 1981, 1991 Digital Boundary Data (Scotland) 1981, 1991 Digital Boundary Data (Northern Ireland) 1991 Lookup Tables 1991, 2001
Census Microdata Unit (1992–2006)	
Angela Dale (Director), Ruth Durrell Cathie Marsh Centre for Census and Survey Research, Faculty of Economic and Social Studies, University of Manchester, Dover Street Manchester M13 9PL web: *http://les.man.ac.uk/ccsr/cmu/* email: *ccsr@man.ac.uk* tel: 0161 275 4721, fax: 0161 275 4722	Individual Sample of Anonymized Records GB, 1991 Household Sample of Anonymized Records GB, 1991 Individual Sample of Anonymized Records NI, 1991 Household Sample of Anonymized Records NI, 1991 Individual Sample of Anonymized Records UK, 2001 Household Sample of Anonymized Records UK, 2001 Small Area Microdata UK, 2001
Longitudinal Study Support Programme (1991–2001)	
Heather Joshi (Director), Kevin Lynch The Centre for Longitudinal Studies, Institute of Education, 20 Bedford Way, London WC1 H 0AL web: *http://www.cls.ioe.ac.uk/Ls/lshomepage.htm* email: *hj@cls.ioe.ac.uk* tel: 0171 612 6875, fax 0171 612 6880	Longitudinal Study, EW 1971-1981-1991 Longitudinal Study, EW 1971-1981-1991-2001

(continued overleaf)

Table 1.1 *(continued)*

Data unit	Data sets and functions supported
Census Registration Unit (2001–2006)	
Kevin Shurer (Director) Data Archive, University of Essex, Wivenhoe Park, Colchester, CO4 3SQ web: *http://www.data-archive.ac.uk/* email: *archive@essex.ac.uk* tel: 01206 872 009, fax: 01206 872 003	Making arrangements for authentication, registration and licensing of users of data sets. Promoting use of the whole suite of census information.
ESRC/JISC Census Programme Coordinator (1992–2002)	
Phil Rees School of Geography, University of Leeds, Leeds LS2 9JT web: *http://www.geog.leeds.ac.uk/staff/p.rees/* email: *p.rees@geog.leeds.ac.uk* tel 0113 233 3341, fax 0113 233 3308	Negotiations with data suppliers, licensing arrangements, programme development and coordination, steering group organization (ESRC/JISC Census Advisory Committee), consultation with users, organizing workshops, preparing annual reports on the programme, disseminating knowledge about the programme

Note: GB: Great Britain, NI: Northern Ireland, UK: United Kingdom, EW: England and Wales, SC: Scotland; Contact details may change over time but up-to-date references can be easily found using a search engine or Web sites *http://www.census.ac.uk/* or *http://census.ac.uk/*

census and vital events longitudinal data, linking 1971, 1981, and 1991 Censuses and which will be extended to the 2001 Census by 2004.

The Data Units perform the following functions: (1) They register users and institutions for access to the data, though this function will be centralized in the 2001 to 2006 period; (2) They receive data sets from the Census Offices, mount them on servers, load them into appropriate software, and carry out quality checks; (3) They liaise with the Census Offices if problems are detected with the data and receive new versions; (4) They create new derived data from that supplied from the Census Offices; (5) They provide comprehensive user on-line documentation for the data sets and interface software; (6) They provide courses and seminars for users on the nature of census data set, on how to access the statistics, and on how to carry out analysis; (7) They field inquiries from users; (8) They carry out some data processing for users who lack the resources and skills to do the processing themselves, or where the resulting products can be captured for general use; (9) They receive from users the results of analysis, particularly derived data, which can be mounted for wider access; and (10) Finally, they carry out surveys of user activity and gather information about publications.

A couple of examples illustrate the value of the support that ESRC/JISC Census Programme Data Units provide. The UK Censuses have not, to date, asked a question about income. The Census Microdata Unit has been able to add to the SARs (Samples of

Anonymized Records) earning scores derived by imputing to SAR individuals the earnings of individuals matching characteristics in the New Earnings Survey and in the Labour Force Survey (Drinkwater and Leslie, 1996). In the 2001 Census, similar procedures may be applied more generally, in the absence of a question on individual income. A second example is the creation of SAS for postal districts and postal areas for general use by the CDU from postal sector SAS supplied by the Census Offices.

1.4 CENSUS DATA: TYPES AND STRUCTURES

Figure 1.3 represents the five forms in which census outputs are currently produced in the UK. These outputs are produced principally on computer media, except for statistics for larger areas, which are also published in printed volumes. The diagram provides a picture of the data type on the left-hand side and a representation of the data structure on the right-hand side. There are five data types: census microdata, CAS, census area boundaries, census lookup tables, and census interaction data, which are discussed in turn.

1.4.1 Census Microdata

The census collects completed questionnaires from the nation's households and its communal establishments (colleges, prisons, military bases, long-stay hospitals) and enters them into a database. From this database are produced the outputs released to researchers. Microdata constitute the data type closest in format to the master database and consist of household and individual records.

There are two microdata sets associated with the 1991 Census: the Longitudinal Study and the Samples of Anonymized Records. The *LS* is a linked sample from the 1971, 1981, and 1991 Censuses together with their associated demographic events (births, deaths, marriages, divorces), which will be linked to the 2001 Census by 2004. The sample is just over 1%, selected on the basis of four birthdays during the year. It is widely used for research on mortality because deaths can be linked to prior census characteristics, and it is also of great value for looking at social transitions of more than 10-, 20-, or 30-year intervals. The raw data are not released *per se* because it would be easy to recognize individuals in the information. Instead, researchers request tables or statistical analyses that are vetted before release. Chapter 17 describes the new information that has been added to the LS since 1991 and outlines the plans for adding the 2001 Census link.

The second microdata set, the *Samples of Anonymized Records (SARs),* was extracted from the 1991 Census in both Great Britain and Northern Ireland, with modified coding of sensitive variables and without any identifying names and addresses. The microdata are released as two public use samples, a 2% sample containing individuals (including those in communal establishments rarely covered in household surveys) and a separate 1% sample of households and their members. Researchers can access these data on-line or can transfer them to their desktop/notebook computers. Note that the geography coding, regions, in the household SAR is much coarser than in the individual SAR, districts or grouped districts, because of the greater risks of disclosure of linked individuals in a household—their combined characteristics are much more likely to be unique within a subnational population. Chapter 15 outlines the SARs data sets and discusses the derived variables that have been added since 1991 and proposals for SARs from the 2001 Census.

Census Microdata

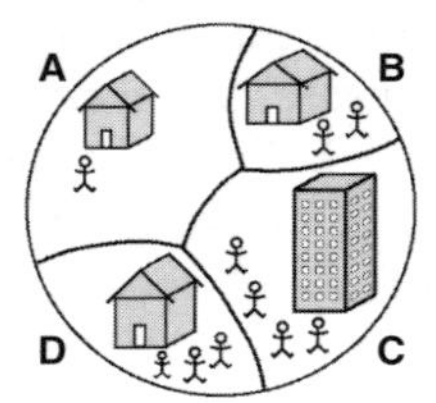

(Individual entity memberships or attributes)

Indexes		Area	Household Size	Individual variable		
House-hold No.	Person No.			Young	Middle Aged	Old
1	1	A	1	0	1	0
2	1	B	2	0	0	1
2	2	B	2	0	0	1
3	1	C	4	0	1	0
3	2	C	4	0	1	0
3	3	C	4	1	0	0
3	4	C	4	1	0	0
4	1	D	3	0	1	0
4	2	D	3	0	1	0
4	3	D	3	1	0	0

Census Area Statistics (counts in classes)

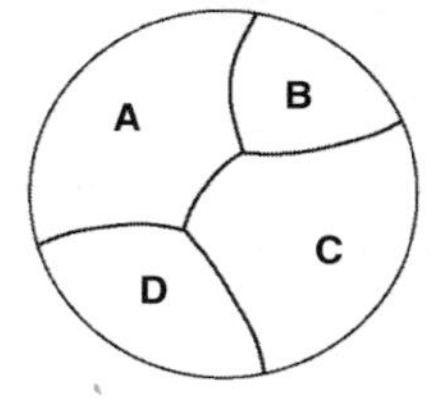

Area	Average Hsld Size	Young	Middle Aged	Old
A	1	0	1	0
B	2	0	0	2
C	4	2	2	0
D	3	1	2	0

Census Area Boundaries (*x*, *y* coordinates)

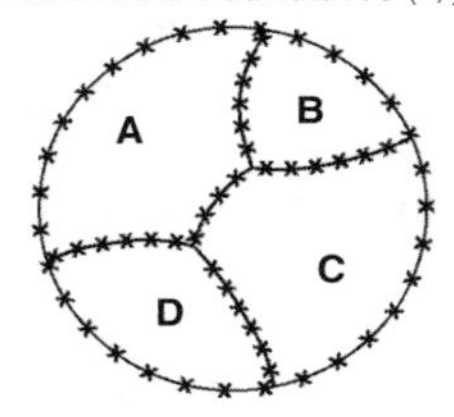

Area	Polygon coordinate pairs
A	(1, 3) (2, 6) (4, 7) (4, 4) (3, 3) (1, 3)
B	(4, 7) (6, 6) (7, 4) (4, 4) (4, 7)
C	(4, 4) (7, 4) (6, 2) (4, 1) (3, 3) (4, 4)
D	(3, 3) (4, 1) (2, 2) (1, 3) (3, 3)

Census Look-up Tables (zone indexes and proportions)

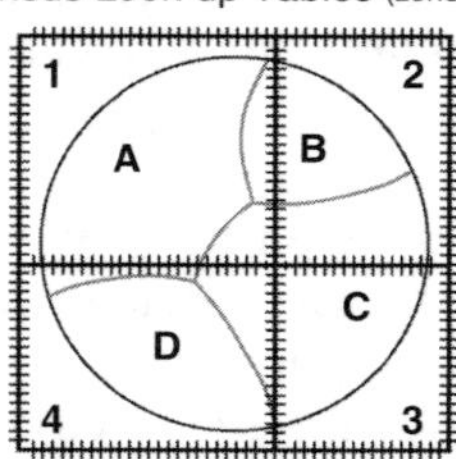

Area	New Zone	Weight
A	1	0.9
A	4	0.1
B	1	0.2
B	2	0.8
C	1	0.1
C	2	0.2
C	3	0.5
C	4	0.2
D	4	1

Census Interaction Data (counts of individuals)

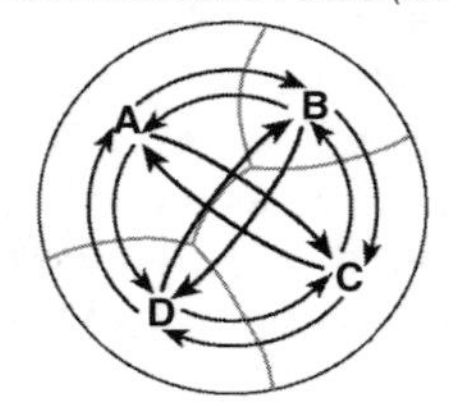

	Destination			
Origin	A	B	C	D
A	1	1	1	0
B	0	1	0	0
C	0	0	1	0
D	0	1	0	1

Figure 1.3 Data types and structures for census outputs.

1.4.2 Census Area Statistics (CAS)

The second type of census data consists of counts for areas organized in tables. Figure 1.3 contains a simple example of a table that counts the people in four areas according to their age. Extraction software (e.g., CASWEB, described in Chapter 8) is designed to reconstitute the cell counts into tables with relevant row and column headings. The content and layout of the tables are agreed in advance between the UK Census Offices and the user communities. The proposals for the 2001 Census have been the subject of consultation during 1999 to 2001 and should be finalized by the end of the year (see Chapter 20 for an account of planned outputs). The area data constitute a fixed set of tables, which are said to be safe from disclosure risk. In many countries the same kind of statistics are produced (for example, Census Tract data in the United States and Canada), although the tables are usually less complex. In other countries, reliance is placed on a bespoke tabulation service for users. The problem with bespoke tabulation is that the national Census Office has to vet each table for disclosure risk in the light of previous requests. This has to be done in order to avoid the 'differencing problem' of subtraction of successive tables, which yield new tables subject to disclosure risk. Publication of a set of fixed tables for areas constitutes a solution to this differencing problem.

Three kinds of area data (tables) were produced from the 1991 Census round in the United Kingdom: (1) National Topic Statistics, (2) LBS, and (3) SAS.

National statistics consist of detailed tables published in Topic Reports (for example, on migration, ethnicity, and country of birth) for the countries making up the UK (England, Wales, Scotland, Northern Ireland). Topic Reports also contain data not available elsewhere (for example, the full 34-group ethnic classification for all districts in Great Britain in Table A of OPCS/GROS, 1993a). These tables cover demographic, housing, occupation, employment, ethnicity, birthplace, migration, car availability, journey to work, and health. From the 2001 Census, a similar set of topic reports are likely to be published, but with fewer tables and more commentary and analysis, with accompanying CDs holding further detailed tables in a variety of computer formats.

The second kind of area data are the Local Base Statistics (LBS) that consist of circa 20,000 counts arranged in 99 tables. LBS tables are published in computer and printed form for UK countries, regions, counties/Scottish regions, local government districts, and in computer readable form for wards (England and Wales) or postal sectors (Scotland). The areas for which LBS are available are set out in Table 1.2. There are slight differences between countries in the areas for which LBS are provided, and none were produced for Northern Ireland. The LBS tables for the smallest areas (wards in England and Wales, and pseudo-sectors in Scotland) differ from those for districts and larger areas in the disclosure control measures applied. The interior cells of most tables in the ward/sector level LBS have been randomly perturbed by the addition of 0, +1, or −1 in secret proportions in two rounds, though Rees and Duke-Williams (1994) have suggested only single blurring be applied.

The equivalent data set to be produced from the 2001 Census will be called the *Standard Tables*. These will be simpler and more numerous than the LBS tables in the 1991 Census, avoiding the concatenation of many cross-tabulations into one table. Chapter 20 provides more detailed information and references to planning documents.

The Small Area Statistics (SAS) constitute the third type of CAS. The SAS consist of about 9,000 counts arranged in 86 tables. The tables are 'junior' versions of those in the LBS with fewer variable categories. Table 1.3 shows that SAS were produced for a

Table 1.2 LBS from the 1991 UK Census: geographical coverage and number of spatial units.

Geographical coverage	Spatial scale	Number of units	Blurring applied
England	Wards	8985	Yes
	Districts	366	No
	Counties	46	No
	Regions	8	No
	Country	1	No
	District health authorities	186	No
	Family health service areas	90	No
	Regional health authorities	14	No
Wales	Wards	945	Yes
	Districts	37	No
	Counties	8	No
	Country	1	No
	District health authorities	9	No
	Family health service areas	8	No
Scotland	Pseudo-sectors	1003	Yes
	Districts	56	No
	Scottish regions	9	No
	Island areas	3	No
	Health board areas	15	No

Source: Denham (1993: 58, Figure 3.1)
Note: Blurring is applied to cell counts by random addition of 0, +1, −1.

very wide range of areas in each UK constituent country. For areas below district scale, the cell counts are blurred in the way described above. The SAS were produced in two tranches: primary and secondary.

Primary SAS were produced first for the main census and administrative areas, followed at varying intervals by the SAS for secondary areas. The secondary area SAS met the needs of smaller groups of users such as political scientists (SAS for Parliamentary and European constituencies) and business users (postal sector SAS in England and Wales). The equivalent data set to be produced from the 2001 Census will be termed the *CAS*. These will contain fewer counts than the 1991 SAS, in simpler tables avoiding concatenation (see Chapter 20 for more details).

1.4.3 Census Boundary Data

In order to map the statistics for geographic areas, using cartographic or geographical information system (GIS) software, researchers require digital files that give the coordinates of the area boundaries. This is the third type of data depicted in Figure 1.3. The data

Table 1.3 SAS from the 1991 UK Census: geographical coverage and number of spatial units.

Geographical coverage		Spatial scale	Number of units	Blurring
England	Primary	Enumeration districts	106 866	Yes
		Wards	8985	Yes
		Districts	366	No
		Counties	46	No
		Regions	8	No
		Country	1	No
	Secondary	Civil parishes	10 262	Yes
		Parliamentary constituencies	524	Yes
		European constituencies	38	Yes
Wales	Primary	Enumeration districts	6330	Yes
		Wards	945	Yes
		Districts	37	No
		Counties	8	No
		Country	1	No
	Secondary	Communities	866	Yes
		Parliamentary constituencies	66	Yes
		European constituencies	4	Yes
England & Wales	Secondary	Postal sectors	7892	Yes
		Urban areas	1859	Yes
Scotland	Primary	Output areas	38 255	Yes
		Pseudo-sectors	1003	Yes
		Districts	56	No
		Scottish regions	9	No
		Island areas	3	No
	Secondary	Wards	1158	Yes
		Regional electoral divisions	524	Yes
		Civil parishes	916	Yes
		Inhabited islands	144	Yes
		Parliamentary constituencies	72	Yes
		European constituencies	8	Yes
		Localities	534	Yes
Northern Ireland	Primary	Enumeration districts	3729	Yes
		Wards	566	Yes
		Districts	26	No
	Secondary	Postal sectors	400	Yes
		Health board areas	4	No
		Education board areas	5	No

Source: Denham (1993: 58, Figure 3.1) and MIMAS Web pages (*http://census.ac.uk/*)

consist of geographic coordinates (grid references or latitude, longitude) representing the boundaries. The boundary data matrix consists of rows of codes and labels for each area together with the grid coordinates describing the polygon or set of polygons that make up the area's boundaries. The exact format in which the data are organized will depend on the software being used. The UK digital boundary data are created and supplied by a variety of organizations. The academic community (through the ESRC/JISC purchase) has acquired 1991 Census boundary data for Scottish Census areas from GROS, and for English and Welsh areas from the ED-line Consortium. This consortium consists of MVA Systematica, the London Research Centre (now part of the Greater London Authority), Ordnance Survey Great Britain, and the Data Consultancy. Boundary data for Northern Ireland are supplied by OSNI direct to users under an annual licence.

1.4.4 Lookup Tables

The fourth type of data associated with the 1991 Census consists of lookup tables (LUTs), which can be used for the conversion of census data from an official source geography to a user target geography. Figure 1.3 shows that the entity in this data structure is the intersection of the source areas for which area statistics have been supplied (areas A, B, C, D) and target geography (zones 1, 2, 3, 4). The simplest lookup table is simply an index that records which other area the basic unit belongs to. However, often the census geography and the alternative geography overlap, creating intersections. These intersections form the rows of the example data matrix shown in Figure 1.3. The third column of the lookup table holds the weights for converting area statistics from the standard census geography to an alternative. In the absence of other information, the weight might be the area of the overlap. In the case of the ED/postcode directory produced by OPCS/ONS, the weights are the number of households in the intersection resident at the 1991 Census. Such a lookup table enables researchers to convert data from census areas to postal areas at a small scale. One recommendation emerging from the current consultation process between Census Offices and the academic community is that LUTs be developed more systematically and accurately in response to user demands for census data for non-standard geographies, and in order to update the standard geographies as they change between censuses. Chapter 5 gives a systematic account of the derived lookup tables generated by census users and made available for general use. A facility for generating user lookup tables from the All Fields Postcode Directory (AFPD) has been developed through an ESRC project (Yu and Simpson, 1999; Simpson and Yu, 2000 and Simpson 2001) and is available via the MIMAS service (*http://census.ac.uk*).

1.4.5 Interaction Data

The fifth data type shown on Figure 1.3 consists of interaction data, which involve the flow of people from origins to destinations. Two kinds of flow are captured by the 1991 Census in the United Kingdom: migrant flows and journey to work flows. The number of persons in each flow can be counted in a variety of categories, which are organized into tables (see Flowerdew and Green, 1993 and Rees and Duke-Williams, 1995a for details). Interaction data are most naturally represented as origin–destination tables as shown in Figure 1.3. Each cell in the table refers to an origin (the row) and a destination (column). These data will be known as Origin–Destination Statistics in the 2001 Census.

The UK Census Offices created software systems for producing migration and journey to work statistics called the Special Migration Statistics (SMS) and the Special Workplace Statistics (SWS), respectively. Users were able to submit their own definitions of origins and destinations, and flow statistics were produced. The tables and counts made available depended on spatial scale in the case of the SMS, with very few counts being provided at ward scale. The ESRC/JISC Census Programme purchased both SMS and SWS data sets for wards/(England and Wales) and sectors (Scotland) and an SMS set for districts. The Census Offices had major difficulties in producing these complex statistics and the academic community had problems in producing suitable software to handle 10 933 by 10 933 flow matrices. ESRC/JISC are funding a new Centre for Interaction Data to provide support for use of these statistics (Table 1.1). Chapters 18 and 19 discuss what has been achieved with the 1991 Census, SMS and SWS, and what is proposed for the 2001 Census.

1.5 CENSUS DATA: MEASURES TO PROTECT CONFIDENTIALITY

The UK Census Offices pledge to their respondents that their individual data will not be disclosed to any person except Census Office staff involved in the initial processing of the census. This pledge is a requirement of the Census Act of 1920 amended by the Census (Confidentiality) Act 1991 (Marsh, 1993a). Names and addresses are recorded on the census form, but will remain there unused for 100 years. Only the unit postcode part of the address is captured in digital form together with the collection area code.

The 2001 Census is being processed initially by scanning the census returns. The scanned images will be used only for checking purposes. Software will convert the ticks and characters, except for names and addresses, into digital data for editing, imputation (where responses are missing) and output processing. All records will be assigned a unique address reference incorporating the grid coordinates of a point within the property parcel. These address references will be made available to the Census Offices by the mapping agencies in England and Wales (Ordnance Survey GB to ONS) and in Northern Ireland (OSNI to NISRA). GROS will be able to generate their own address references from their own in-house GIS system.

Two strategies are adopted to protect data from the risk of disclosure (Marsh *et al.*, 1994). The first is to keep the data within the Census Offices in a secure computer environment and to release outputs, such as tables, which have been carefully vetted for unique and sensitive classifications. The second strategy makes sure that the data is safe by applying one or more of a number of protection devices and then releasing the data to the users.

The census master database is classified as too risky to release *in toto* and so remains on Census Office computers, protected by both physical security and by the absence of physical connection to computer networks. From the master database are prepared the standard outputs of published tables and published data sets. Approved users are allowed to request tables and analyses from the LS. A service is offered to users for the production of bespoke, special tables, but these must be vetted in detail before release.

Census data are made safe in a variety of ways and then released for use by researchers or the public. Table 1.4 gives details of the nine protection devices used for the different data sets. The final column refers to record modification, which is likely to be used in the 2001 Census instead of random perturbation or suppression, as it makes for outputs that are more consistent.

Table 1.4 Protection devices used with census data sets.

Dataset	Vetting	Tables	Broad coding	Samp-ling	Thresholds	Random changes	Suppression	Licence agreement	Record modification
Longitudinal data									
LS 1971–1981–1991–2001	✓		✓	✓				✓	
Microdata									
SARs 1991			✓	✓	✓			✓	
SARs 2001			✓	✓	✓				✓
Special tabulations 1991	✓	✓	✓					✓	
Area statistics									
SAS 1991 for EDs, wards		✓	✓	✓	✓	✓		✓	
SAS 1991 for districts		✓	✓	✓				✓	
CAS 2001		✓	✓		✓				✓
LBS 1991 for wards, sectors		✓	✓	✓	✓	✓		✓	
LBS for districts		✓	✓	✓				✓	
ST 2001		✓	✓		✓				✓
Interaction data									
SMS 1991, wards/sectors		✓	✓					✓	
SMS 1991 districts		✓	✓		✓		✓	✓	
SWS 1991		✓	✓	✓				✓	
ODS 2001		✓	✓		✓				✓

1.6 CENSUS DATA: OUTPUT FORMATS AND SOFTWARE

The forms in which census data are delivered to the user are under rapid evolution. For most of the history of the census the main output format has been in books of printed tables, still an important resource for research, and purchased by libraries. Typeset output is, however, expensive to produce and does not have a great market beyond research institution libraries. Therefore, Census Offices in the 1960s and 1970s offered computer printouts. In the 1980s, with the development of desktop computers, computer printouts were largely superseded by files of digital counts that were the numbers in the tables, supplied initially on magnetic tape (reels or cassettes), but increasingly on easier-to-handle media such as diskettes, DAT tapes, and CD-ROMs.

The organization of supply of data from a few sources to users at many locations is a problem faced by many manufacturing, wholesale, retail, and financial organizations. Two models of dissemination coexist: the packet and the on-line network. The former was invented in 1840 by Roland Hill as the prepaid post at uniform rates, and the latter by Alexander Graham Bell in 1876 as the telephone. There is a strong case to be made for each of these two models. The convenience of packet data dissemination has vastly improved with the development of CD and associated technologies (DVD), while the convenience of on-line access to information resources has improved considerably with the adoption of the Web standard for interfaces on the internet.

The model of Web access to central resources is likely to become more important over time: the user can keep up-to-date with new versions of data products and can learn about and use new derived data sets deposited at central data units. However, the speed of access to the Web via normal telephone lines and modems, although improving, is currently slow and relatively expensive for most users, which is why there still remains a market for packet delivery via CD or DVD. In planning a dissemination strategy for the 2001 Census, the Census Offices and user communities will need to consider carefully the balance between these two dissemination strategies.

The Census Offices delivered 1991 data to the academic community very largely as ASCII (plain text) files in fixed format or comma-delimited form, together with documentation on its meaning, usually in paper form. Essentially, the files were just streams of numbers. The only exception was the LS, which could only be released in the form of vetted tables. Three kinds of software were needed to make sense of the numbers: (1) extraction software, (2) analysis software, and (3) display software.

Extraction software allows the user to make choices about filtering a large data set, selecting the table or tables and formatting the outputs. Filtering might involve choosing a set of geographic areas or types of household; selecting tables might involve choosing the row variables and column variables, and how each was organized into categories; and formatting might involve selecting a data structure, which could be used with a spreadsheet. Extraction software is specific to the census data sets from which the extractions are being done, whereas analysis and display software can handle a wide variety of data input by the user.

To read the files of digital counts or records required specialist software written in higher level programming code. This was beyond the skills of most researchers, so there was a demand for software for extracting the data required from the large digital file and presenting it in an easy-to-understand format. At its simplest, this software simply marries the digital counts in a table with the metadata describing the data—that is, the row labels and the column headings. Local authorities in Great Britain commissioned a University team of researchers to develop software to access the Small Area Statistics from the 1981 Census (Rhind, 1983). This software, SASPAC, was further developed to handle the 1991 Census (Davies, 1995) for UNIX, DOS, and Windows environments. Licences for SASPAC were purchased by the academic community for use at individual universities or via the MIMAS service at Manchester Computing. An alternative but similar product called C91 was developed for PC use by Powys County Council, which has also been used within the academic community. More recently, a new system for Web-based access to the SAS/LBS, CASWEB, has been developed. These issues are discussed more fully in Chapter 8.

These extraction packages provide rather limited analysis capabilities, but they do contain facilities for exporting extracted data in formats compatible with widely used spreadsheet, statistics, and mapping/GIS software. Integrated extraction and graphical presentation software has been prepared by CD-ROM designers such as Chadwyck–Healey. An academic team (Unwin and Dykes, 1996) has developed a software system called CDV (Census Data Visualizer) with JISC funding to facilitate exploratory visual analysis of integrated SAS-DBD data sets (see Chapter 7). A similar package, DESCARTES, developed by the German national GIS centre, GMD, has recently become available for use via the Web.

Table 1.5 sets out the main software packages made available by the ESRC/JISC Census Data Units. The LS is held in database package MODEL 204, not widely used and for which the ONS had to write specific interface software, FLEXTRACT, to help both in-house and LS support programme staff to produce tabulations and prepare database extracts for use with statistical packages. The SARs are held in a variety of statistical

Table 1.5 Interface and analysis software for census data sets.

Dataset	Database data format	Extraction software	Analysis/display software (examples)
LS	MODEL 204	FLEXTRACT	SPSS, SAS, MLN, GLIM
SARs	ASCII	SPSS, SAS, QUANVERT, SIR, QUICKTAB, *USAR*, NDSTAT	SPSS, SAS, GLIM, NDSTAT
KS, SAS, CAS, LBS, ST	ASCII, DATABASE	SASPAC, *CASWEB*	SPSS, SAS, GLIM, *CDV*, DESCARTES
SMS, SWS, ODS	ASCII, DATABASE	QUANVERT, *SMSTAB, WICID*	SPSS, SIM, GLIM
DBD	ASCII, ARC-INFO	*UKBORDERS, DBD91, CASWEB*	ARC-INFO, MAP-INFO, *CDV*, DESCARTES

Note: ASCII = American Standard Code for Information Interchange; *SMSTAB* = italics indicates academic community software.

package system file formats that cover most user needs. Users can also download the files in ASCII format along with full data dictionaries and glossaries for use with other packages. A project funded by the ESRC/JISC Census Programme has produced a very fast interactive tabulation package called USAR (Turton and Openshaw, 1995) for use on computers running the UNIX and DOS operating systems (described in Chapter 15).

Bespoke software written by projects funded by the ESRC/JISC Census Programme are important for providing access to the more complex interaction data sets and DBD. SMSTAB was originally written by Duke-Williams (see Rees and Duke-Williams, 1995a) as a program to read and check the SMS data set for wards/sectors, but proved useful as a general extraction tool. Proprietary software, QUANVERT, is also used to access the interaction statistics. SMSTAB code was adapted very quickly by the Census Dissemination Unit to access the interaction (journey to work) statistics in the SWS. An ESRC/JISC Census Programme project has developed a more user-friendly Web Interface to Census Interaction Data (WICID).

A project at the Data library of the University of Edinburgh has developed an interactive software interface called UKBORDERS, described in Chapter 6, to access the digital boundary data for Great Britain held as polygon descriptions of the smallest Census units (enumeration districts, EDs, and output areas, OAs) in ARC/INFO format. The user is able to create subsets of the data at small or larger scales. An equivalent interface called DBD91 carries out similar functions for England and Wales ED boundaries within counties (see *http://census.ac.uk*). Users carrying out interaction or DBD extractions need considerable support because the data sets are both large and complex. Chapter 6 describes the UKBORDERS interface.

1.7 WEB RESOURCES AND THE COMPACT DISK

All these data and software products are housed in computer files on various university servers. The information about the files is freely accessible via the Web, while most of the data are available on-line to registered users on university servers or government computers. It is likely that much of the 2001 Census outputs will become freely available via the Web. However, as with many other Web resources, it takes a great deal of time and knowledge to find these resources. The chapters in this book provide Unique Resource Locators (URLs), which are the addresses that users need to track down the Census Data System. URLs may change over time, so the editors of the book have established a set of Web pages. This can be found at *http://census.ac.uk/censusdatasystem/*, which will be maintained over the print life of the book. For those readers who do not have an internet connection but have a suitable disk drive on their personal computer, the editors have assembled the contents of the Web resources referenced in the book that are in the public domain (free to users) on a CD. The website and CD are referred to collectively in the following pages as the *CDS Resources*.

1.8 ORGANIZATION OF THE BOOK

The book is divided into six parts. Part I discusses the *Geography of the Census and Lookup Tables*. In the 1991 Census the small areas used for publishing data from the census coincided with the EDs used to collect the census in England, Wales, and Northern Ireland. In Scotland, the use of OAs separate from collection areas was introduced. Chapter 2,

The Debate about Census Geography, describes the discussions that took place during a two-year period, 1995 to 1997, on the question of how small areas for the 2001 Census should be designed. Chapter 3 describes the solution to be adopted in England and Wales, *Output Areas for 2001* built up from unit postcodes, which adapts and improves on the Scottish model for 1991. In Chapter 4, the authors describe the software *Tools for designing geographical zones* for a variety of purposes. Part I of the book concludes with Chapter 5, *Lookup tables and new area statistics for the 1991 Census,* which gives an account of the lookup tables available to convert 1991 Census data from the geography current at the time of the census to new geographies.

Part II continues the geography theme by reviewing *Boundary Data and Visualization*. Chapter 6 describes *Handling and accessing boundary data of the census*, while Chapter 7 introduces innovative software systems for *Visualizing census data* on maps, graphs, and cartograms.

Part III takes the reader through the most widely used census data set: the *Area statistics*. Until recently, the range of software for extracting data items from the area statistics was dominated by one software package, SASPAC, originally developed for local government use. In Chapter 8, *Disseminating census area statistics over the Web,* a new interface is described, which makes possible Web-based access to area statistics. This is the way in which many data products from the 2001 Census will be delivered. The next four chapters describe products, based on CAS, developed by the Census Offices or researchers. Chapter 9 gives an account of *Deprivation indicators* based wholly or partly on census data. Chapter 10 describes how data for irregular reporting areas has been converted to a regular grid and expressed as *Census population surfaces*. Chapter 11, on *ONS classifications and GB profiles*, shows how an enormous amount of census data for areas can be summarized into a set of intuitively meaningful *census typologies for researchers*, at a variety of spatial scales. Chapter 12 tackles the problem of *Dealing with the undercount* that mars all censuses and shows how revised estimates of population statistics after the census were prepared for general use. Chapter 13 summarizes the research carried out since the 1991 Census on approaches to *Updating the census*, which aim to overcome the decay over time in statistical relevance of census data.

Part IV of the book provides reviews of census *Microdata*. Chapter 14 describes the enhancements that have been effected to the *Microdata data from the census: samples of anonymized records*. Chapter 15 provides a guide to the software for *On-line tabulation for the SARs*. Chapter 16 covers the main features of the post-1991 *ONS Longitudinal Study*. Chapter 17 reviews methods that have been employed to generate *Synthetic microdata* needed in a variety of research applications.

Part V tackles *Interaction Data* (origin-destination flows). Chapter 18 summarizes the variety of machine-readable census *Migration data* now accessible, together with an account of the derived data that fill crucial gaps in the district flows created by data suppression, and which corrects for the undercount. The chapter also discusses how the improved migration information to be generated from the 2001 Census. Chapter 19 describes the *Workplace data*, its structure, derived additions, and methods of analysis.

Part VI, the concluding part of the book, looks forward to the *outputs from the 2001 Census*. Chapter 20 discusses the new questions for the 2001 Census and comments on the research opportunities in prospect. Chapter 21 reviews the careful plans being put in place to overcome anticipated underenumeration to produce *The one number census*. Chapter 22 outlines *An outputs strategy* for the 2001 Census, as currently planned. Users will need guidance about how to find particular topics and variables in the census outputs, and Chapter 23 on *Metadata*, describes ways in which this guidance will be provided. The last two chapters introduce the topic of disclosure control measures, designed to provide protection of confidentiality for census data with minimum loss of information to the potential user. The final Chapter 24, introduces new methods for *Testing user-requested geographies*, the first brick in a plan to provide much more information from the 2001 Census to users on request. Readers of this sixth section should make use of the book's website to trace developments concerning the 2001 Census data.

This chapter describes how the extensive outputs of the 1991 Census of Population in the United Kingdom have been made available to researchers in the country's universities. This has been possible because of a productive partnership between government (the Census Offices), higher education (ESRC/JISC and the Universities involved in the Census Programme), business (software suppliers, digitizing companies) and local government. The academic community has performed an important quality control function, has developed critical pieces of extraction software, and has filled in critical lacunae. Its most important role, however, has been to use the data to provide reports on the state of nation and to draw socioeconomic patterns, trends, and problems to the attention of policy makers. The ESRC/JISC Census Programme has provided part of the information infrastructure needed for the academic community to play this role. Many lessons have been learnt in the process of building this information infrastructure that has informed the consultative process leading up to the UK Census in 2001 and the dissemination strategy for the new census outputs.

Box 1.1 Key web links for Chapter 1.

http://www.hmso.gov.uk/	A Web-based licensing facility for value-added use of official data, administered by Her Majesty's Stationery Office (HMSO)
http://census.ac.uk/	Web page giving access to census data resources supported by the ESRC/JISC Census Programme for academic use
http://census.ac.uk/censusdatasystem/	The book Web pages in which the links referenced in the text will be kept up-to-date
http://census.ac.uk/cdu/	CDU home page
http://www.geog.leeds.ac.uk/	CIDS, home department web page
http://edina.ac.uk/ukborders/	UKBORDERS home page
http://les.man.ac.uk/ccsr/cmu/	Census Microdata Unit home page
http://www.cls.ioe.ac.uk/Ls/lshomepage.html	Longitudinal Study Support Programme home page
http://www.data-archive.ac.uk/	Census Registration Unit home department web page
http://www.geog.leeds.ac.uk/staff/p.rees/	ESRC/JISC Census Programme Coordinator staff page
http://www.statistics.gov.uk/	National Statistics and ONS web site
http://www.london-research.gov.uk/	London Research Centre (now Greater London Authority) home page

(continued overleaf)

http://www.open.gov.uk/gros/	General Register Office Scotland home page
http://www.nisra.gov.uk/census/start.html	Northern Ireland Statistics and Research Agency Census web page
http://www.ordsvy.gov.uk/	Ordnance Survey Great Britain Web site
http://www.osni.gov.uk/	Ordnance Survey Northern Ireland web site
http://www.open.gov.uk	UK Government web site
http://www.ukonline.gov.uk	UK Government web site

Part I

Geography and Lookup Tables

2

The Debate about Census Geography

Phil Rees and David Martin

SYNOPSIS

This chapter reviews the debate that has taken place during the 1990s about the geographical organization of the 2001 Census. It presents a range of relevant data sets and some general principles that have come to be widely used in the context of this debate. A range of alternative options for census geography is presented, and the strengths and weaknesses of each reviewed. Although it is possible to define the desirable characteristics of geographical divisions of the country for census purposes, it is apparent that no single geographical system can meet all the objectives.

2.1 INTRODUCTION

This chapter sets out to explain why the geography of the census is so important, and to review the key issues in the debate about choice of geography. Undertaking an exercise as large as a population census requires that an appropriate geographical base is drawn up, both to organize the census operation, and to present the results. The quality of the geographical base therefore has enormous impacts on the success of the data collection exercise, and on the ease with which users of the data can access the findings. While the organization of the census is largely an internal consideration for the Census Offices, the way in which the output data are organized is a subject of intense debate among users, who frequently have conflicting requirements of the geographic base. These requirements are not static, but develop and change between censuses, as does the actual distribution of population that it is sought to capture. The UK Census Offices have engaged in extensive consultation with users across all sectors concerning the geography that should be used for the 2001 Census. The discussion presented here draws on a range of contributions to the debate about census geography (such as those assembled by Rees, 1997a), while seeking to update the commentary initially developed in Rees and Martin (1997).

It is important to recognize that the geography of the census concerns not only the areas used for the release of area-based statistics but also provides the structure for the compilation of interaction data such as migration and commuting flows and the building blocks from which all other more specialized forms of census output are constructed.

Technological change has led to a shift in emphasis from a geographical base that was used primarily to specify and map census results to one that forms part of the digital geographic information infrastructure, used for many analytical and management purposes far beyond the census itself. As geographic information technologies have become more sophisticated, and particularly with the widespread availability of geographic information systems (GIS) since the late 1980s, the expectations of users, and the operating environment within Census Offices have changed rapidly. Independently of the census, large amounts of geographically referenced data about individuals are routinely collected by many organizations, and there is increasing pressure for the census to be provided in a way that allows integration with these non-census sources.

In the second and third sections of this chapter, a number of important definitions and principles are set out before moving on in the fourth section to examine various aspects of the ongoing debate about census geography. The material presented here should be read alongside Chapters 3 to 7. In Chapter 3, David Martin explores the development of the output area system to be used in 2001.

2.2 SOME DEFINITIONS

In this section, we introduce a number of important definitions that must be understood in order to follow the ongoing debate about census geography. In our introductory remarks, it was noted that census geography serves two major purposes: organization of the census data collection and organization of the output results. Conceptually, it is helpful to separate these two aspects, which can be termed the *collection geography* and the *output geography*. Traditionally, these two systems have in fact been the same, but there is no underlying reason for this to be the case, and the development of new data handling technologies makes their separation a real possibility, as discussed in the following section.

The collection geography is a geographical division of the area to be enumerated, defining those areas that are to be the responsibility of individual census employees, known as *enumerators*. In the United Kingdom, these have become known as enumeration districts (EDs), each of which is covered by one enumerator. EDs in 1991 typically contained around 200 households and 400 persons. In census geography terms, a household has a special definition concerning the relationships between a group of people (OPCS/GROS, 1992a) and is not necessarily coterminous with an address, a geographical code that identifies a particular front door—usually for the purposes of mail delivery. One address may contain many households, such as a house subdivided into bed-sitting rooms, whereas some households may have more than one address, such as a main residence and a holiday home. EDs are defined by the Census Offices according to specific criteria intended to provide approximately even workloads for enumerators, taking into account factors such as the distances to be travelled between addresses, the numbers of households expected to be found at each address, non-native language speakers, and so on. This is achieved by varying ED size, such that enumerators in 'difficult' areas have smaller EDs. The conventional way of indicating the area to be covered by each enumerator has been to provide them with a map, on which the boundaries of their ED are marked. For management purposes, the collection geography is usually hierarchical in nature, with more senior census managers responsible for the conduct of enumeration in successively larger groups of EDs.

One of the basic mechanisms of census output is the aggregation of results to standard areas. The output geography is a geographical division of the enumerated area, to which the results of the census are aggregated. In recent decades, the collection geography has been used as an output geography, with census results tabulated for EDs, although in fact there may be more than one output geography. In 1971, for example, census results were tabulated for both EDs and regular grid squares. An important prerequisite of the standard output geography is that it should nest within the statutory geography of the country in force at the time of the census. The statutory geography represents the administrative and political division of the country, and one of the basic requirements of the census is that it provides information on the characteristics of the populations falling within each of these areas. This is necessary so that aggregated results can be provided for parliamentary constituencies, local government areas, health and education authorities, and so on. This information frequently forms the basis for resource allocation from central to local government. In practice, EDs have been designed to nest within wards, the smallest of the statutory areas, and aggregated census results have been produced for the full hierarchy of administrative divisions above this.

2.3 SOME PRINCIPLES

The organization of the census, as depicted through our definitions in the preceding section, has come to embody a number of ways of working that have become accepted principles of the census. As discussed in the following, some of these (such as the confidentiality requirement) are enshrined in census legislation, and others (such as the trade-off principle) are practical devices used in the operationalization of the census, which help to ensure the legal requirements are met.

The motivation for data aggregation to standard output areas is one of the guiding principles of the census, that of confidentiality. The confidentiality principle seeks to prevent the inadvertent disclosure of information about any identifiable individual in the published census results. This has necessitated the use of a number of confidentiality mechanisms, including thresholding and data modification. No aggregated results may be released for areas falling below defined thresholds. In 1991, these were 16 households and 50 persons for the ED-level small area statistics (SAS), and 320 households and 1000 persons for the local base statistics (LBS). If a census output area is found to be below threshold, then it is known as a restricted area, and its results must be exported to one or more neighbouring areas, until the population total exceeds the threshold. As a further confidentiality measure, a degree of data modification is undertaken by pseudo-random modification of the tabulated counts for each area before publication. An understanding of the confidentiality principle allows us to understand the issue of differencing output geographies. Although great care may be taken over the confidentiality of a single geography, problems occur when more than one output geography is produced because it is theoretically possible for the two geographies to be overlaid in order to deduce data for areas that fall below the thresholds and hence violate the confidentiality principle (Duke-Williams and Rees, 1998). This issue has become more pertinent with the widespread use of GIS, allowing users to manipulate alternative boundary sets with ease.

The trade-off principle is that the amount of published socioeconomic detail increases with the size of the geographical area under consideration. At the very smallest geographical level, the directory of EDs and postcodes (OPCS/GROS, 1992b) provided a household count for each unique intersection of an ED and a unit postcode (the smallest division of the postal geography). At the next geographical level, the extensive data set available for

EDs in 1991 was known as the SAS. For wards, local authority districts, and above, a still more detailed set of tabulations was released, known as the LBS. The rationale for this trade-off is that it helps to ensure census confidentiality by 'hiding' individuals and households with unusual socioeconomic characteristics within larger aggregations of population.

In Scotland, in 1991, practice differed somewhat from that in the rest of the United Kingdom in that output areas were created on the basis of aggregations of unit postcodes, which differed from EDs, allowing a greater degree of continuity with the EDs used for data publication in 1981 (Clark and Thomas, 1990). Postal geography throughout the United Kingdom is designed independently of the census and statutory boundaries, and even unit postcodes may therefore straddle statutory boundaries (for a full discussion of postal geography, see Raper *et al.*, 1992). The aggregation of unit postcodes to wards therefore involved a degree of approximation because the published figures for each ward represent the totals for the best-fitting aggregation of postcodes and not the precise count of population falling within the ward boundary. Outside Scotland, the approximation principle has not been considered acceptable for statutory purposes, and this is relevant to our subsequent discussion of output area design. These observations lead to another principle of concern to many users, that of consistency in census geography definition across all parts of the United Kingdom.

A lookup table is a device that helps users to overcome the approximation problem and these are discussed more fully in Chapter 5. Lookup tables such as the 1991 directory of EDs and postcodes indicate the overlap between non-matching geographical units, and can be used even where one of the units (as in the case of postcodes) does not have defined geographical boundaries. For each postcode, its ED membership is given, together with an indication of the size of the overlap expressed as a number of households. This information may be used in order to provide weighting for the estimation of ED-level socioeconomic characteristics down to the postcode level, or to assign each postcode to an ED on a 'best fit' basis. Lookup tables also provide a means to assist users with non-matching geographic problems such as dealing with changes in units between censuses. Atkins *et al.* (1993) describe a lookup table that provides best-fit linkages between 1991 EDs and 1981 wards. Barr (1993) notes that many GIS operations involving census data can be reduced to the processing of lookup tables if the relevant information is available in this form.

A final principle that has emerged during the 1990s is that of delivering the census as a GIS. Openshaw (1995b) is particularly direct in his criticism of the 1991 outputs in their 'neglect of GIS'. 1991 Census geography in England and Wales was created by manual cartography, and only subsequently digitized by commercial organizations external to the Census Offices. During the intercensal period, there has been a growing number of applications that deliver integrated census and digital map data (Martin *et al.*, 1998). For 2001, with the widespread availability of desktop GIS and web-mapping applications, it seems essential that users have the option of obtaining integrated census statistical and boundary data directly in digital form.

2.4 ELEMENTS OF THE DEBATE

2.4.1 Separation of Collection and Output Geography

Historically, the practicalities of census taking have restricted the census output geography to be either the same as or a simple aggregation of the collection geography. The basic enumeration area in England and Wales was the parish before 1841, and data were published at the level of the ancient county, thus making use of areas originally defined for ecclesiastical and administrative purposes. From 1841, census-specific EDs were used, and the size of the areas for which data were published was reduced to the local authority level (Mills, 1987). Computer processing of the results since the 1960s has allowed reduction of the smallest output area to that of the ED itself. Nevertheless, the use of the same ED-based geography for data collection and output represents a trade-off of conflicting priorities, in which the considerations of the data collection exercise have been predominant (Denham, 1980). The use of GIS for census geography management (Martin, 1998a) has for the first time made the complete separation of collection input and output geographies an entirely viable technical option, drawing both geographies from the same underlying GIS database, but designing them for different purposes.

2.4.2 Basic Spatial Units

An issue that follows from the potential separation of collection and output geographies is 'which output areas should be used?' Previously, with output based on EDs, this was not a relevant consideration, but the opportunity to design a new set of output areas has provoked much discussion, in particular as to the smallest geographical areas for which census data should be published. The key difficulty here is that different users require aggregations that are based on completely different criteria. Many census users are now heavily committed to other georeferencing systems and integration with these non-census applications is at least as important as the establishment of another new (incompatible) geography for 2001 data output. Growth in the use of postcode- and address-based information has played a significant role in the development of the present situation (Raper *et al.*, 1992; Martin and Higgs, 1997).

For many purposes the most attractive option will be to build areally aggregated data sets that 'fit' into widely used standard geographies, and it is therefore necessary to consider the most appropriate geographical objects from which such output geographies can be defined (termed basic spatial units, or BSUs by Openshaw, 1990). Candidates clearly include the individual address, the unit postcode, aggregations of postcodes, 1991 or 2001 EDs, grid squares, and others, including hybrid objects constructed from the intersection of these basic building blocks. Examples of the main contenders are given in Figure 2.1(b)–(f), each based on the same (hypothetical) residential area shown in Figure 2.1(a).

Individual address references such as those provided by Ordnance Survey's ADDRESS-POINT are illustrated in Figure 2.1(b). The confidentiality principle prevents the use of individual addresses as output units, but the address provides the most flexible basic unit for aggregation to larger areas. Address referencing forms the basis for the functioning of many organizations who are current or potential census users, and a difficulty with the 1991 output has been the lack of any readily available means of placing individual addresses unambiguously within EDs for the majority of users without access to ADDRESS-POINT and digital ED boundaries. Not only has the use of address-level applications increased during the 1990s, but may be expected to grow still further beyond 2001, as organizations increasingly enhance their address databases to be consistent with British Standard 7666 (British Standards Institute, 1994), defining a standard for address and street referencing that offers the potential for far easier address matching than at

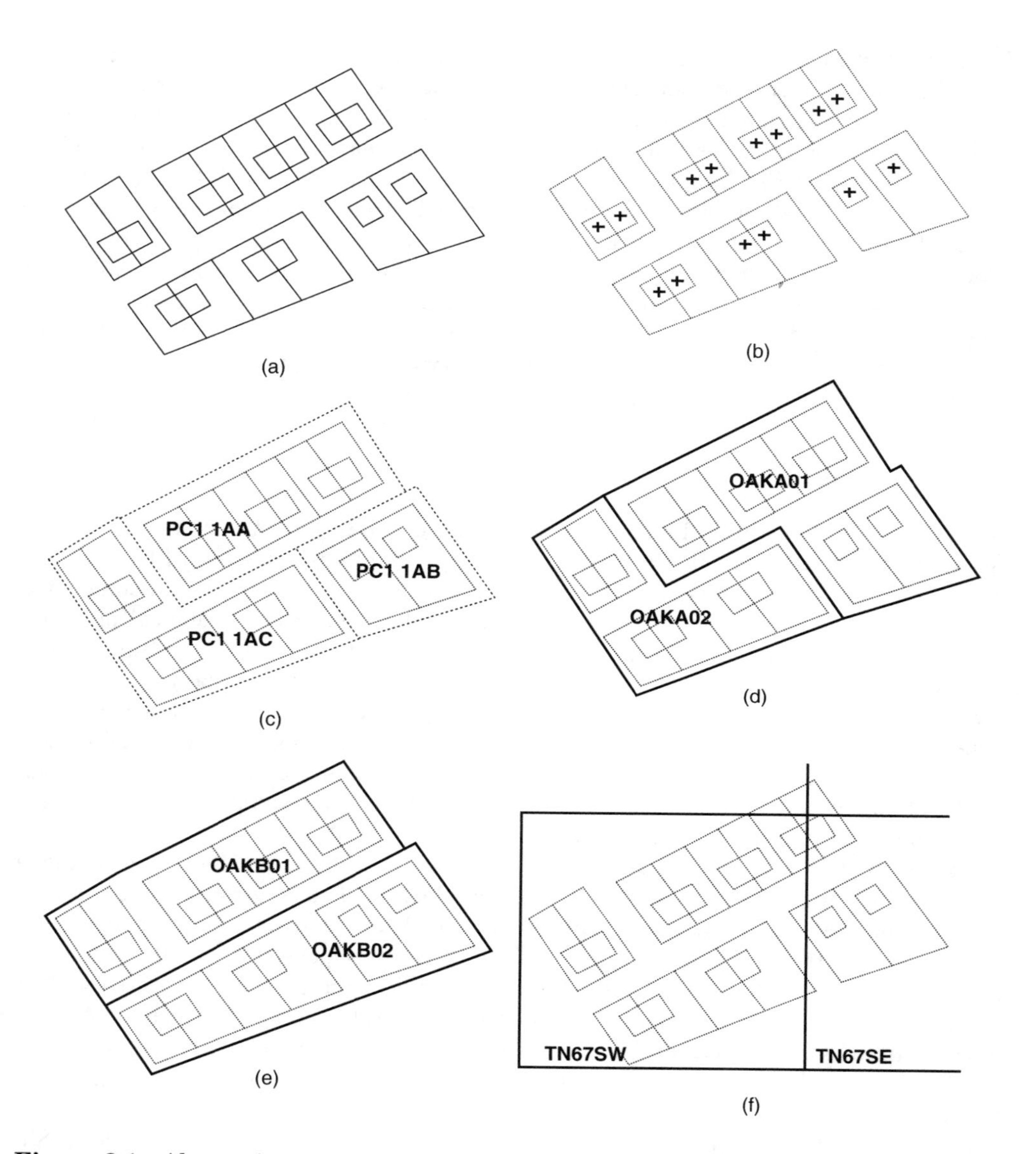

Figure 2.1 Alternative census geography models: (a) residential geography, showing property and building outlines; (b) individual address referencing; (c) unit postcode boundaries (undefined at present); (d) census output areas based on aggregated postal geography; (e) census output areas independent of postal geography; (f) output based on grid squares.

present. There would therefore be considerable demand for any product that allowed the automatic assignment of addresses to one or more of the standard output geographies.

There are around 1.7 million unit postcodes in the United Kingdom, typically containing 14 postal addresses, and these are the most widely used and understood small area references. The postcode forms the basis for the operations of many census-using organizations, and it is therefore important that there should be some clear relationship between census output and the postcode system. In 1991, this has been provided by the ED/postcode directory, a lookup table, which gives the count of households falling within each unique ED/postcode intersection, and allocates each postcode into a 'pseudo-ED', representing the ED in which the majority of its population lives. There is strong user demand for a geographical system based on postcodes (Dugmore, 1996; Rees, 1998a). There is still no single definitive set of areal boundaries, such as those illustrated by Figure 2.1(c) for unit postcodes in England and Wales, although these are available in Scotland. The creation of such boundaries by automated means is now being undertaken both by Office for National Statistics (ONS) for census purposes and by commercial organizations, although some important design considerations remain unresolved, such as: 'should boundaries incorporate higher level administrative boundaries or major physical features?' and 'should they cover the entire land area?' The adoption of slightly different rules by different agencies means, for example, that the unit postcode boundaries created by the Census Offices will not be coincident with those generated by others. As noted in the preceding text, the use of any postcode-based system requires some degree of approximation of counts to statutory areas if the unit postcodes are not to be split. Most fundamentally, it is unclear whether the requirements of statistical data collection should be allowed to have an impact on the geographical definition of unit postcodes, which are created and operated by Royal Mail for the smooth working of the postal system, and who have no direct interest in census applications as discussed here.

The unit postcode has some considerable attractions as a basic building block for other output geographies, such as that illustrated in Figure 2.1(d). Postcodes provide a convenient intermediate zone for which household data may be aggregated, and then further aggregation rules may be applied in order to form groups of postcodes that correspond to higher level geographies. The trade-off principle would suggest that it should be possible to provide a small number of basic counts for the postcodes themselves, with increasing detail at higher levels of aggregation. Unit postcodes form the basis for the 2001 output area design proposals described in Chapter 3.

A further output option, illustrated in Figure 2.1(e) involves the use of census-specific small areas that are not based on postcodes but which aggregate neatly to wards. These might either be entirely new areas, or based heavily on the 1991 ED boundaries to permit analysis of change and to avoid unnecessary effort for users in moving between one essentially arbitrary output geography and another. Where extensive new development has taken place, 1991 EDs would need to be subdivided to provide smaller geographical units meeting 2001 confidentiality constraints. This 'new' geography would therefore be known in advance to users, and a small 'update' of modified boundaries would allow existing users to convert to the new geography immediately. Implicitly, such a route would set up the 1991 boundaries as a 'frozen' census geography, but a major difficulty is again the requirement to nest within statutory areas. Because the statutory geography is itself subject to constant revision, this forces a large number of changes to lower level boundaries, making the continuity of small areas between censuses rather illusory.

An alternative frozen geography is aggregation to regular grid squares such as those used in 1971 (CRU/OPCS/GROS, 1980), and illustrated by Figure 2.1(f). These have the advantages of permanence and the absence of a digital boundary requirement, but suffer from high rates of suppression as a result of the wide range in population densities, and the inability to recombine them into any other standard zonal geography. Such an aggregate geography would be very easy to produce from an ADDRESS-POINT database, and should perhaps be reinstated as a geography for intercensal comparison, avoiding many of the interpretation problems inherent in the irregular zonal systems. Grid square geographies have many attractions but suffer from being completely incompatible with the needs of many users, and are therefore unlikely to ever emerge as the single compromise system preferred by the majority. Various approaches to the spatial remodelling of census data, such as those proposed by Martin (1989) and Langford and Unwin (1994) offer the possibility of constructing grid square systems from the small area outputs without some of the accompanying constraints of direct data aggregation, and these are discussed more fully in Chapter 10.

2.4.3 Multiple and Flexible Output Geographies

In the past, the manual boundary definition process used in census geography design has been a major obstacle to the preparation of multiple output geographies, but the existence of address-level referencing in 2001 offers the potential to create multiple aggregated geographies automatically. Whereas time and personnel constraints made it impossible to consider the creation of more than one geographic base, the computation of further geographies, once the elemental units exist in digital form, is entirely feasible. The constraints are therefore no longer primarily technical ones, but are concerned with the societal and administrative issues of confidentiality, production costs, and user demand. For the first time, it is technically possible to create a number of different census output geographies at relatively low cost, each of which is well suited to a particular purpose—an idea developed further by Martin (2000). For example, there would be no technical obstacle to the creation of area statistics for 1991 EDs; 2001 EDs; output areas comprising groups of postcodes, and for grid cells simultaneously.

Looking towards the future, perhaps the most fundamental issue concerns the possibility of creating flexible output geographies. Flexible geographies are those that might be defined by users, for which custom data aggregation is required. An example would be a set of sales areas defined by a company without reference to the existing census geography. Discussion of this issue requires a distinction to be drawn between safe data and data in a safe setting. Safe data, such as the 1991 SAS and LBS are released only after data protection measures have been applied to ensure confidentiality. Data in a safe setting include the full census database before aggregation, held by the Census Offices but not released to users. Safe data for publication are derived from data in a safe setting. In the past, this has been a one-off procedure in which standard data sets have been prepared for standard geographies. Flexible geographies would require a data aggregation and control system that could assess the safety of any requested geography and derive safe data for output. Some initial steps towards such development are presented in Rhind *et al.* (1990).

The outstanding obstacle to the production of multiple and flexible output geographies is the problem of differencing, discussed in detail by Duke-Williams and Rees (1998) in Chapter 24. They present a series of experiments that allow the comparison of pairs of geographies in order to test whether it is indeed possible to determine counts for subthreshold areas. Their recommendation is that only one small area geography be produced, perhaps constructed from small groups of postcodes, and separate aggregations of statistics to larger administrative areas, and to a large mesh grid (5 km) would also be acceptable. They consider a second set of small areas to be unsafe due to the differencing problem. The implication of this work is that multiple small area geographies and flexible geographies are unsafe unless all combinations could be evaluated and assessed and declared safe by the aggregation system.

2.5 CONCLUSION

It should now be apparent that the debate about census geography is complex and far-reaching. It concerns both fundamental and poorly understood statistical issues such as the differencing problem, set in continually evolving data and computing environments, in which products and procedures that were impractical for a previous census can become viable and practical solutions at the next. Further, as with many aspects of census design, decisions made in the years preceding the census itself will have to bear scrutiny from the very different perspectives of the following 15 years, until data from a subsequent census are released. Coombes (1995) provides a helpful set of seven tests against which any census geography should be evaluated:

1. Are the building blocks the smallest that the confidentiality restrictions deem to be possible (to allow maximum flexibility of aggregation)?
2. Is each of the areas in a set of building blocks, or other set of areas, defined on a consistent basis across the country?
3. Does this set of areas represent (parts of) 'real world' entities, such as settlements, which can be recognized using these boundaries?
4. Does the set of areas allow comparison with previous census(es) at this level, or for some minimal grouping of areas to create consistent boundaries?
5. Does the set of areas cover the whole country without leaving any locations whose data are too sparse to allow them to be published?
6. Are the boundaries of these areas available in digital form?
7. Can these areas be readily and accurately linked by their location coding to the areas used in many non-census data sets?

It will be apparent from the preceding discussion that it is not possible to devise any single geographical division of the country that can meet all the conflicting requirements of the census. Even for the purpose of data collection taken in isolation, there is no single solution that meets all objectives, and user requirements of output geographies are far more complex. We have seen how all census geographies therefore represent a degree of compromise. Nevertheless, new tools and technologies are available for the creation of 2001 Census geographies that far exceed the capacities of those available for previous censuses, and which offer a reasonable expectation of passing or closely approaching Coombes' tests 1, 2, 5, 6, and 7. Aspects of the 2001 developments mentioned here are discussed further in the following chapters. It can be observed that a number of the conceptual and technical developments first initiated in the discussions preceding past censuses are likely to come to fruition in the implementation of the census in 2001. It is therefore perhaps a reasonable expectation that some of the more exciting possibilities discussed in 2001 but currently infeasible for various reasons will be implemented in the early censuses of the new millennium.

Box 2.1 Additional CDS resources for Chapter 2.

WP9701.pdf	Text of Rees (1997a), an edited collection of papers on the geography of the 2001 Census, which influenced Census Offices' thinking on designing output areas for the 2001 Census

3

Output Areas for 2001

David Martin

SYNOPSIS

This chapter describes the plans for 2001 Census geography and geography-related products. Following a brief discussion of the way in which census geography has been managed in the recent past, the case for the separation of collection and output geographies is presented. Each is then discussed in some detail, with reference to the computer systems being put into place for their creation and management. The chapter concludes by identifying the principal implications for census users of the developing 2001 approach to census geography.

3.1 INTRODUCTION

As discussed by Rees and Martin in Chapter 2, the basic geographical unit of the 1991 Census was the enumeration district (ED). EDs were the elemental units used in England, Wales, and Northern Ireland, both for data collection and publication, and may be aggregated directly to form all higher level statutory boundaries. In Scotland, EDs were used for data collection, but a separate geographical division of the country into output areas (OAs) was used for data publication. The Scottish geographies were both constructed from a digital boundary set representing unit postcodes, the smallest geographical units in the postcode geography. In the rest of the United Kingdom at the time of 1991 Census planning, not only were there no digital boundaries available for EDs or unit postcodes, but unit postcodes themselves were defined only as lists of addresses and had no definitive geographical extents. 1991 ED planning was undertaken manually on the basis of paper Ordnance Survey maps and a variety of ancillary data sources in order to produce the 1991 census geography as hand-drawn lines on paper base maps, which were photocopied for the organization of enumeration. These maps were subsequently digitized commercially, forming the basis for the digital boundary products that have been widely used throughout the 1990s.

All this must be seen in the context of growing calls for a closer integration between census and postcode geographies since the late 1980s, embodied in the report of the Chorley Committee on the handling of geographical information (DoE, 1987), and subsequently echoed by many users (Raper *et al.*, 1992; Dugmore, 1996), particularly because postcodes have become an increasingly widely used georeferencing system for other sources of socioeconomic information. Other aspects of the 1991 geography model have proved problematic to users: the separate creation of the digital geographical base and small area statistics (SAS) allows some mismatches to occur between the two data sets, and presents the user with a substantial data manipulation task in simply integrating the census data with the corresponding boundaries (illustrated by Charlton *et al.*, 1995). The design of EDs in order to even out enumerator workloads leads to significant variations in population size and geographical area and shape, which can make direct interpretation of the tabulated values difficult. In the 1991 data, EDs are suppressed, in which the population or household totals fall below the thresholds of 16 households and 50 persons, requiring exporting of the counts to another nearby ED.

This chapter describes the approach that is being taken to address these problems for the 2001 Census. Perhaps the most important aspect is the separation of the collection and output geographies and the use of computer-assisted design procedures in a geographical information system (GIS) context. These developments have only been made possible by the emergence of a number of important new digital geographical products during the early 1990s that provide the necessary data infrastructure to support such an approach. As will be seen, the proposed approach is closer to the model adopted in Scotland in 1991, but, in addition, requires the creation of unit postcode boundaries. The following sections deal, respectively, with collection geography and output geography. Having introduced the two components of the geography system in this way, we go on to discuss some of the more indirect implications of the proposed 2001 geographical base for census users.

3.2 COLLECTION GEOGRAPHY

In this section, we address the characteristics that are required of a census collection geography, initially by looking at those aspects of the design task that are common to the 1991 and 2001 Censuses. We then focus in more detail on how the 2001 collection geography has been designed. The objective of census collection geography is to provide a framework for the organization and management of enumeration, in such a way that enumerators' workloads are equalized as far as possible. Enumeration difficulty varies considerably by type of area. For example, suburban housing tends to be relatively easy to enumerate, with properties of conventional design and layout, with obvious means of access and little difficulty in identifying addresses. There will usually be only one household per address, and there is a tendency for these populations to be compliant in the completion of census questionnaires. Some types of neighbourhood, however, such as many inner city areas and remote rural areas, present particular obstacles to enumeration. In the inner city case, there may be many subdivided properties, with idiosyncratic layouts, unconventional numbering, concealed access, and many households behind each front door. In remote rural areas, enumerators may have to cover extremely long distances between addresses. The Census Offices' approach to the management of this wide variation in neighbourhood types is to grade EDs according to anticipated enumeration difficulty, and then to adjust ED size so that enumerators of difficult EDs have to cover fewer households (Clark, 1992). In order to assist enumerators in the identification of their areas, boundaries are drawn wherever possible along obvious topographic features such as roads, rivers, and railway lines. This complex task makes use of information from the previous census, estimates of household numbers and geographical area, and additional information on new residential development or demolition provided by local authorities.

If the collection geography is also to be used as the output geography (as in 1991), then the areas created must, in addition, respect all parish and ward boundaries in force at the time of the census. In 1991 this complex job was undertaken by Census Office staff working from 77 000 paper base maps (usually at 1:10 000 scale), and 400 000 copies were required for use by field staff (Clark and Thomas, 1990).

For the 1997 Census test, an automated approach to collection geography design was adopted by the Office for National Statistics (ONS), making use of digital data sets that had been created in the early 1990s, and using a GIS as the design environment. The system, known as the geographical area planning system (GAPS) is an ARC/INFO and ORACLE application, commercially customized for use by ONS. The system is outlined schematically in Figure 3.1. Extensive use is made of digital geographical data in replacing the paper resources that formed part of 1991 ED planning. The Ordnance Survey raster 1:10 000 images provide background information for the planner, which is the direct digital equivalent of the paper base mapping used previously. 1991 ED geography is provided by ED-Line, one of the products commercially digitized from the original Office of Population Censuses and Surveys (OPCS) maps. This information is useful because it is desired to change the enumeration geography as little as possible, and ED-Line thus provides a ready-made digital solution for 2001 in areas where no changes

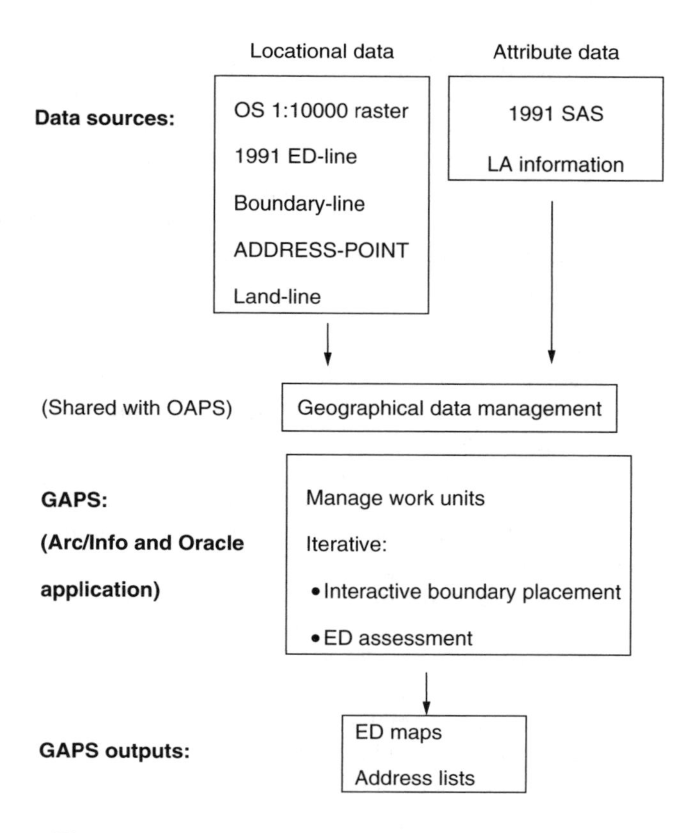

Figure 3.1 The geographical area planning system.

are necessary. Unfortunately, statutory boundaries are subject to continual revision, and all levels, including the latest ward and parish boundaries are provided by Ordnance Survey's Boundary-Line product. These are the boundaries to which census outputs must aggregate neatly, and it is therefore simplest (although not strictly necessary) if EDs are neat subdivisions of these areas, as in 1991. The locations of individual postal addresses are given to 0.1 m by ADDRESS-POINT, which also serves to help identify areas where significant residential development has taken place since 1991. If these changes are not reflected in the 1:10 000 mapping, then more up-to-date Land-Line information is used to provide additional topographic detail. Selected variables from the 1991 SAS are held in the system so as to enable estimates of enumeration difficulty to be made. These include indicators of multi-occupancy and a language other than English as the main language.

GAPS provides tools for the management of the national ED planning database, and many of the geographical data sources are shared by the output planning system described in the following text. Processing begins by automatic assessment of a work area (typically a ward), to ascertain whether any 1991 EDs need to be redesigned. Redesign is required if any statutory boundary changes have taken place, or if an ED falls outside the acceptable size range for its estimated degree of enumeration difficulty. The actual target population sizes to be used for EDs in 2001 will also be affected by the use of the postal system for the return of census questionnaires, with many enumerators responsible for the equivalent of more than one 1991 ED. The estimate of enumeration difficulty takes account of the selected SAS variables and the ED's area and estimated number of households. Areas that may have experienced major population increase are identified by comparing the address counts that may be derived from the 1991 SAS count and ADDRESS-POINT. Where adjustment is required, planning proceeds by interactive boundary placement and editing by the ED planner, with the ED boundaries as the edit coverage and raster mapping, statutory boundaries and address locations displayed as a background. ED boundaries are automatically clipped to statutory boundaries, and new estimates of enumeration difficulty computed, which may indicate that further cycles of boundary redesign are necessary. Only when all EDs fall within the acceptable ranges can the work area be checked into the 2001 database.

For the purposes of enumeration, GAPS offers two important new outputs that were not directly available in 1991. The first of these is the provision of single-sheet customized maps for enumerators with ED boundaries clearly indicated against background mapping at an appropriate scale. The ORACLE database containing the full address information from ADDRESS-POINT is then used to produce enumerators' record books with all known addresses pre-printed. In 1991, enumerators were provided with blank record books and had to write in each address located within their ED. The use of GAPS cannot guarantee that no addresses will be missed, but serves to improve the resources available to enumerators and thereby decrease the likelihood that addresses will be missed or visited twice. A potential disadvantage is the existence of situations in which the map and address list given to an enumerator are not in complete agreement because of inaccuracies in the ADDRESS-POINT database.

The use of GIS in collection geography design in this way represents the automation of the 1991-style design process without any particular innovation in methodology, but nevertheless brings important advantages to the Census Offices. A broader model of this geographical data handling as part of the increasing automation of census processing tasks is presented in Martin (1998b), but this theme will not be pursued further here. Most importantly for census users, the development of GAPS has provided the geographical data management infrastructure within which the separation of collection and output geographies has become possible, and from which a wide range of digital geographical products may potentially be derived. Indeed, an increasing number of different 2001 census 'geographies' are emerging for specific purposes (Martin, 2000).

3.3 OUTPUT GEOGRAPHY

In 1991 and previous censuses, the collection geography has been used as the standard output geography in the United Kingdom, with the exception of the Scottish case noted earlier. This has occurred for a number of reasons, not least the enormity of the task of creating an alternative geography by manual means and of cross-referencing the census results to the alternative geographical base. In 1971, each census return was related to a 1-km grid square, permitting output both by EDs and on a regular grid. Although much data suppression was necessary, the grid square mapping found in CRU/OPCS/GROS (1980) was a unique and innovative output. Unfortunately, the same convention has not been adopted in subsequent censuses and it has not therefore been possible to take advantage of the grid square structure in order to map and monitor population change. In 1981, the only geographical identifier associated with each census return was the ED. 1991 saw the addition of the postcode, and subsequent to the publication of the small area statistics and local base statistics, an additional area data set was created for postcode sectors, which contain on average 2700 addresses. The output of data at the unit postcode level was prevented by their small size (on average 14 addresses), and these could not be combined into larger areas because of the lack of any definitive boundaries that define which unit is adjacent to which.

The key difficulty associated with multiple output geographies created from small base areas is the fear that it may be possible to overcome census confidentiality measures by differencing, an issue dealt with in detail by Duke-Williams and Rees (1998; Chapter 24 of this volume). It is therefore necessary to adopt a single standard output geography, and for 2001 this will more closely follow the 1991 Scottish model, comprising specially constructed OAs that match as far as possible to postcode geography and that are separately designed according to a different set of objectives to those set for collection geography.

The ideal characteristics of an output geography are rather different to those of a data collection geography, discussed in the preceding text. For output, the difficulty of enumeration is not a relevant consideration, and it would be more useful to have OAs that are of approximately equal population size. Uniformity of population size aids the interpretation of rates and proportions by equalizing the population denominator. Further, if population size is to be controlled, then below-threshold areas should be avoided, so that there are no areas for which statistical data are suppressed. For many purposes, it is still desirable that boundaries should fall along recognizable features on the ground, and for mapping purposes uniformity of geographical area and shape would be desirable. Possibly more important than coincidence between OA boundaries and physical features is coincidence with important social divisions, that is, it is very helpful to many analyses if distinct social neighbourhoods are contained within, rather than split between, OAs. Morphet (1993) demonstrates the extent to which this is not achieved by EDs. The demand from many census users for ease of integration between the census and other geographical

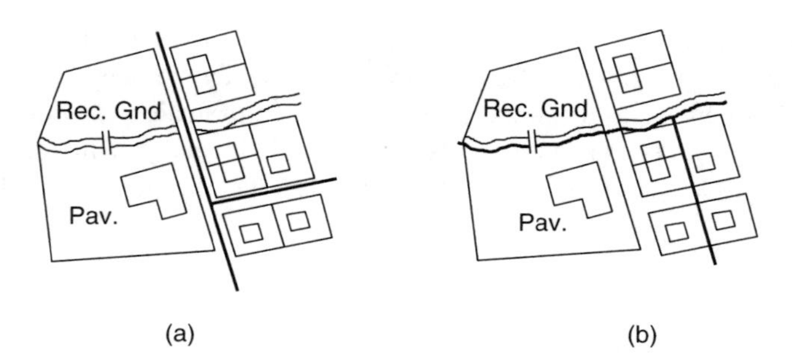

Figure 3.2 Illustrations of possible (a) ED and (b) OA boundary placement for the same small urban area.

systems, particularly the postcode system, has already been noted, and the requirement that census OAs should aggregate neatly to all statutory areas also remains. It can be seen therefore that the ideal specification of a census output geography is just as complex, although different in many important respects, as that of a census collection geography. Figure 3.2 illustrates the extent to which this may lead to differences in boundary placement in a small (hypothetical but typical) urban area comprising a recreation ground on one side of a street with a residential area on the other. The area is crossed by a small stream. For enumeration purposes it may be simplest to divide the area as shown in Figure 3.2(a), in which the road centrelines are used as the boundaries between different EDs, and each street block on the map falls into a different ED. For output purposes, however, the stream may form part of a ward or parish boundary, which must therefore be followed by OA boundaries, and the houses facing onto the main road are likely to have a different unit postcode to those in the minor road. The boundaries may therefore most appropriately fall along the stream and rear fences as shown in Figure 3.2(b). In this simple example, there are no common boundary components at all between the collection and output geographies. We shall now consider the system that is being implemented to address these requirements in the creation of 2001 output geography.

The prototype 2001 Output Area Production System (OAPS) has been built using the geographical data infrastructure already provided by the collection geography design system, GAPS. OAPS is illustrated in Figure 3.3. A number of the data sources for OAPS are shared with GAPS and they have been constructed as different AML applications with common ARC/INFO workspaces. The basic methodology for OA design is to combine unit postcodes in such a way as to produce geographically compact OAs that are of similar population sizes and socially homogeneous. In order to ensure that the best available population counts are used, this process will be undertaken after 2001 enumeration. The unit postcode is thus the building block of the output geography (as already the case in Scotland), and it is necessary for unit postcode boundaries to be created in the first instance. Unit postcode boundaries will be created by ONS in advance of 2001 OA planning, as a first stage of the OAPS process. This stage is outlined by dotted lines in Figure 3.3.

The absence of unit postcode boundaries was the major obstacle to the integration of census and postcode geographies for the 1991 Census (Wrigley, 1990), but the digital geographical products now available make possible the direct derivation of a set of unit postcode boundaries from existing products, albeit subject to a number of important

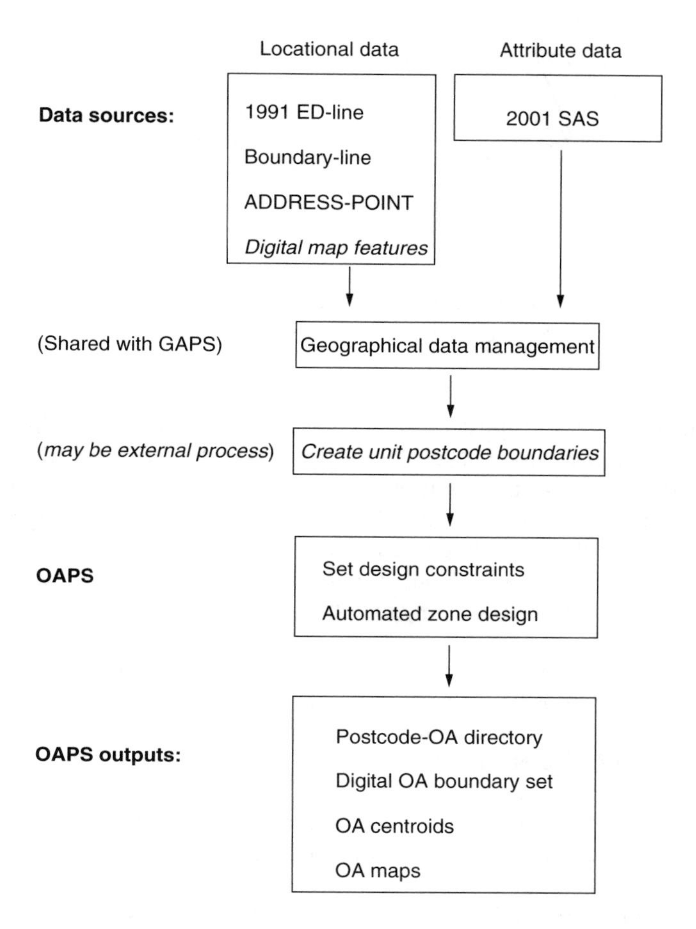

Figure 3.3 The Output Area Planning System (OAPS).

design criteria. At present there are no areal extents associated with addresses, and there is no national coverage of basic land and property units (BLPUs) (Pearman, 1993). It is therefore necessary to create synthetic parcels that may be aggregated to form unit postcode polygons. Thiessen polygons (Boots, 1986) are readily constructed around each ADDRESS-POINT location, and adjacent polygons sharing the same postcode may be merged to form postcode polygons. If conducted without constraints, this process results in highly irregular polygon boundaries in many types of neighbourhood, although in regularly laid-out urban areas the results may be quite acceptable. Nevertheless, it is necessary to constrain the process to respect statutory boundaries, as these must be followed precisely in the census outputs. A clip coverage is therefore created comprising ward and parish boundaries, and such additional digital map features (such as road centrelines and railway lines) as it is considered appropriate to incorporate into OA boundaries. Thiessen polygons around individual addresses are then intersected with the polygons of the clip coverage, and merged if they share a postcode and fall within the same statutory area.

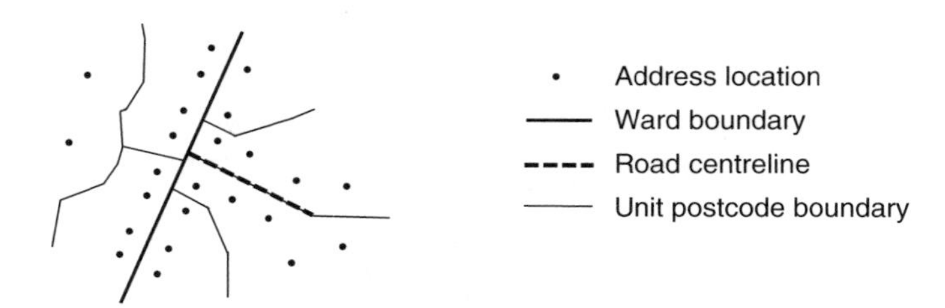

Figure 3.4 Unit postcode boundary design, illustrating clipping to ward boundaries and road centrelines.

Spatially coincident 'stacked' postcodes (such as those occurring in multilevel buildings) will be assigned into the same polygons, and other specific technical 'fixes' are required to accommodate other local irregularities in the postcode system. The solutions adopted in each case are the result of user consultation and experimentation. Figure 3.4 demonstrates this process applied to a hypothetical urban area in which Thiessen polygons are clipped along the line of a ward boundary, and also to a minor street centreline, and then merged to form unit postcode polygons that fall neatly on either side of the clipping features.

In Scotland, a small degree of approximation has been considered acceptable at ward boundaries, so that unit postcodes need not be split, but in England and Wales it will be necessary to create two separate polygons for any postcode falling across a ward boundary. On the basis of the 1991 directory of EDs and postcodes, 2.6% of unit postcodes are split across ward boundaries, compared with 14.9% split across ED boundaries, so the degree of approximation in the census to postcode geographies will be much less than in 1991, using this approach. In Scotland, a rather different approach to the creation of 2001 OAs has been taken to that described here, more strongly focused on maintaining consistency with the OAs used in 1991. Clusters of adjacent postcodes will be created that share the same membership of higher level areal units. If these clusters are large enough, they will be subdivided, otherwise they may remain as OAs in their own right or be merged, introducing increasing levels of approximation to the higher level boundaries. Northern Ireland will follow the same general procedure as England and Wales, but with postcode polygons created manually as opposed to the Thiessen-based procedure described in the preceding text.

The core of OAPS is the automated design of 2001 OAs from the unit postcode (and part unit postcode) building blocks provided, as discussed earlier. The unit postcode boundaries will be prepared so as to match as far as possible those in place at the time of the census, and immediately provisional area statistics are available for these building blocks, OA design can begin. This operation will proceed once all one number census (ONC) processing, described in Chapter 21, is complete. The adjacency matrix, showing the building blocks that have common boundaries, together with their boundary lengths, centroids, and areas are all provided by the GIS, and combined with the enumerated counts, allow iterative recombination of building blocks into OAs in an implementation of Openshaw's automated zoning procedure (AZP). The original AZP is described by Openshaw (1977a), and an application to the grouping of 1991 EDs is discussed by Openshaw and Rao (1995), who also explore different algorithms governing the iteration process.

Zone design proceeds by the initial random grouping of adjacent building blocks into approximately the desired number of OAs within a work area. A number of summary measures describing the overall differences from a target population size, irregularity of shape, and homogeneity within each OA may be computed. In particular, more sophisticated measures of homogeneity than those initially used in the 1997 census test experiments (Martin, 1997) are discussed by Tranmer and Steel (1998), and more sophisticated mechanisms for the combination of different criteria are considered in Chapter 4.

3.4 CONCLUSION: IMPLICATIONS OF THE 2001 GEOGRAPHY MODEL

For census users, a number of important implications may be identified from the new methods described here for handling the geographical bases of the 2001 Census. Firstly, the 2001 output geography will be closely related to unit postcode geography, making it much easier to relate socioeconomic data georeferenced by census and postcode systems. Even where postcodes are split, there is a greater likelihood that the splitting boundary will correspond to a major topographic feature existing in another key digital geographical data set. Further, the smallest census OAs themselves will be more numerous than 1991 EDs, and display greater equality of population size, and greater internal homogeneity. Conversely, boundaries in unpopulated regions will be more likely to divide space in abstract geometric figures rather than following some arbitrarily chosen ground features that are unrelated to population distribution. With the possible exception of sub-threshold parishes and ward areas, which will need to be combined, there is no reason for there to be any sub-threshold OAs created. Digital boundaries for OAs will be supplied directly in digital form by the Census Offices, and will share a single source with the various directory products and (SAS) themselves, resulting in a lower rate of mismatch between the various data resources. The actual costs of these new geographical products will be related not only to the costs of the census but also to the extent to which existing digital data sets are incorporated within them.

In the longer term, the existence of all the relevant geographical bases within a GIS at the Census Offices provides the starting point from which the management of all subsequent boundary changes may proceed. Effectively, GAPS and OAPS provide the basis for a national population GIS. Although the detailed information in such a system would remain confidential, it provides the Census Offices with the capacity to manage all subsequent changes in directories and lookup tables, and to provide 2001-based population estimates for the new geographical areas created by future boundary change. It is extremely difficult to make predictions concerning the likely conduct of censuses beyond 2001, and their probable geographical base. Nevertheless, in an environment in which all the detailed geographical information is held within a GIS, and in which the OAs are part of an entirely separate system from the collection geography, it is a reasonable expectation that the task of managing change between future geographies may be considerably easier than it has been till date. The adoption of address-referencing as a central component of ONS' geographical referencing strategy, and increasing demand for neighbourhood statistics should seek to ensure that the use and quality of georeferenced data within the Census Offices will continue to increase rapidly post-2001. As in Scotland in 1991, the necessary changes to collection geography do not inevitably force corresponding changes in output geography, and where these must be made, their relationships with past geographical areas may be directly calculated. The future of census geography rests in increasing geographical flexibility and in value-added digital products, each incorporating a major component derived from existing geographical bases.

Acknowledgements

Various aspects of the developments described in this chapter have been undertaken by ONS, and by the author in collaboration with ONS, and the work has been further supported by economic and social research council (ESRC) Award H507255154. Grateful thanks are extended to the various ONS staff whose input is reflected here. Responsibility for all views expressed here remains with the author.

Box 3.1 Key web link for Chapter 3.

http://www.geog.soton.ac.uk/research/oa2001	Website of the Output Area design project (funded under the 1999–2001 ESRC 2001 Census Development Programme) with 2001 Census geography demonstrator data

4

Designing Your Own Geographies

Seraphim Alvanides, Stan Openshaw, and Philip Rees

SYNOPSIS

This chapter briefly reviews zone design concept and discusses the methods used to design new zones from smaller areal building bricks. Zone design is a field pioneered by Stan Openshaw, and this chapter draws extensively on Openshaw's work. It is based on material from Alvanides's doctoral dissertation, which provides a comprehensive account of zone design theory and method. Results from the 2001 UK Census are to be published for about a quarter of a million output areas (OAs), which were designed using an algorithm developed by Openshaw (see Chapter 3). However, zone design methods also have enormous potential for use in grouping together 2001 Census OAs into zones for analysis purposes. Zone design methods are outlined in the chapter and reference is made to work on software for zone design, the developers of which can be contacted for the code and advice.

4.1 INTRODUCTION

Since 1991, higher education institutions, local authorities, and government departments have benefitted from direct access to census information. There has also been a massive growth in the commercial use of digital information. Developments such as the Census Access Project (CAP) (see Chapter 1) mean that following the 2001 Census, all users with access to a Web browser will be able to produce their own census statistics and maps.

This chapter informs census data practitioners and users alike on the problems faced in using areal units for reporting census research findings. Section 4.2 introduces the modifiable areal unit problem (MAUP) and briefly reviews the way it has affected census-related literature. Section 4.3 suggests the use of zone design as a way around the MAUP and outlines the zone design method. Section 4.4 describes the software resources available for zone design. Use of zone design will provide census users with the means to develop geographies specifically related to their research, rather than having to rely on official census geography. Zone design can help users cope with the huge quantity of information that will be available at OA scale.

4.2 FROM AREAL UNITS TO ZONES

In the discussion that follows, we adopt the following definitions: by *areal units* are meant the small territorial building bricks that are aggregated into larger entities, by *zones*, we mean those larger spatial entities that are the product of the zone design process.

4.2.1 The Areas for Which Census Data are Produced

Areal units are widely used by official agencies in most countries for reporting statistics and disseminating census information. Areal aggregation is a confidentiality protection device, as cross-tabulations do not reveal information about the full profiles of the individuals whose data are being summarized. Statistics from the 1991 Census in the United Kingdom were initially published for small areas used in collection, enumeration districts (EDs), and the administrative areas they fit into—electoral wards and local government areas (LGA). During the last ten years, increasing use has been made of geographical entities not directly used in the standard products such as postcode sectors, for which small area statistics were published in a second round. Shortly after the 1991 Census, Raper *et al.* (1992) published their groundbreaking book, advocating the use of postcode geography as a means of georeferencing data. Much administrative and most private sector data are referenced by postcode and address, and it makes sense to be able to aggregate census data to postal geographies. Longley and Clarke (1995) assemble geographical information system (GIS) applications for business and service use that are dominated by postal geographies. The conversion of 1991 Census Small Area Statistics to postal geography (other than units based on postal sectors) is accomplished through use of the ED/postcode directories, the Central Postcode Directory (CPD), and the All Fields Postcode Directory (AFPD). Census users have thus been engaged in the creation of new geographies for census statistics.

4.2.2 The MAUP

Clearly, as soon as areal units are used in reporting data and their analysis, the user faces the MAUP. The MAUP was defined by Openshaw (1984) as two distinct, but interrelated, problems that emerge when space is subdivided into areal units. The aggregation problem is associated with uncertainty about the correct number of zones needed to represent a phenomenon. Relationships at one geographical scale may disappear at another. The aggregation problem is related to the configuration of the zonal boundaries in space: there are many different ways to define the boundaries of zones, holding the number defined within a study region constant.

The diagrams in Figure 4.1 illustrate both the aggregation and zoning effects for a small set of areal units. The areal units are represented by a set of cells, each of which contain a dot that represents the areal unit centroid. The areal units and their centroids are regularly distributed over an imaginary study area in a very simple configuration and the underlying population values are assumed to be equal. In this demonstration of the MAUP, 16 areal units are arbitrarily aggregated into two different scales and four zonings per scale level. The aggregation effect can be followed vertically, expressed here as the number of zones. As we move from the scale of 4 output zones to the scale of 2, the average size of the zones increases. The zoning effect can be followed horizontally and it

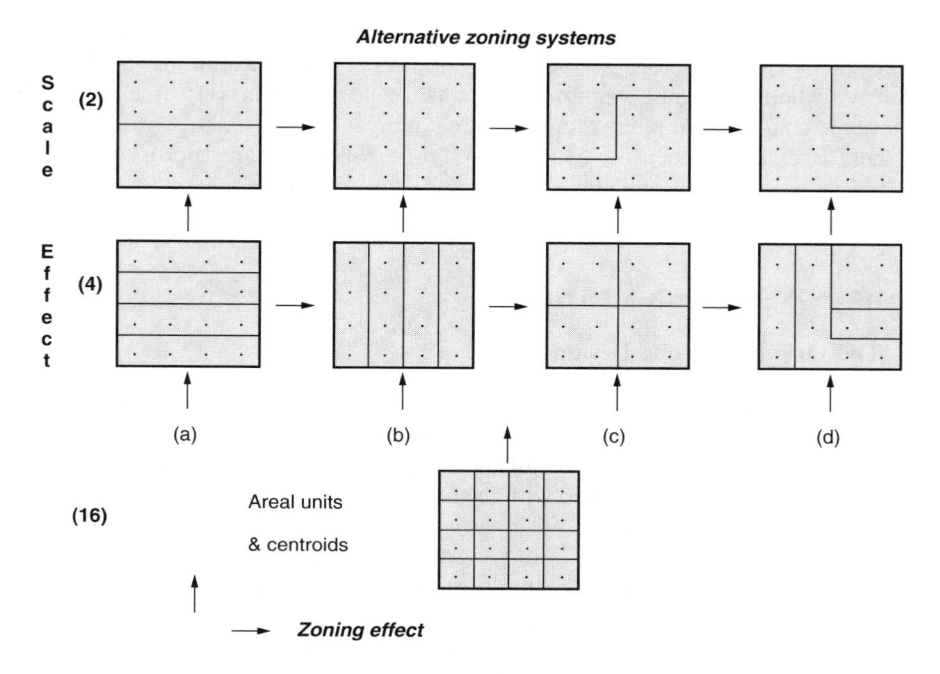

Figure 4.1 The aggregation and zoning effects of the MAUP.

is shown as different configurations consisting of the same number of zones (either 2 or 4). It is also worth noting here that for zonings (a), (b), and (c), the zones are comparable because they have exactly the same size and shape. However, this is not the case for the last zoning system (d), in which the zones have unique sizes and shapes. In reality, it is very unlikely that data can be found at exactly the same scale, but even for similar scales, the results may vary considerably because of the MAUP. This example serves only as a simple illustration of the MAUP effects, but gives some idea of the complexity of the problem. Even for such a small example, it would be difficult to exhaust the different zonings for all the aggregations to which the 16 areal units could be aggregated.

4.2.3 It Matters Which Areas You Use

It has become clear that the areal units for which statistics are reported can have a considerable influence on the message conveyed. Academics and practitioners alike are concerned by the arithmetic of politics and the way statistics can be deliberately manipulated to produce desirable effects (Dorling and Simpson, 1999). Similarly, maps can be designed to mislead and Monmonier (1991) presents some excellent examples of 'cartocontroversy', resulting from deliberate manipulation of cartographic information to support political agendas. The arbitrary definition of zone boundaries, by excluding or including certain areas, is one of the most controversial issues in map propaganda. For example, electoral zones can be created for partisan political purposes so that a political party retains a majority of votes in an area. This is known as gerrymandering, and is, in fact, a controlled exploitation of the MAUP effects. One solution to the zoning effects might be to use only the smallest areal units available in any analysis. However, the reliability of a census statistic decreases as the size of the reporting area diminishes. Users may also be overwhelmed by the weight of information available in very fine geographical aggregations, consisting of many thousands areal units (Openshaw and Alvanides, 1999a). Thus, there is clearly a need for methods to define zones that are appropriate for study or analysis purposes, using objective criteria for determining the suitability of these new zoning schemes.

4.3 THE ZONE DESIGN METHOD

4.3.1 The Origins of Zone Design

Although the use of arbitrary collection areas for reporting census data has been widely criticized (Morphet, 1992), the role of small areal units in a zone design context has been surprisingly overlooked. The zone design concept has its origins in attempts to address the MAUP and the difficulties associated with the modifiable nature of zones. Zone design was originally suggested by Openshaw (1977a) as a means for tackling aggregation uncertainty associated with the use of areal units in geographical research. The modifiable nature of areal units may indeed provide the solution to the problem: if there is uncertainty in the number and configuration of zones, then why not try different aggregations and zonings of the areal units? The equivalent statistical problem would be to group observations using different intervals so that the statistical properties of the data are communicated clearly. The equivalent cartographic problem is the classification of observations by quantiles, equal intervals, and standard deviations, so that underlying patterns emerge. These facilities exist in all statistical and cartographic packages, and although the results of their misuse can be highly deceptive, the tools themselves are hardly ever questioned. Zone design has the potential to become the geographer's tool to encounter the MAUP, and census areal units are the most suitable context for demonstrating it.

4.3.2 Zone Design Procedures

Zone Design is based on a simple concept: aggregate *A* areal units into *Z* larger zones so that the output zones are contiguous and cover the whole area. In other words, the output zones should not form 'islands' or be disconnected. In addition to what the output zones look like, they also possess some imposed characteristics. These are described by the objective function optimized in each case. For example, the output zones might have equal (or approximately equal) population; this is the simplest function one might want to use. An explanation of the available optimization functions is given later in the chapter. Initially, this challenge was named the automated zoning problem and the solution was provided through the automated zoning procedure (AZP) (Openshaw, 1977a). The solution involved the reconfiguration of *A* areas into *Z* larger zones in a way that optimizes an objective function imposed by the user.

Several tasks have to be accomplished to operationalize AZP. First, contiguity matrices for the areal units employed as building bricks for the output zones have to be developed from GIS data for the areas. Second, the nature of an objective function appropriate for the zone design task has to be determined. Third, one or more optimization algorithms implementing the AZP idea need to be developed and tested to identify the most suitable one, given the size of the problem and the objective function. Finally, it is sometimes necessary to build in fixed constraints on zone characteristics. These issues are discussed in turn.

4.3.3 Creating Contiguity Matrices from GIS Data

It was necessary to develop a method for defining areal unit contiguities from map information. The idea of a contiguity matrix is easy to grasp: given a set of areal units A, an $A \times A$ contiguity matrix is defined so that any element is 1 if the two areal units are spatially connected and 0 if they are not. In order for an output zone to be contiguous, it is necessary for its member areal units to be connected in the contiguity matrix. Back in the 1970s, contiguities had to be identified on paper maps and typed in manually! Implementing contiguity matrices today is much more straightforward, because most GIS packages store topological information about map elements or offer the means to retrieve such information. However, depending on the way the geographical features are represented in a particular GIS, the topological information needs to be retrieved in a form suitable for zone design. For example, if topology is stored for the borders of the areal units rather than the areal units themselves, then some coding in a computer language outside the GIS is necessary to create the contiguity matrix. As GIS software becomes more sophisticated, it will be possible to retrieve contiguity information for areal units automatically. However, for the systems described in the next section, simple programs for creating contiguity matrices from standard GIS were written and incorporated into zone design.

4.3.4 Zone Design Objective Functions

In zone design, the user will want to optimize certain properties of the zones produced. These properties are combined into an objective function, which is evaluated at each iteration of the zone design process to see if its value has improved. Most objective functions involve sums of the distances in property space among all the zones, so that optimization means seeking a minimum value of the objective function. Among zone attributes that have been found useful in zone design are the following (Alvanides 2000, for the relevant formulae):

- *Size equality*. The aim is to make the zones as similar in size as possible. Normally, size is defined in terms of the resident population of interest in the zone, but this concept could refer to numbers of people, households, addresses, retail outlets, schoolchildren, men aged 40 and above, and so on, depending on the aim of the research.
- *Attribute similarity*. The aim is to make the zones as similar as possible in terms of one or more ratio variables that measure zone character. For example, attributes such as the percentage of households by tenure and by housing type have been used in the design of OAs for the 2001 Census (see Chapter 3), with the aim of maximizing the socio-economic homogeneity of zones.
- *Accessibility/compactness*. The aim is to achieve maximum accessibility of building brick populations to zone centres or to achieve the most compact shapes of areas. The former goal can be reached by minimizing the aggregate population-weighted distances that building brick populations have to travel to zone centres. The latter goal can be achieved by minimizing the ratio of zone perimeter to zone area.

- *Correlation*. The aim is to achieve a given level of correlation between two variables computed for zones. An objective function incorporating correlation was used by Openshaw and Taylor (1979) and Openshaw and Rao (1995) to demonstrate the serious effect of zone design in statistical studies using areal data.

4.3.5 Zone Design Algorithms

How does a zone design algorithm work? Here we describe the AZP algorithm developed by Openshaw (1977a, 1978). The AZP algorithm has been enhanced by Openshaw and Alvanides through use of Tabu Search and Simulating Annealing techniques (Alvanides, 2000). The AZP algorithm involves the following steps: The formal mathematics are given in Alvanides (2000), in an exposition that fleshes out the compressed account in the Openshaw papers).

Step	Description
0.	Create an initial random aggregation (IRA) of the A areal units into Z zones represented by a classification vector and compute the value of the objective function for the IRA. The method for IRA is given in Openshaw (1977b).
1.	Randomly select any zone z and make a list of areal units bordering z, but not members of it.
2.	Randomly select and remove any areal unit a from the list.
3.	Identify the zone that areal unit a belongs to and update the classification vector.
4.	If the remaining areal units in the donor zone q are not contiguous then restore the zoning system and return to *step 2*.
5.	Accept the move of areal unit a to zone z, calculate the model parameters (if necessary), and calculate the new value of the objective function.
6.	If the new value of the objective function is not an improvement on the previous value then restore the previous classification vector; return to *step 2*.
7.	Update the current best value of the objective function. Extend the bordering vector by adding areal units connected to a, which is now a member of zone z and return to *step 2*.
8.	When the list of bordering areal units to zone z is exhausted, then return to *step 1*.
9.	Repeat *steps 1 to 8* until the algorithm converges, that is, the change in the objective function is less than a specified tolerance.
10.	The algorithm ends with a nearly optimum system of Z zones in the classification vector with the current value of the objective function considered as optimal.

Alvanides (2000) describes an adaptation of the AZP algorithm that uses a "simulated annealing" (SA) approach, which is employed in a series of zone design case studies.

4.3.6 Zone Design Constraints

In many cases, it is insufficient to merely optimize an objective function. The user may have limits or constraints that must also be considered, so that constrained optimization methods must be used. The concept of optimization constraints in zone design is relatively difficult to grasp, and the best way to explain it is by using three real-world examples. First, a minimum size threshold is often imposed to help protect the confidentiality of statistics released for zones (Martin, 1998a). In Chapter 3, for example, OAs for the 2001 Census (England and Wales) will be specified with a minimum threshold size (e.g., at least 100 households in every OA). A second example relating to confidentiality thresholds is the release of incident data from hospital admissions and health registries, subject to a minimum population at risk within an area (e.g., at least 1000 males above age 40 for prostate cancer). A final example that requires the implementation of an optimization constraint is the use of zone design for electoral redistricting, where a minimum number of citizens eligible to vote are necessary to ensure fair representation of the electorate.

The difficulty with such constraints is that they are problem specific and hard to implement, both conceptually and algorithmically. For example, in the case of electoral redistricting, it is possible to define an objective function so that the electorate is comparable in each constituency, or define a constraint that ensures equal numbers of people eligible to vote alongside a minimum electorate constraint. In addition, it might be desirable to add a constraint that ensures fair political representation of people from ethnic minorities. The problem then becomes one of explicitly defining all these constraints and implementing them so that they have a reasonable effect on the zone design process. Even more complicated is the interaction between the objective function and the optimization constraints so that the output zones are the outcome of a balanced combination. It is outside the purpose of this chapter to cover in detail the optimization constraints, and the reader may wish to consult the relevant operational research literature on constrained optimization. Suffice to say that geographers have so far suggested problem-specific constrained methods for optimizing census OAs (Martin, 1998a; Alvanides *et al.* 2002) and for performing exploratory spatial analysis with health data (Wise *et al.* 1997; Alvanides *et al.* 2001).

4.4 CARRYING OUT ZONE DESIGN

At time of writing, the ingredients for a general-purpose zone design tool exist, but are not available as an easy-to-use package. One reason for this is the need to tailor the zone design objective function to the aims of the research being undertaken. To alter the objective function requires alteration of the AZP routines and testing of the modifications. Methods such as the SA version of AZP also involve skilled parameter setting. So the resources listed in Box 4.1 reference the source code for two versions of zone design systems (ZDES) and the software designers who can help develop applications with users as collaborators.

4.4.1 Zone Design Within a GIS: The ZDES3b Resource

ZDES3b (ZDES, version 3b) is a system loosely coupled to a GIS that automatically performs zone design with minimum requirements from the user. The main components of ZDES3b are: AZP procedures, a graphical user interface (GUI), and a proprietary GIS. The GIS is necessary for handling the spatial data, linking the geography (zones) with the attributes (census variables) and creating the essential contiguity matrix needed for

ZDES3b to run. In addition, the GIS offers the tools for developing the GUI and handling the resulting regionalizations from the AZP procedures. The GIS selected for developing ZDES3b was the well-known ARC/INFO package by ESRI. The choice of GIS package can be justified by the accessibility of ARC/INFO through Manchester Information and Associated Services (MIMAS) for registered users, the affordable CHEST (Combined Higher Education Software) licensing agreement for the academic community, and the continuity with earlier research by Openshaw and Rao (1995).

The first step involves downloading boundary data for the study area, together with a number of census variables relevant to the zone design problem. Boundary data down to the ED level can be downloaded either from UKBORDERS (see Chapter 6) or from MIMAS, or purchased through licensed providers. Census data can be accessed on-line by registered census users through the CASWEB interface described in Chapter 8 or by running a SASPAC program on MIMAS. Once the boundary and census data are in place, it is necessary to combine them so that the census variables become polygon attributes. This can be achieved by turning the census data into an INFO table and subsequently using the JOINITEM command in ARC/INFO (for a detailed description of this process, see Charlton *et al.* 1995). Alternatively, both boundary and census data can be downloaded together through the new CASWEB interface (see Chapter 8). The new CASWEB produces files that combine census text data with digital boundary data (DBD) for the selected census zones, thus saving the user from the join operation in ARC/INFO. At the moment, it is only possible to retrieve two types of GIS outputs from CASWEB: MapInfo interchange format and ArcView shape format. In order for the ArcView shape files to be converted to an ARC/INFO coverage for use with ZDES3b, the user is simply required to run the SHAPEARC command in ARC/INFO. As soon as the coverage contains a polygon topology with the census variables properly attached to polygons as attributes, it is ready for input in ZDES3b.

Figure 4.2 shows the contents of the ZDES3b system in terms of files and programs that are embedded in an ARC/INFO shell. The AZP tools are seen as foreign programs by ARC/INFO. AZP requires various data input files and a text parameter file (.zdset), which contains user choices and parameters set via the interface described later. As mentioned earlier, the zone design procedures operate on the topological relationships, stored as a contiguity list, between the input areal units. In order for the contiguity information matrix to be constructed, a file of segment data (.seg) needs to be prepared. This is a list of segments consisting of the two areal unit identification numbers or codes defining the segment and the length of the boundary between the two areas. The segment file is used as input to the program SEG2CON, which outputs two files: the contiguities, stored as lists of neighbour area IDs for all areas in file .con, and the boundary lengths for all areas in file .len. In addition to topological data, zone design needs the input of population weights (file .pop) used when the aim is to achieve equal population zones, and additional areal unit variables (file .mvr) used when the aim is socio-economic homogeneity. Two more data sets with geometric characteristics of areal units need to be prepared: the physical extents of areal units (file .are) when area compactness is optimized, and the centroid coordinates of areal units (file .cxy) for optimizing accessibility to zone centres. These input files are entered into the system and after the AZP code has been run, three output files are produced: (1) a list of zones and their membership in terms of the areal unit building bricks (file .zon), (2) aggregated values of the variables used in the objective function (file .agg), and (3) information on the performance of each AZP run (file .sum).

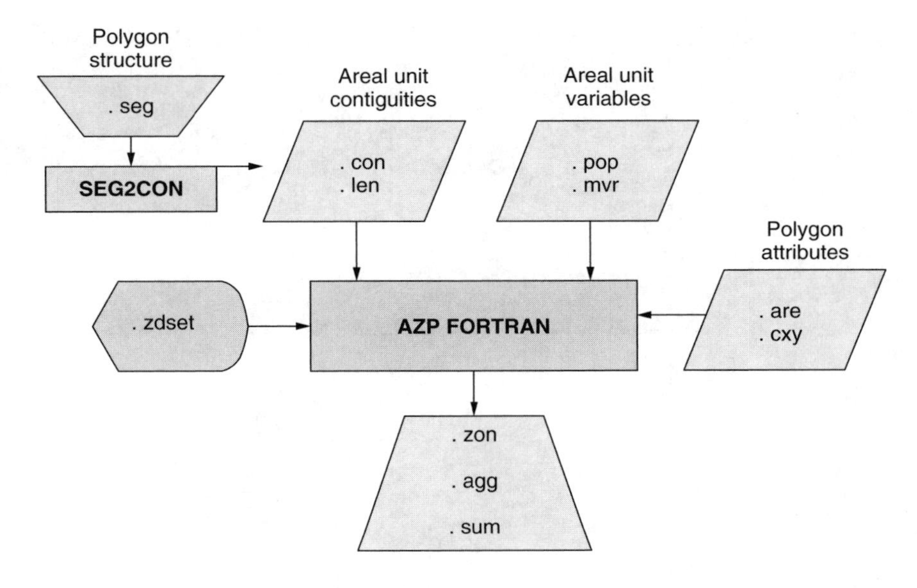

	Files to prepare for ZDES3b run	
.zdset	Parameter setting file	File names and zoning parameters
.seg	Areal unit segment list	Left area id, right area id, length
	Files for input to ZDES3b	
.con	Neighbour lists	Area id, neighbour area ids
.len	Boundary lengths	Area id, boundary lengths
.pop	Areal unit populations	Area id, population/weighting
.mvr	Areal unit other variables	Area id, variable(s)
.are	Areal unit area size	Area id, physical area
.cxy	Areal unit centroids	Area id, centroid coordinates
	Output files by ZDES3b	
.zon	Output membership list	Area id, zone id
.agg	Zone aggregated values	Zone id, aggregated value
.sum	Summary of function	Run number, objective function performance

Figure 4.2 Data files and contents for ZDES3b.

The main components comprising ZDES3b are connected through the GUI so that the system requires minimal GIS knowledge from the user. The ZDES3b interface, shown in Figure 4.3, has been designed so that the user does not have to deal with the more advanced GIS functions offered by a powerful GIS such as ARC/INFO. The interface assists the user through a couple of menu windows in the following tasks:

- geographical and attribute data preparation
- selection of AZP function and parameters
- running the AZP program
- reading the AZP results
- display of the output zones
- preparation of output coverage

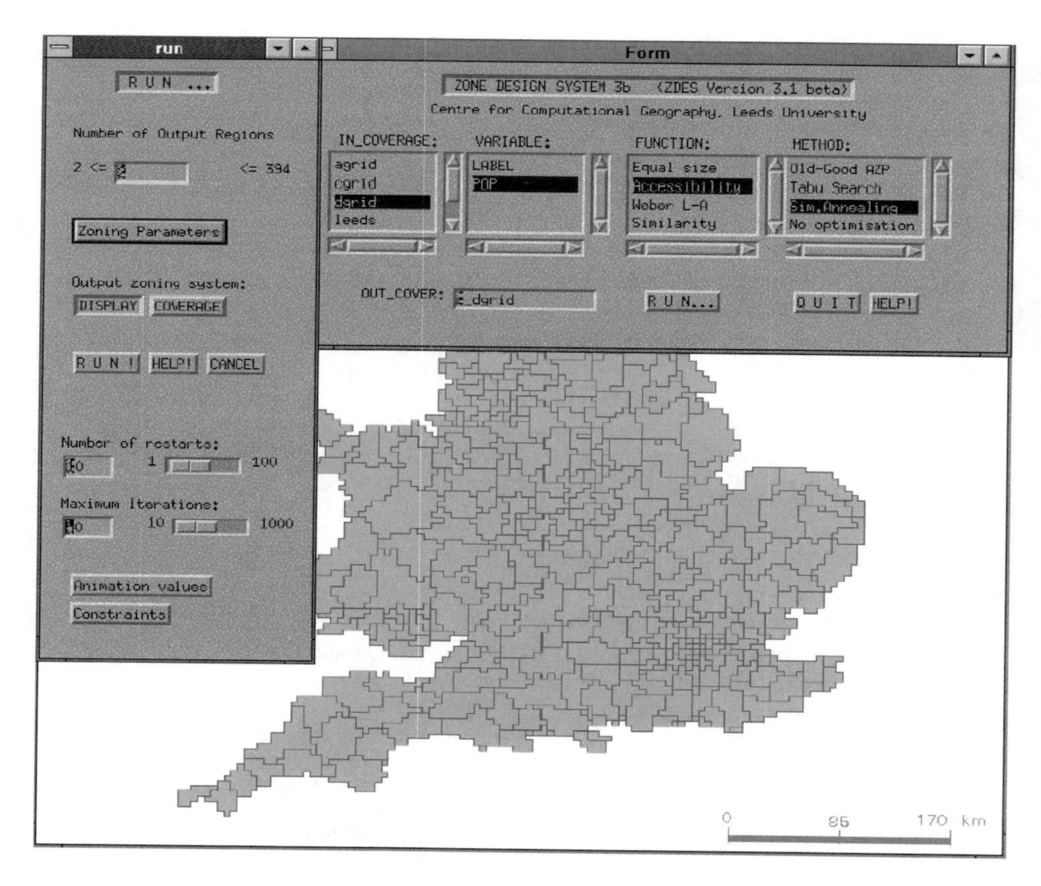

Figure 4.3 The ZDES3b graphical user interface in ARC/INFO.

4.4.2 Using ZDES3b for Exploratory Analysis

In this section, ZDES3b is demonstrated with two examples of very basic exploratory analysis. First, zones of optimal accessibility are designed in England and Wales, using Local Authority District (LAD) boundaries from the 1991 UK Census as building blocks. The second example is a simple zone design classification using the attribute similarity objective function on a synthetic data set. The formulae for these objective functions is not presented here, but the models can be found in most geographical analysis textbooks (for example, see Robinson, 1998, under *similarity* and *accessibility*).

The interface between the AZP and the GIS is written in simple AML, and as a result, all that is needed to run ZDES is the usual ARC/INFO command for running programs in batch mode: `&run zdes3b`. The system is then initiated and the ZDES screen together with the graphical menus appears. First, the user is prompted to select an input coverage from the input coverage list. If a coverage has already been downloaded and prepared, as described in the previous section, then it needs to be stored under the ~/zdes3b/data/ workspace in order to appear in the menu. For the purposes of this demonstration, readers are asked to use the sample coverages that come together with ZDES3b. The three sample coverages are:

- agrid: a simple synthetic coverage with cells containing 3 values (10, 50, 100) as attributes. This coverage is provided for basic experimentation on understanding the objective functions.
- cgrid: a modified coverage of counties in England and Wales. The county boundaries have been heavily generalized and the population counts have been rounded, so that the data can be used for demonstrating the system to unregistered users.
- dgrid: a modified coverage of LADs in England and Wales. As with the previous coverage, the district boundaries have been heavily generalized and the counts have been rounded.

As the census data in the last two coverages (cgrid & dgrid) have been modified to speed up the zone design process, users are advised to use these data sets only for zone design training. Registered users can download the full data sets and prepare them as described earlier for analysis. For demonstrative purposes, the 'district' coverage (dgrid) is selected here, as shown in the main menu of Figure 4.3. As soon as the coverage name is highlighted, the system retrieves and displays the boundaries in the graphics window and updates the list of *variables* (*label* and *population*, in this case). In this example, we want to group together LADs so that their population-weighted distance from the zone centres is minimized. The population count for each areal unit will be used as a weighting factor (so the population variable is selected) and accessibility is the *objective function* that needs to be highlighted in the menu.

Three *optimization methods* are listed in the final menu: old AZP (some times referred to as 'Monte Carlo' optimization), tabu search and SA. The optimization quality usually increases as we go down the list at the expense of computational time, with SA providing the best results, but taking up more time to converge than the other two methods. As a rule of thumb, it is advisable to start with the old AZP for experimentation and progress to SA for better-quality analysis. It may also be useful to run different optimization methods to compare the optimization function performance (stored in file .sum) with computational time. In this case, the problem is relatively small (394 areal units), so computational time is not an issue and, therefore, the SA method is highlighted. Also, note that an *output coverage* name is displayed by default, but the user can give a more comprehensive name. Finally, the *Help!* button invokes a new window with advice on preparing the input data, a brief description of optimization methods, and parameters and contact details.

After the basic optimization choices have been made, the *Run...* button invokes the final window with more specific parameters. The number of *output regions* is the most important parameter and the system indicates the minimum and maximum values, the maximum being the number of areal units in the coverage less one. It is also possible to set the number of algorithm *restarts* and the maximum number of *iterations* for each run, so that the system does not run indefinitely. Depending on the nature of the problem, it is advisable for the number of runs to be set to at least 10, so that there is a pool of solutions to choose the best one from. However, it is noted that if a large number of restarts is combined with a time-consuming optimization algorithm, such as SA, then the user should be prepared for long processing times. The number of iterations can be set at a high value, since the algorithm converges after a few idle iterations, that is, when subsequent swaps of areal units between zones do not improve the performance of the objective function dramatically.

The final choice to note in this menu is the *output zoning system* mode, which can be set to *display*, for mapping the results only on the graphics window, or *coverage* for consequent storage of the resulting zones to the defined output coverage. Finally, it is possible to see the contents of the *zoning parameters* file (.zdset) before running (or cancelling) the AZP algorithms. If the parameters shown in Figure 4.3 are used (2 output zones; 10 restarts; 50 iterations), then the areal units in the dgrid coverage will be redesigned to form compact zones with high internal accessibility.

The second example involves some basic zone design with a simple synthetic data set consisting of a of a 14 × 26 cell grid. The 364 areal units have been given the arbitrary values 10 (low), 50 (medium), and high (100), so that two clusters are hidden in the data shown with different shades of grey in Figure 4.4(a). If the data set is classified in any desktop GIS, using an in-built classification algorithm such as *natural breaks*, then the output is Figure 4.4(a) for three classes and Figure 4.4(b) for two classes. Although the medium and high values are correctly classified together in Figure 4.4(b), the classification cannot be used as a two-zone system because the two grey areas are not contiguous. The difference between unconstrained classification and contiguity constrained zone design can be explored further using ZDES3b. The coverage agrid contains the data for this example and can be loaded in ZDES3b by highlighting the name in the input coverage list of the main menu. The variable containing the values is again named *population*, and we need to optimize the *similarity* function since we want to identify zones with similar values. The optimization method makes no difference in this simple example, so the traditional AZP should be able to cope with such a simple contiguity arrangement.

In the Run menu, we start with two output zones to compare the zoning system with the classification in Figure 4.4(b), and increase the number of restarts to 50 and the number of iterations to 100, so that the AZP has sufficient time to converge. The difference between the output zoning system with two zones in Figure 4.4(c) and the cartographic classification with two classes in Figure 4.4(b) is that the two distinct areas with the higher values are connected to form a single zone. Although there are two clusters in the data set, the algorithm was forced to produce two zones, thus misallocating a few low value cells to 'bridge' the two areas, the values of which are higher than the rest of the cells. Although potentially useful for pattern analysis, the two-zone solution does not reflect the underlying patterns accurately. By increasing the number of zones to three, the output zoning system becomes similar to the cartographic classification with three classes, as shown in Figure 4.4(d). Their underlying difference is that while the zoning solution is achieved using a combination of attribute and spatial data (contiguities), the classification operates solely on the values of the cells (attributes).

Consider now what happens if the system is asked to design more zones than required by the needs of the analysis. In the case of the similarity function, zone design is forced to partition space, by splitting one of the zones, as shown in Figure 4.4(e). As there are no other size or shape constraints imposed by the user, the algorithm proceeds by partitioning space in a 'random' fashion, as long as the output zones are contiguous. The result is usually one or more zones, with irregular shapes that do not reflect any underlying patterns, other than the randomness of the zoning process. However, it is interesting to note that even in this situation, the algorithm still seeks to balance the objective function by allocating the same number of cells in each of the low value zones in Figure 4.4(e).

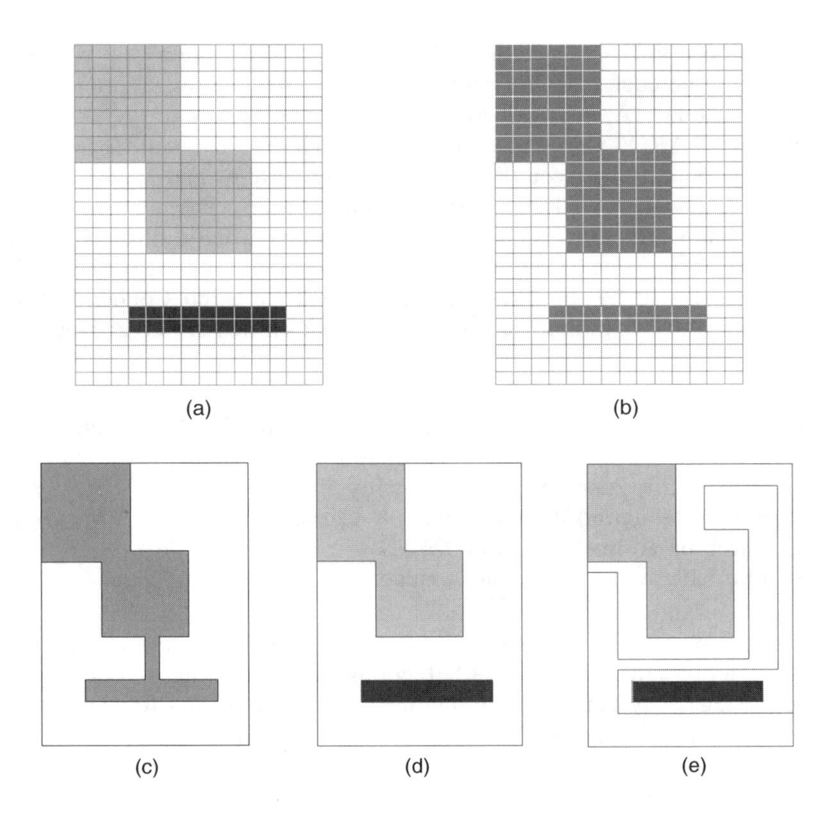

Figure 4.4 Cartographic classification and zone design outputs: (a) input areal units with low (white cells), medium (grey cells), and high (black cells) values; (b) cartographic classification with two classes: low (white cells) and medium-high (grey cells); (c) zone design with two output zones: low (white) and medium-high (dark grey). *Note the shapes of the two zones, especially the 'bridge' connecting the medium and high-value areas in the darker zone;* (d) zone design with three output zones: low (white), medium (light grey), and high (black). *Note the similarity with the three-class cartographic classification in (b);* (e) zone design with four output zones: low (white), medium (2 light grey zones), and high (black). *Note the similarity with the three-zone system in (c) and the shapes of the two low value zones in white.*

This simple example clearly demonstrates the aggregation component of the MAUP, when geographical data are mapped using the 'wrong' aggregation (number of zones); it is then inevitable that underlying patterns are lost. One solution is the use of the disaggregated data in Figure 4.4(a), but since this is not always possible, it is important to find the aggregation with the minimum loss of information.

4.4.3 Zone Design Outside a GIS: The ZD2k Resource

One of the drawbacks of ZDES3b is the need for users to have some familiarity with the UNIX operating system and the ARC/INFO package. In addition, users wishing to extend the capabilities of the AZP algorithms by adding optimization methods, objective functions and constraints need to have an in-depth knowledge of the system as a whole; they are also required to write code in FORTRAN, while more flexible object-oriented languages are widely used today. A final consideration with ZDES3b is the reliance on both a specific version of ARC/INFO and a specific FORTRAN 70 compiler and library set, thus impeding portability between different operating systems and platforms. To resolve these problems and make zone design widely available to the academic community and the census users, a major exercise was undertaken converting the algorithms and extending the system in Java code capable of running through an internet browser interface (Box 4.2). The project was funded by the Economic and Social Research Council (ESRC) 2001 Census Development Programme and the initial proposal had six objectives (Openshaw and Alvanides, 1999b):

1. Provide enhanced zone design functions for use with the UK 2001 Census
2. Provide new tools for designing and aggregating census OAs
3. Allow users to define zone design functionality and constraints
4. Integrate flow (interaction) data from the UK Census
5. Develop a Web-based interface to the ZDES
6. Develop a portable tool that operates independent of any commercial GIS

The first step in responding to the needs of access and flexibility involved placing a Web-based interface around the existing ZDES3b components, allowing users to communicate and control the legacy system by using familiar Web-based forms. To achieve this, some form of server side scripting that was capable of communicating with ARC/INFO and the AML scripts was required. A Java servlet was designed and built to perform this role. The Java solution was preferred to conventional CGI scripts because Java offers superior performance, flexibility, and security. The initial system architecture, shown in Figure 4.5(a), made it possible to create sets of Web pages that prompt users for all the information required to drive ZDES3b components from start to finish. This solution was very similar to the ZDES3b interface, but could be accessed remotely, thus achieving the accessibility objectives of the project. However, it quickly emerged that performance was a major drawback with this solution because each individual run required multiple ARC/INFO sessions to be executed simultaneously on a single server, making the process both slow and resource hungry. It was clear that a move away from ARC/INFO (or any other commercial GIS) was needed for increased flexibility and complete portability.

The next stage in development focused on removing the ARC/INFO bottleneck; this strategy freed up resources and considerably improved the portability of the system. Both ZDES3b and the early version of ZD2k made use of the internal data architecture of ARC/INFO that stores topological information about map elements. However, more commonly used GIS storage structures, such as the MapInfo interchange format and the ArcView shape format offered by CASWEB, do not have any concept of contiguity. To solve this problem, a Java module was written that automatically generates contiguity information from ArcView shape files. The Java module also performs area and perimeter calculations and exports the information in the same format as the original ZDES3b. Being Java based makes this section of the system much easier to interface with the servlet engine on the server, and it directly allows users to perform geometry calculations without the need for a commercial GIS package. In addition to the contiguity generator, a Java module was written to reproduce the ARC/VIEW boundary dissolve process used to

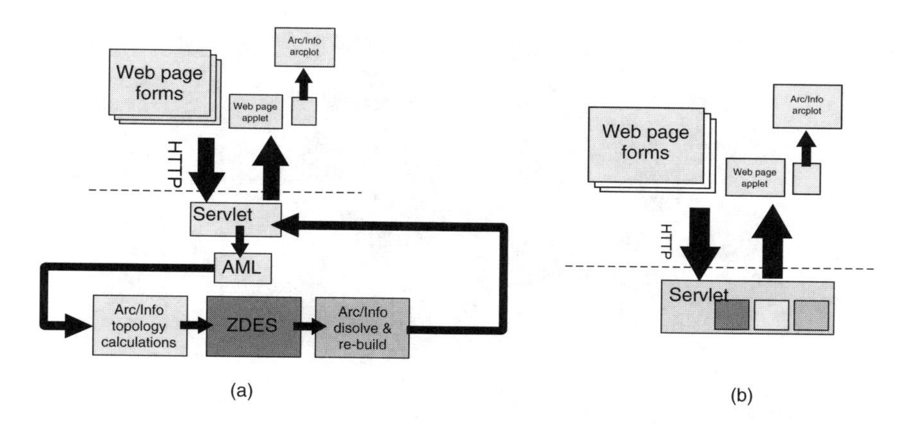

Figure 4.5 ZD2k development: (a) replicating ZDES3b functionality by using a Java servlet that accesses ARC/INFO functions; (b) GIS independent geometric and optimization ZD2k modules controlled by the Java servlet.

merge the input areas into the output zones. With this final stage of geometric operations completed, the performance of the system as a whole was much improved.

However, the legacy FORTRAN core running the AZP routines was still a significant hindrance to portability. The final development task called for a Java-only solution, based initially on a subset of the methods and functions provided by ZDES3b. Specifically, a limited number of 'most frequently used' options were converted from FORTRAN to Java. The resulting ZD2k system has been developed with expandability in mind, allowing for additional methods, functions, and optimizations to be added as components without the need to edit or recompile the system core (as was the case with ZDES3b). The ZD2k system now consists of a number of separate Java components, which can either be used independently (the two geometry phases are useful in their own right) or together within any system, as shown in Figure 4.5(b). Thus, as it stands, ZD2k is not a replacement for ZDES3b, but a set of components that can add zone design capabilities to other systems, or that can be quickly added to simple user interfaces for specific purposes.

4.4.4 Using ZD2k for Electoral Redistricting

The new ZD2k components are designed so that they can be easily combined within a user's computer environment to tackle a user-defined problem. For demonstrative purposes, the basic ZD2k components have been grouped together in an interesting example of electoral redistricting in Great Britain, adapted from Macgill *et al.* (2001). It has long been recognized that the shape and size of spatial zoning systems could make a significant difference to the patterns of economic and social variables displayed in choropleth maps (for reviews and examples see Alvanides and Openshaw, 1999; Openshaw and Alvanides, 2001). When the variables are continuously defined, it may not make a significant difference. However, when the variables are discrete (1/0 or yes/no variables), then a different configuration of size or shape can make a crucial difference to the value that a specific variable takes within the zone. ZD2k illustrates this aspect of zone design in political

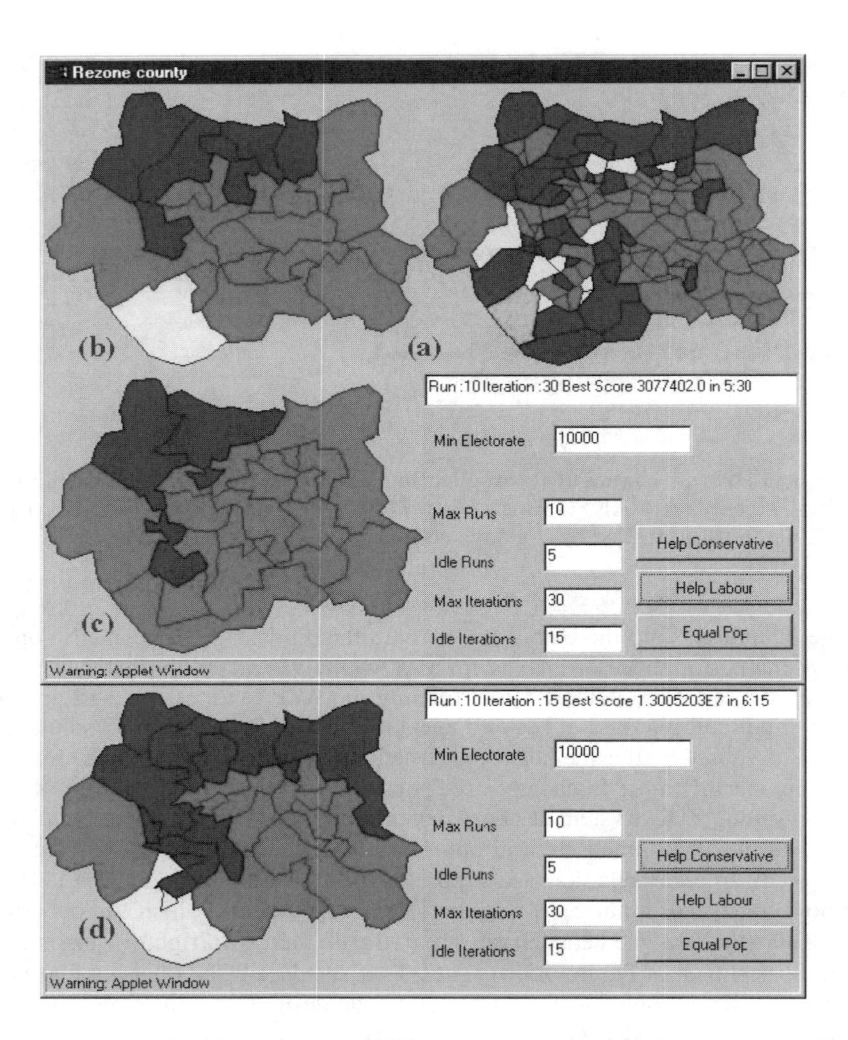

Figure 4.6 Electoral redistricting in ZD2k: (a) electoral wards for West Yorkshire with voting data used as input areal units: Liberal Democrats are shown in yellow (white), the Labour party is represented by bright red (light grey), and the Conservatives with dark blue (dark grey). *Note the domination of Conservative voters in the north and south and the relatively isolated clusters of Liberal Democrats. The single green ward in the south-west represents the Green party;* (b) electoral constituencies: the Labour party has the majority of seats, with the Conservatives dominating the north, whereas the Liberal Democrats gain one seat in the south-west. *Note how the isolated Conservative wards in the east have been 'lost' in Labour party constituencies and the elimination of the green party ward in the south-west;* (c) Gerrymandering in favour of the Labour party. *Note the contraction of the Conservative seats to the north-west, where most voters are clustered, and how a relatively modest rearrangement of the boundaries in the south-west eliminates the Liberal Democrat seat;* (d) Gerrymandering in favour of the Conservative party. In an attempt to assist the Conservatives, the Liberal Democrats also benefit, by gaining an extra seat in the south-west. *Note the decreasing size of the constituencies in the centre in order to satisfy the objective function by reaching the minimum electorate constraint.*

geography, taking to the extreme the process of biased electoral redistricting (gerrymandering). Gerrymandering has long been a concern in political geography, but how easy is it to create optimal zoning systems for one political party or another? ZD2k first allows the user to choose on screen a particular part of the country for analysis. The rest of this section focusses on West Yorkshire County.

Once the desired county is highlighted and the Rezone button selected, the system retrieves the ward level boundaries and the results of various elections over the last ten years, as shown in Figure 4.6(a) for West Yorkshire. In addition, the voting patterns for electoral constituencies are shown in Figure 4.6(b), so that the current distribution of party seats becomes clear. The shades reflect political parties: mid-grey for Labour, dark grey for Conservative, white for Liberal Democrats, and light grey for the green party. It is clear from Figure 4.6(b) that the Conservative seats are concentrated in the north, with the rest of the region dominated by the Labour party. As expected, there is much more variation in local pockets of good performance for each of the political parties at the ward level, as shown in Figure 4.6(a), which demonstrates the considerable effects of aggregation on discrete variables.

The key question: can these electoral wards be aggregated in such a way as to help either the Conservative or the Labour party to win more parliamentary seats (that is, redesign the constituencies to ensure greater success)? Leaving the parameters at default values (minimum electorate 10 000 voters, 10 maximum runs of 30 iterations), select the *Help Labour* option. Note that, unlike the static ZDES3b, results here are constantly mapped to the screen and the user can view the zone design process in action as the algorithm swaps wards between zones, trying alternative arrangements of size and shape. Figure 4.6(c) shows the final results for a new zoning system that maximizes the number of Labour seats that could be created on the basis of the initial electoral ward results. Thus, in each of the zones shaded red, the Labour party would have the majority votes. A close inspection shows how the shapes of the new zones have been manipulated to encompass wards with low votes from other parties. In the same time, the Conservative wards are grouped together in larger and fewer Conservative 'constituencies', resulting in the lowest possible seats. The Liberal Democrat seat is also lost by successfully spreading the supporting wards into zones dominated by Labour voters.

For the final example, we run ZD2k with the same starting conditions, but try and *Help Conservative*, this time. After a few runs, the patterns in Figure 4.6(d) emerge, showing a very different picture of the political landscape. Surprisingly, by attempting to help the Conservatives, the Liberal/Democrats also benefit with an additional seat. The Conservative zones are expanded this time to touch all wards with a significant Conservative representation. It is also worth noting the decreasing size of the Conservative zones in order to satisfy the objective function by reaching the minimum electorate constraint. The power of zone design as an exploratory tool becomes apparent in this example, but it also serves to demonstrate how ZD2k can assist our understanding of spatial processes and models with the use of visualization.

4.5 CONCLUSION

The problem researchers face in using official boundaries with social data is that the boundaries of the areas have been defined without knowledge of the purpose of the research. Geographical boundaries are usually drawn by official bodies to define administrative areas, such as EDs, police beats or health districts. The last 30 years, with the realization of the importance of space in human behaviour and the adoption of geographical models by businesses and service providers, have seen the emergence of even more boundaries partitioning geographical space into divisions such as catchment areas, sales territories, and service provision domains. Every set of boundaries drawn by a governmental or non-governmental body results in a more or less formally established new geography. More often than not, these new geographies consist of areal units defined by arbitrarily drawn boundaries, following convenient natural or technical features, but bearing little relevance to the underlying socio-economic patterns (Morphet, 1992).

Zone design can be seen as a scientific way of manipulating aggregation effects in order to produce optimal zoning systems for specific purposes. The novelty of the approach is that while previous studies examined these effects in isolation, zone design is concerned with both the zoning and aggregation concurrently. This chapter attempted an up-to-date explanation of the zone design method and demonstrated two ZDES using three comprehensive case studies. In his widely cited study on the MAUP, Openshaw (1984) confidently stated: 'Very soon it will be possible to routinely apply the same methods to any spatial data set and any model or function, no matter how complex. When this happens often enough then a new geographical revolution will surely have occurred.' It took a quarter of a century for technology to catch up with Openshaw's groundbreaking ideas on tackling the MAUP. We have come a long way towards realizing the zone design method, but the technological advancements in hardware and software and the increasing data availability have opened up new windows to the study of the MAUP. We are presented with opportunities to improve optimization techniques by employing unprecedented computing power, developing new visualization tools by utilizing exciting software, and exploring data sharing by exploiting the cyberspace.

Acknowledgements

All census data used in this chapter are Crown copyright and supplied through the ESRC/Joint Information Systems Committee (JISC) Census Programme via the Census Dissemination Unit (CDU) at the University of Manchester. Boundary data are copyright of the ED-Line Consortium and are supplied through the ESRC/JISC Census Programme via the UKBORDERS service at the EDINA Data Centre, University of Edinburgh. Electoral data were kindly made available by Professor Danny Dorling of the School of Geography, University of Leeds. The ZD2k version of the zone design software was developed with funding under the ESRC 2001 Census Development Programme, award H507 25 5153 and was written by James Macgill of the Centre for Computational Geography, University of Leeds. Dr. Adrian Bailey provided valuable advice and comments on zone design as a participant in the ESRC project.

Box 4.1 Key web links for Chapter 4.

http://www.ncl.ac.uk/sarg/zonedesign	ZDES3b: ZDES, version 3b, for UNIX ARC/INFO.
http://www.ccg.leeds.ac.uk/census/zdes2001	ZD2k: ZDES, Web-based version.

Box 4.2 Additional CDS resources for Chapter 4.

ZDES3b Guide	Instructions on how to download fully functional version of ZDES from CD (UNIX version)
ZDES Demo	Demonstration version of ZDES showing how rezoning political constituencies can impact upon election results

5

Lookup Tables and New Area Statistics for the 1971, 1981, and 1991 Censuses

James Harris, Danny Dorling, David Owen, Mike Coombes, and Tom Wilson

SYNOPSIS

This chapter reviews the principal lookup tables that have been assembled to aid the comparison of data between the different zonal systems used for different censuses and also between census zones and alternative geographical systems. The construction and structure of the tables are reviewed, and extensive resources, including many of the lookup tables themselves, are provided in the CDS Resources that accompany this book.

5.1 INTRODUCTION

This chapter discusses the use of lookup tables to re-aggregate census statistics to different spatial units. Lookup tables are poorly understood and their value to research is not widely appreciated. However, these data sets provide the crucial link between the fixed geography of the census and the spatial units for which many research applications require data.

It is helpful, at the outset, to explicitly state what is meant by a lookup table. A lookup table can be defined simply as a data structure that links one set of data to another. The focus of this chapter is specifically on geographical lookup tables and the 1991 Census geography. There are, of course, many other types of lookup tables in common use (e.g., telephone directories linking names and addresses to telephone numbers, cinema listings giving cinema names and films showing, and book content pages listing chapters and page numbers). There are many ways in which the two data sets in a lookup table can be arranged, though in a lookup table consisting of data sets A and B, a simple distinction can be made between one in which data set A is in some kind of order and one in which B is in order (e.g., 1981 Census areas listed in census Office area code order in terms of 1991 Census areas or 1991 Census areas in terms of 1981 Census areas).

The plan of this chapter is as follows. In Section 5.2 the official lookup tables produced by the Census Offices are discussed. The next section describes the creation of a lookup table linking the 1971, 1981, and 1991 Censuses on the basis of the 1981 geography, followed by Section 5.4 discussing the production of a lookup table linking the 1981 and 1991 Censuses on the 1991 geography. Lookup tables linking census areas to functional regions form the basis of Section 5.6, while the next section examines the late 1990s local government reorganization in Great Britain and describes a lookup table that links the 1991 Census to the 1998 geography. The final section looks forward to the 2001 Census and includes some lookup table suggestions.

5.2 CENSUS OFFICES' LOOKUP TABLES

With the exception of the Samples of Anonymized Records (Chapters 14 and 15) and the Longitudinal Study (LS) (Chapter 16), all of the statistical outputs from the 1991 Census of population consist of areally aggregated data. That is to say, rather than providing information about individuals, the census data sets comprise counts of persons and households by area. Aside from the issue of confidentiality, it would be impractical to analyse the voluminous individual-level data, and therefore one of the objectives of aggregation is to achieve a level of statistical parsimony to reduce the large number of observations to a manageable set of descriptive statistics (Barr, 1993b). The smallest areas to which census data are aggregated are enumeration districts (EDs) in England and Wales (typically 200 households) and (Output Areas (OAs) in Scotland (50 households)). These areas form the building blocks of the census geography and nest into higher level administrative units, such as electoral wards and local authority (LA) districts. In 1991, this meant that the local authorities' statutory obligation to report statistics at ward level was met without need for further manipulation of the data. In addition to the postal address, two geographical codes were recorded on the census returns: the ED/OA code and the full postcode. This enabled the data to be aggregated to postcode sectors, districts, and areas, allowing comparison with other postcode-based survey data. Over time, both the administrative and postal geographies have been subject to continual change in response to demographic, political, and operational factors. This constant redefinition of geographical units presents a significant hurdle to researchers who wish to use the 1991 Census statistics in conjunction with many other data resources, including previous censuses.

The ideal solution to this problem would be to create new aggregations from the individual-level census database to reflect each new alteration to the boundaries, be they postcode sectors, wards, or any other administrative or user-defined zones. Indeed, with the advent of Ordnance Survey's ADDRESS-POINT product it is now technically feasible to aggregate the data to any arbitrary geography using the address as the atomic unit. Unfortunately this could potentially compromise the anonymity of the data, a risk that arises from the potential of identifying individual records from marginally different aggregations using a technique called *differencing*, an issue further addressed in Chapter 24. As a result, the Census Offices do not supply new aggregations and the only method of rezoning the data is to re-aggregate the existing published outputs. Re-aggregation is the process of assigning data values from one set of zones to another using geographical intelligence about the intersection of the two geographies. This geographical intelligence may come from detailed boundary data or a grid reference for the centroid of each zone or it may simply be descriptive information about the composition of the zones.

In the context of the census, the objective of re-aggregation is to define units composed of census zones that approximate as closely as possible to the target areas for which we require data. In the most straightforward scenario the target geography may consist of areas into which the census units nest perfectly, but in reality a degree of fuzziness will be introduced by the inevitable mismatch between the two geographies. Various methods can be applied to resolve the best approximate match between the two sets of areas, often involving an areal or population-based weighting factor. In most cases this will be achieved by overlaying the two geographies in a geographical information system (GIS). It would be highly inefficient, however, to repeat such a resource-intensive process each time we want to perform a new re-aggregation task. Instead, it is preferable to create a link file recording the allocation of census zones to the target zones. These files are variously called lookup tables (Cole, 1993; Australian Bureau of Statistics, 1996), constitutions (OPCS, 1993a), geographic equivalency tables, or correlation lists (Blodgett and Meij, 1997). They are produced by a number of organizations in the United Kingdom including the Census Offices, Ordnance Survey, and Royal Mail, as well as local authorities, research groups, and private sector organizations. Lookup tables are essential for census data re-aggregation as well as a range of other data manipulation and geographic analysis tasks, and yet they are poorly understood and inadequately documented.

The Economic and Social Research Council (ESRC)/JISC Census Programme has acquired a number of lookup tables that relate the 1991 Census geography to other spatial units, which, due in part to lack of suitable software for accessing and manipulating the data, are presently underexploited by academics. With the release of the 1991 area statistics, the Area Master File (AMF) was made available by Office of Population Censuses and Surveys (OPCS) and provided a 'universal' census geography-to-administrative geography lookup table for England and Wales (the Scottish equivalent was the Output Area/Higher Area Index (OAHA) produced by General Register Office Scotland (GROS). A sample of records from the top of the AMF is illustrated as Figure 5.1. These tables are available from the census site (*http://census.ac.uk*), and are linked from the enclosed *CDS Resources* CD. This made possible re-aggregation of the area-based data to a variety of spatial units including Regional and District Health Authorities, FHSAs, Urban Development Corporation zones, and Parliamentary and European constituencies. As each of these geographies has changed over time, the usefulness of the AMF has diminished, and it has largely been replaced by one-off lookup tables for individual geographies. An example of this is the new local government geography lookup table described later in this chapter.

Perhaps the most important lookup tables for census research are the postcode link files produced jointly by the Census Offices and Royal Mail: the Postcode/Enumeration District Directory (PCED) for England and Wales and the Scottish postcode index, again linked from *CDS Resources* and the census Web site. Annual releases of these files have enabled users to re-aggregate census counts to reflect the most up-to-date reorganization of postcodes, thereby allowing comparability with other survey data and the large resources of data collected by health authorities and local and central government for operational purposes. A welcome development in this area has been the All Fields Postcode Directory (AFPD) that relates unit postcodes to a range of administrative and statutory units and incorporates the PCED link to 1991 EDs for England and Wales. However, this file does not replace the AMF because it requires further processing to produce a usable link from the census areas directly to geographical units other than the postcode. Software has been developed by Yu and Simpson (2000) and Simpson (2001) and is accessible via *http://convert.mimas.ac.uk/afpd/*.

```
101      05AA0140032050Inner London
201AA    060260000274City Of London                                          0501
301AAFA  0100000013Aldersgate                                             28E 01
401AAFA01AAAA01TQ321381911     203    350  FF10   01  110   69K60227 F840501
401AAFA02AAAA02TQ322681861      75    156  FF10   01  110   69K60227 F840501
401AAFA03AAAA03TQ322181841     139    228  FF10   01  110   69K60227 F840501
401AAFA04AAAA04TQ322481761       1      1  FF10   01  110   69K60227 F840501
401AAFA05AAAA05TQ321781731     195    309  FF10   01  110   69K60227 F840501
401AAFA06AAAA06TQ322581681      43     68  FF10   01  110   69K60227 F840501
401AAFA07AAAA07TQ320981891       1      4  FF10   01  110   69K60227 F840501
401AAFA08AAAA08TQ321681451       6     14  FF10   01  110   69K60227 F840501
401AAFA09AAAA09TQ320681481                 FF10   01  110   69K60227 F840501
401AAFA10      TQ320381562       0    116  FF10   01  110   69K60227 F840501
301AAFB  0050000014Aldgate                                                28E 02
401AAFB01AAAB01TQ333881321                 FF10   01  110   69K60227 F840501
401AAFB02AAAB02TQ333781211       9     21  FF10   01  110   69K60227 F840501
401AAFB03AAAB03TQ332581111                 FF10   01  110   69K60227 F840501
401AAFB04AAAB04TQ335681131       1      2  FF10   01  110   69K60227 F840501
401AAFB05AAAB05TQ335480911                 FF10   01  110   69K60227 F840501
```

Notes:

Field	Variable
1	Area type (1=county; 2=district; 3=ward; 4=ED)
2–3	County code
4–5	District code
6–7	Ward code
8–9	ED code
10–15	ED code in 1981
16–25	Centroid grid reference
26	ED type (1=standard; 2=special; 3=shipping)
27–35	Number of private households in ED
36–44	Number of persons in ED
45–46	New town
47	Regional health authority
48–50	District health authority
51–55	Parish
56–57	Central statistical area
58–59	National park
60–64	County electoral ward
65–67	Parliamentary constituency
68–69	Urban development corporation
70–71	Housing action trust
72–73	European parliamentary constituency
74–80	Urban area

Figure 5.1 The first few records of the AMF.

As each successive census adds to the corpus of digital demographic data, an increasingly important research activity involves comparing data from different censuses. Incompatibilities between population bases and cross-tabulations are only one of the problems that present themselves with change over time. The lack of a common geography between the 1981 and 1991 Censuses meant that comparative studies were all but impossible at a fine spatial resolution, and this is another area where lookup tables have made an important contribution to census research. The construction of 1981 to 1991 intercensal lookup tables is discussed next in this chapter.

5.3 THE 1991 CENSUS IN TERMS OF THE 1981 CENSUS GEOGRAPHY

This section describes the work undertaken to produce a lookup table of 1981 Census wards in terms of 1991 EDs and OAs covering Great Britain. It includes a description of what has been done, together with details of, and justification for, the methodology employed. Dorling and Atkins (1995) and Atkins *et al.* (1993) give full reports—including statistical and graphical analysis of the resulting data, pseudo-code of some of the algorithms used, and tables highlighting the more interesting spatial changes.

A number of factors, including the availability of boundary data for 1981, but not at the time for 1991, and the existence of a lookup table of the 1971 Census in terms of 1981 Census wards, led to the selection of 1981 Census wards as the consistent areal base to be adopted for this study.

It became apparent before work had progressed very far that the initial geographical referencing of the 1991 Census data was not as reliable as had been hoped, and consequently a large amount of manual checking was needed to facilitate the production of an allocation table linking each 1991 ED to its best fitting 1981 ward. Digitized boundaries did not become available for academic research until after the work had been completed. Four independent allocations of a 1981 ward were made for each 1991 ED. This was done despite knowing that a large number of wards had unchanging boundaries so that the method as a whole could be checked. The data sources used include the 1981 and 1991 ED level Small Area Statistics (SAS) files, the 1981 and 1991 AMFs, the 1981 digitized ward boundaries and the 1991 PCED file. The four allocations made were then combined into one robust lookup of a best-fitting 1981 ward for every 1991 ED. This process was largely automated, although the programs used highlighted a large number of cases in which further manual investigation was necessary. When the available data had been combined into a single lookup table, there still existed an unacceptable number of misallocations of EDs to wards. To remedy this, a number of additional checks were performed, cross-checking the data in the lookup file with various other sources, paper-based maps of census geography and documentation on boundary changes.

The last stage of the work reported here was the production of 1991 SAS files aggregated to the 1981 ward level. The SAS files produced, along with the final version of the lookup file, have been deposited at the ESRC Data Archive at Essex University and at the Census Dissemination Unit at Manchester so that they can be of benefit to other researchers. The lookup table is also available as part of *CDS Resources*, which provides a linked set of 1971–1981–1991 SAS counts aggregated to common 1981 ward boundaries together with a program for the user to extract data selections.

Attention should be drawn to the following possible sources of error in the data sets:

- The process of aggregating blurred counts has the potential to introduce significant errors (especially where the total counts are low, which is likely to be the case when performing analysis at ward level). This does not affect the population totals used in this study, as they were never blurred
- Where 1991 EDs are split between 1981 wards, misallocation is inherent to the process of assigning them solely to one or other of the wards they span. The misallocations will 'cancel out' locally, however, and so should have almost no effect on, for instance, district level analysis of ward-based statistics
- Census data for residents in special EDs that contain very small communal establishments are suppressed in the ED level SAS. Consequently these people cannot be allocated to a 1981 ward using the methods outline here. This affected approximately 0.06% of the 1991 population. The influence of this effect should have been insignificant for this study, and at higher geographical levels these people are included in census output
- Where the census variables for an ED were imported or exported by OPCS to other 1991 EDs (to maintain confidentiality) that were across 1981 ward boundaries, there will be local misallocation of the population. Again, this does not affect the population counts used in this study but would be important in particular areas for studies based on statistics other than population counts
- Visitors who did not provide an accurate address of their usual residence were not included in the 1991 ED SAS, and hence could not be allocated. This affects approximately 0.4% of the population, and would have been a similar problem in 1981 and 1971 so is unlikely to bias the results shown here
- All the errors associated with the undercount of 2.2% of the population in 1991 remain. This is by far the most significant source of error and eclipses the aforementioned problems. The remedy used for this study was the choice of a changing population definition that behaved similarly to the national level changes

This work resulted in the production of a lookup file of 1991 EDs to 1981 wards, covering all of England, Wales, and Scotland, and the creation of the full set of 1991 10 and 100% SAS tables aggregated to 1981 ward level. The lookup file allocates a best fitting 1981 ward to each 1991 ED, and the grid reference of the ED centroid is given along with an indication of the source from which the 1981 ward code was derived. The 10 and 100% SAS files are in a similar format to the OPCS ward level SAS tables.

Following the 1981 Census, a great deal of work was done to link together the 1971 and 1981 Censuses, in terms of both spatial equivalence and comparability of the variables. For a criticism of the census tract method of linking censuses see McKee (1989). The 1971 census data were aggregated to 1981 ward level at Newcastle University (Dorling 1991). As a consequence of this, the vast majority of change-over-time analysis at Newcastle, covering the 1971 to 1981 period, makes use of the 1981 ward level (or aggregates of 1981 wards) as its areal base. Because 1981 is also the mid-point in the three census periods from 1971 to 1991, the 1981 areal units are a sensible base for studies covering this period. The alternative approach, aggregation of 1981 EDs to 1991 wards, would not facilitate comparisons to the 1971 Census (unless a similar exhausting task was undertaken with the 1971 Census).

The development of the lookup file was a three-stage process, following which the transformed SAS files were derived from the refined version of the lookup file. The first stage made use of OPCS ED level data, in particular, the ED centroids (for both 1981 and 1991 EDs), together with the digitized 1981 ward boundaries. These data were used to produce three alternative lookup files from 1991 EDs to 1981 wards. The grid references of the 1991 EDs and 1981 ward boundary data were then incorporated within a GIS database. With this visual aid a large number of EDs, suspected to be in the wrong positions, were checked interactively using on-screen maps (see Atkins *et al.*, 1993). The second stage made use of the PCED file, which contains details of individual unit postcode centroids and population counts and the 1991 EDs within which they fall, for all of England and Wales. Data from this file were used to produce population weighted ED centroids, which were allocated to 1981 wards by the point-in-polygon algorithm. This produced a further independent lookup file of 1991 EDs to 1981 wards. This was not necessary in Scotland where the OA naming scheme contains an intrinsic link to 1981 Census postcode geography. In the third stage, the lookup files created in the previous two stages were combined to produce what became, after further scrutiny and revision, the final lookup file. This was achieved by choosing the 1981 ward indicated by one or other of the lookup files, the choice of which to use being determined by the matching between the 1981 wards suggested by the various lookup files for each 1991 ED.

Numerous checks were then carried out on this penultimate lookup file and corrections made accordingly. Ensuring one to one matching of wards in districts that have not had ward boundary changes between the censuses, checking the distances between EDs within each ward, and drawing 'ward polygon' diagrams (see Dorling and Atkins, 1995: Figure 19) were among the checks carried out. When all the checks had been carried out, the lookup table was used to produce 1991 SAS data files aggregated to the 1981 ward level, from 1991 ED level SAS data.

5.4 THE 1981 CENSUS IN TERMS OF THE 1991 CENSUS GEOGRAPHY

One of the most serious omissions of census planning in England and Wales is the failure to provide a definitive link between the geographical units used by successive censuses, in sharp contrast to Scotland and Northern Ireland. After the 1981 Census, an attempt to address this problem was made by the definition of a set of census 'tracts' forming spatial common denominators between 1971 and 1981 EDs. The availability of these spatial units plus OPCS/GROS definitions of variables in terms of the 1971 and 1981 Censuses greatly facilitated the analysis of change over the decade 1971 to 1981 (Champion *et al.*, 1987). Despite the dramatic improvement in computer power and data-handling techniques, the ability to analyse change over the decade 1981 to 1991 was made harder by the simultaneous changes in census geography, census base and variables. However, the one advance was in the availability of digitized boundaries for census areas.

The work of Atkins *et al.* (1993) in using a variety of techniques to identify the way in which 1991 EDs were related to 1981 Census wards, provided a resource for spatially linking data from the 1971, 1981, and 1991 Censuses in the form of a lookup table (which they made available to the academic community on the Manchester Information and Associated Services (MIMAS) computer service). This data set has been used to create two files that describe the spatial relationship between the wards used in 1981 and 1991. There is a large amount of socioeconomic data for which the ward is the smallest spatial unit (e.g., monthly unemployment data and annual vital statistics data). Some of these data sets are collected using 'current' wards (i.e., the electoral wards that exist at the time the data are collected), whereas others use 'frozen' wards, usually those defined by the most recent census. Through the creation of estimates for one set of ward definitions by converting data collected for a different set of wards, it becomes possible to create longer spatial time-series of data or estimate rates using denominators collected for a different set of wards than the numerator.

Two converters were created; one for translating data collected on a 1991 ward base to a 1981 ward base; and the other to convert data collected for 1981 wards to a 1991 ward base. An example application of the former would be the creation of a 1991 economic activity data set for 1981 wards, enabling ward-level unemployment rates to be calculated from JUVOS unemployment data. The latter converter was used in a project to estimate ward-level population change by ethnic group, converting 1981 Census data on country of birth to a 1991 ward basis, in order to enable comparison with 1991 Census data on population by ethnic group (Owen and Ratcliffe, 1996). This form of conversion is useful where 1991 data are richer and more complex than 1981 data, and thus the possible loss of information incurred through the approximation involved in the conversion is less great for 1981 Census data.

Each converter takes the form of a set of 'factors' that distribute the population resident in a ward as defined by one census to the wards defined by another census that cover the same territory. The method relies on having population data for smaller spatial units that can be located within the each set of wards, in this case 1991 EDs with 1981 ward codes.

The first step is to read the lookup file and create two linked lists, one for 1981 wards and one for 1991 wards (identified from the first six characters of the 1991 ED code), which identify for each 1991 ED the sequence number of the next ED within the same ward. The sequence number of the first ED in the ward is also noted. During this first pass, 1991 population totals for each 1981 and each 1991 Census ward are calculated.

In the second step, the program works though each ward in turn, identifying all combinations of wards from the two censuses, and summing the 1991 population for each combination. In the third step, the program works through each ward in turn, converting population sums into proportions of the 1991 population of the ward. The same basic methodology works in both directions, yielding factors that split 1981 wards into 1991 wards and vice versa. The 1981 to 1991 lookup table can be found in *CDS Resources*.

5.5 LINKING THE 1991 CENSUS TO FUNCTIONAL REGIONS

This section of the chapter outlines the reasons—and the methods—that led to the creation of lookup tables linking 1991 Census data to:

- the CURDS functional regionalization (Coombes *et al.*, 1995), and
- the localities and city regions (Coombes *et al.*, 1996).

These two sets of areas illustrate how lookup tables are essential to sets of boundaries that have been created purely for research purposes.

Analyzing data from the census almost always provokes the question of which areas to use for the analysis (Openshaw, 1984). This question arises because census data are collected, assembled, modified, and disseminated within a hierarchy of areas (Coombes,

1995). The fact that there is a choice of areas for which data will be available allows researchers to decide which set of areas is in principle the most relevant to the research questions being examined. If the research is not explicitly dealing with local politics or administration, then there is no reason to assume that local authority areas—which figure so largely among the census published data sets—are indeed the appropriate areas to study. For example, the local authority area of Nottingham includes only a part of that urban area, whereas the nearby city of Sheffield includes not only the urban core but also some of the surrounding rural areas. As a result, data on the two cities' local authority areas cannot be reliably compared in studies of the many social and economic factors—such as ethnicity and unemployment—that tend to vary between any city's inner areas and its more suburban and rural surroundings (Coombes, 1997). As recognized in many countries (e.g., Dahmann and Fitzsimmons, 1995), the need is to group together each city with those nearby rural areas that are closely linked with it: when this grouping is undertaken on a consistent and meaningful basis, the sets of boundaries that are created provide meaningful and comparable definitions for all cities. Areas defined in this way will then be much more appropriate for many research purposes than are local authority areas that are the standard areas for most published census outputs.

No doubt, one reason for lookup tables to be considered 'essential or highly desirable' by 64% of the researchers surveyed by Rees (1998a) was that without them analysis of change over time is scarcely possible. The reason is that each census is tied to the sets of local authority areas that are in existence at the time. Thus, a lookup table becomes essential to provide a link between the areas used in one census and those used for another. Researchers who choose to use non-standard areas will in fact require a separate lookup table to group data to their areas of interest from each census and its specific set of building block areas. One consequence here is that a slightly different pattern of errors results each time building blocks from different censuses are 'best fitted' to non-standard areas. In practice, these errors in lookup tables are only likely to be significant for areas with small total populations—or to be more precise, sets of areas with a low number of building block areas per 'output' area (Coombes 1995)—but this is not the case for either the functional regions or the localities that are of interest here. In addition, both these sets of areas tend to group together adjacent urban and rural areas, so their boundaries rarely run through the urban fringe areas that are the most dynamic in terms of new development and which, as a result, are also where building block areas are most likely to be changed from one census to the next. In short, both the sets of non-standard areas of particular interest here are largely immune to the two main sources of error to which lookup tables can be prone.

Researchers wishing to analyse change through time need to choose which time-point to use as the basis for the definitions of the units within which the change will be traced. For example, a study of the growth of new towns will almost certainly need to use boundaries that are defined in terms of the built-up areas at the latest time-point so that the data can show the new towns 'growing into' their current state. In contrast, an analysis of counter-urbanization might well use boundaries that reflect the limits of the built-up areas at the start of the period so as to monitor the growth taking place in those areas that were initially rural. The two sets of boundaries examined here provide for both perspectives:

- the functional regions reflect British towns and cities as they were in 1971, and
- the localities' definitions are primarily based on 1991 Census data.

With their emphasis on grouping cities with their hinterlands, both these sets of boundary definitions aim to identify areas that are 'functional' in the sense that most patterns of flows and interactions tend not to cross their boundaries. The earlier—functional region—definitions relied exclusively on commuting patterns for evidence on flow patterns, with the method of definition an advance upon those that had evolved over several decades in the United States (Dahmann and Fitzsimmons, 1995). By contrast, the locality definitions were far more ambitious because they drew upon migration and other flow data sets in addition to commuting; an entirely new form of analysis had then to be devised to collate the evidence on patterns of interaction that are present in a variety of very different sources of information (Coombes and Openshaw, 2001).

In terms of lookup tables, the Functional Regions, which were defined specifically to support 1971 to 1981 Census change-over-time analysis, benefited from the innovation in 1981 of widely accessible computerized ward-level data. The original boundary definitions had been derived from the 1930 areas in the 1971 Census commuting data set. The remaining process was a straightforward but time-consuming inspection of maps of the 9267 1981 wards to create a lookup table to the Functional Region Zone boundaries. The subsequent allocation, of the 1991 Census wards to the 1981-based boundaries, used the Atkins *et al.* (1993) lookup table of 1991 EDs to 1981 wards, following the computerized approach described earlier in this chapter. Figure 5.2 shows how the 1991 wards were allocated to Functional Region Zones according to the assignment of the majority of their 1991 population, as indicated by the allocation of their constituent EDs (which was derived via the Atkins *et al.* link to 1981 wards).

In the case of the localities that were defined more recently, lookup tables were inherent to the process of definition itself. The main building blocks for the definitions were to be 1991 wards, but the aim was to draw upon a very wide range of information, which in turn led to relying upon source material based on many different building block areas.

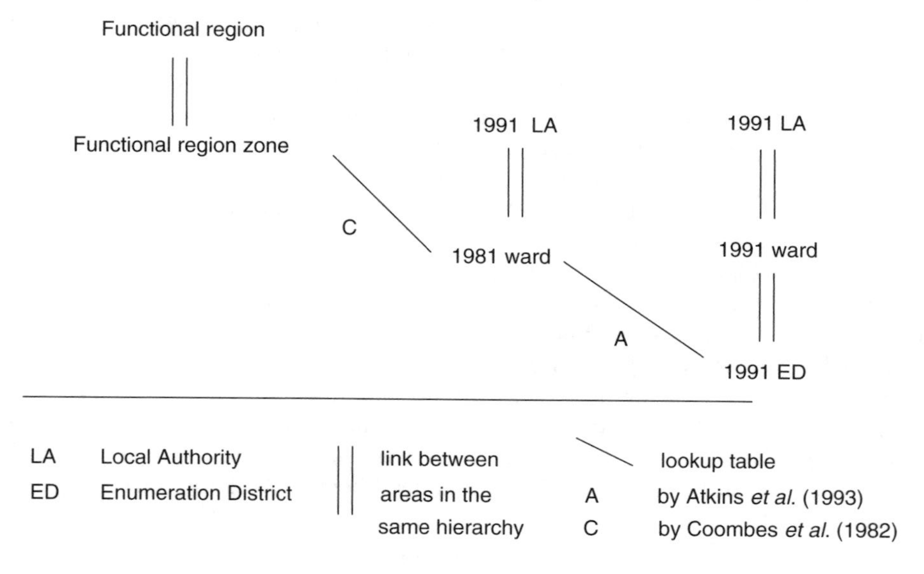

Figure 5.2 Deriving a lookup table for functional region zones.

Some of the information was in the form of boundaries that had been linked to 1981 wards by Owen *et al.* (1986), so these could be linked to 1991 wards through the same method as was used for Functional Region Zones (Figure 5.2). In other cases, a point-in-polygon operation was carried out to identify into which of the areas of interest each 1991 ward's centroid fell. Coombes and Openshaw (2001) describe how the innovative next step was to create 'synthetic data', which, in effect, answer the question 'within how many of the sets of areas examined was this pair of wards grouped together within the same area?' The matrix of these values was then input to the Coombes *et al.* (1986) regionalization algorithm and the localities were defined accordingly. Thus, the building blocks for defining localities were 1991 wards, but the definitions were based on 'synthetic data' whose very creation depended fundamentally upon lookup tables. The functional regions lookup table, localities and city regions lookup table, list of the localities and city regions, and list of functional regions are all provided in the accompanying *CDS Resources*.

5.6 THE 1991 CENSUS IN TERMS OF THE NEW LOCAL GOVERNMENT GEOGRAPHY

The local government structure and geography of the United Kingdom changed substantially between the time of the 1991 Census and April 1998. Many two-tier local authorities were replaced by a single tier, and a substantial number of boundaries were revised. In order for researchers to obtain 1991 Census data, and other data based on the census geography, for the 1998 set of local authorities, and to facilitate a straightforward comparison with the results of the 2001 Census (assuming no more boundary changes occur), a set of lookup tables were constructed. This section begins with a brief description of the need to create the lookup tables—the introduction of a new geography—and then moves on to discuss the lookup tables themselves, and the SASPAC91 programs and data sets based on them that are available to download from the Web.

The change in geography resulted primarily from the wholesale reorganization of local government that was implemented over the period April 1995 to April 1998. A very brief description of the changes follows. More details can be found in Chisholm (1995), Jackson and Lewis (1996), Johnston and Pattie (1996a, 1996b), Leach (1994), and Wilson and Rees (1998, 1999). Maps showing the local government geography of Great Britain in both 1991 and 1998 are available as part of *CDS Resources* (Figures 5.3, 5.4, and 5.5). The idea of reorganization originated with the Conservative government of the early 1990s. It believed that there was a strong argument for replacing the two-tier structure of local government that existed outside the English metropolitan areas with single-tier authorities. This, it argued, would reduce the duplication of services between upper and lower tier councils, and would eliminate the public's confusion as to which council was responsible for which particular service, thus increasing accountability. In Scotland and Wales, proposals were drawn up; little public consultation took place; minor amendments were made, and the altered proposals became law in the Local Government etc. (Scotland) Act 1994 and the Local Government (Wales) Act 1994. The new local government organization of Scotland and Wales came into effect on 1st April 1996. In Scotland, the nine regions and 53 districts and island areas were replaced by 32 single-tier council areas; in Wales the eight counties and 53 districts were replaced by 22 unitary authorities. In England, however, matters were not so simple. The government established a Local Government Commission to review local authority boundaries in England and discuss options with

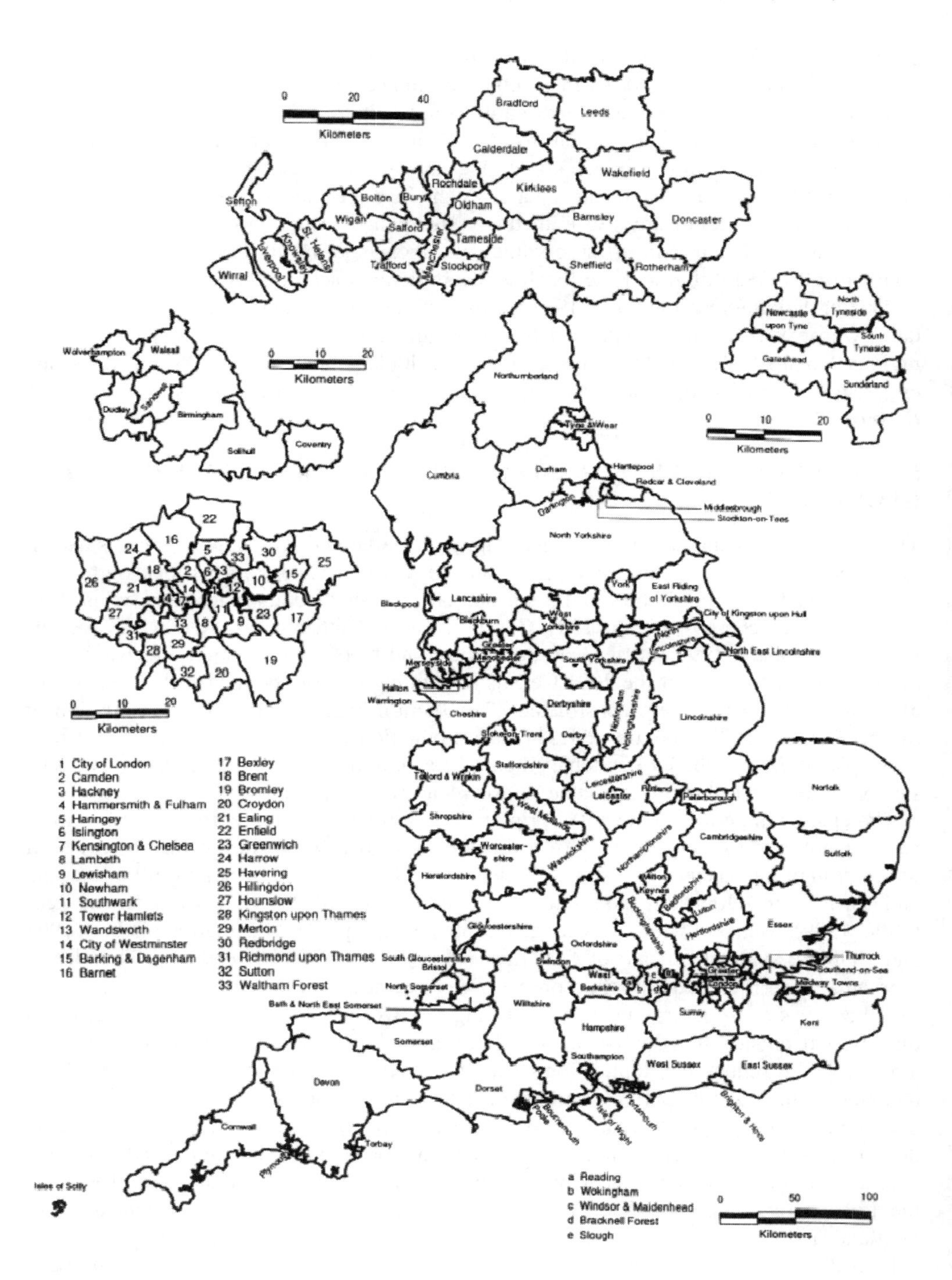

Figure 5.3 Local government authorities, 1998: England.

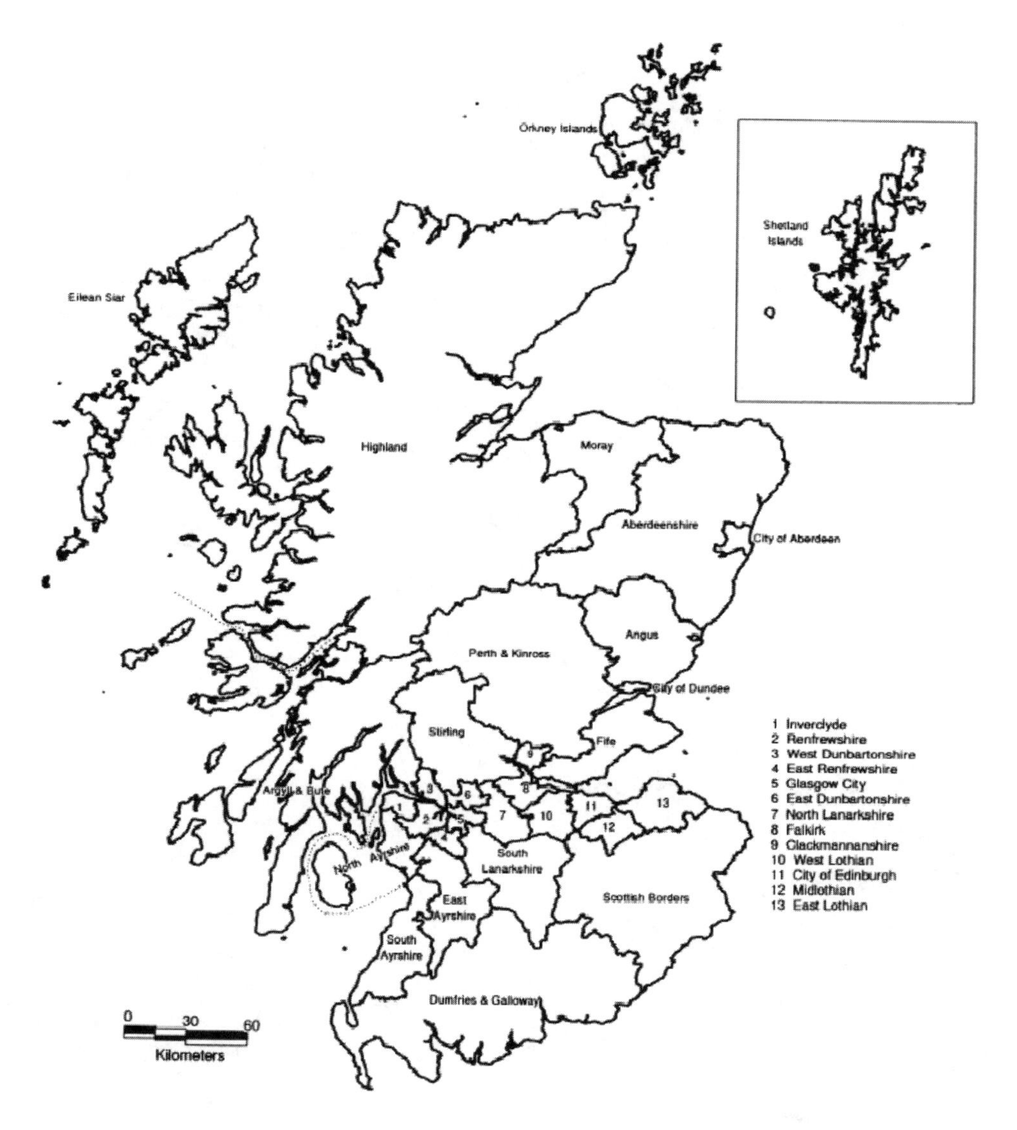

Figure 5.4 Local government authorities, 1998: Scotland.

councils, local people, and other interest groups. Not surprisingly, conflicts of interest emerged between different parties, and the review process took longer than originally envisaged. The eventual outcome was considerably different to the all-unitary authority option the government had hoped for. Eighteen two-tier counties were left unchanged; a further 16 underwent changes in their boundaries, but not their structure, four were abolished, and one (the Isle of Wight) underwent a change to its structure (but obviously not its boundaries!). Unitary authorities were created only for the main towns and cities. Thus, pre-reorganization non-metropolitan England consisted of 39 counties, and 296

Figure 5.5 Local government authorities, 1998: Wales.

districts within those counties, while the new reorganized local government geography comprised 46 unitary authorities and 34 counties, these counties being made up of 239 districts. The new organization for England was introduced in four phases on 1st April 1995, 1996, 1997, and 1998. Local government boundaries were also reviewed in Northern Ireland in the early 1990s. Some small changes were made in the Local Government (Boundaries) Order (Northern Ireland) 1992.

However, the 1998 local authority geography is not only the result of this reorganization. A significant number of small boundary changes took place in the period between the 1991 Census and reorganization. These minor boundary changes may seem rather insignificant compared to the wholesale redrawing of boundaries, but it is still important to account for them. This is especially the case when local level time-series population data are being analysed: changes in population size that appear to be fairly large may be due less to demographic change and more to boundary change. For example, the London borough of Brent's mid-1991 population on 1991 boundaries was 248 600 while on 1998 boundaries it was 242 300 thousand. The estimated population of the borough in mid-1997, on the new boundaries, was 249 600. The real increase in population between 1991 and 1997 was 3.0%, but if the change was incorrectly calculated using the different boundaries of 1991 and 1998 then the increase would appear to be 0.4%.

Perhaps the most useful lookup table linking the 1991 Census with the new local government geography is one that describes the 1998 local authorities in terms of 1991 Census areas (rather than the other way round). Such a lookup table has been compiled using information obtained from Office for National Statistics (ONS) (for England and Wales), the General Register Office for Scotland, and the Northern Ireland Statistics and Research Agency, and is available in *CDS Resources*. This lookup table has also been embedded in a number of simple SASPAC91 programs that can be used to extract user-defined 1991 Census data for the new geography, and are included in *CDS Resources*. Details of how to use SASPAC91 can be found in the *SASPAC User Manual* (London Research Centre, 1992), and on the MIMAS Web site, *http://www.mimas.ac.uk/*. These SASPAC91 programs comprise two parts: the first part reads in the original 1991 Census system files for the 1991 geography and then creates a new set of system files for the 1998 local authorities; the second part then reads in the new system files as would any ordinary SASPAC91 program and extracts whatever census data has been requested by the user. The programs are initially set up to extract just the total resident population. One set of these programs extracts the SAS, the other set extracts the local base statistics (LBS).

The lookup tables mentioned in this section have been used to produce a variety of data, and two data sets are available to users of *CDS Resources*. The first is a re-worked version of the set of tables included in the 1991 Census monitors and at the beginning of the published reports, produced using the lookup tables in the SASPAC91 programs for the LBS. The second data set is a file of mid-1991 population estimates for the 1998 local authorities (Great Britain only). As described in detail in Wilson and Rees (1998) this was produced using the ED and OA mid-1991 estimates produced by the Estimating with Confidence project. Details of this project are given in Simpson *et al.* (1995), and Simpson *et al.* (1997a), and the new geography population estimates are also available in *CDS Resources*. Further, a number of the lookup tables described here have been used in the creation of a new Web interface for linking censuses through time (LCT), which provides access to 1971, 1981, and 1991 data referenced to 1981 wards and re-aggregations to many other geographies. This project has also included extensive recalculation of 1991 area statistics to 'correct' for the 1991 undercount, discussed in Chapter 12.

5.7 CONCLUSION: LOOKUP TABLES FOR THE 2001 CENSUS AND BEYOND

Plans for the 2001 Census outputs include a number of important developments for lookup tables, including the availability of such tables under common licensing arrangements, and the production of a new table identifying the relationship between unit postcodes and 2001 OAs (see Chapter 3). Cole (1998) has suggested a number of further features these lookup tables should possess, and progress is being made towards these objectives:

- they should link 2001 Census areas to previous Censuses;
- they should be updated as boundary changes occur;
- lookup tables should cover the whole United Kingdom; and
- lookup tables should be integrated with digitized boundary data.

Box 5.1 Key web links for Chapter 5.

http://census.ac.uk/cdu/lct/	Interface to the Linking Censuses through Time data sets
http://census.ac.uk/cdu/Datasets/Lookup_tables/Area_Master_File.htm	Area Master File. A list of EDs and the larger areas in which they nest (England and Wales)
http://www.census.ac.uk/cdu/Datasets/Lookup_tables/Output_Area_Higher_Area_Index.htm	Output Area Higher Area Index. A list of OAs and the larger areas in which they nest (Scotland)
http://census.ac.uk/cdu/Datasets/Lookup_tables/Postal/Central_Postcode_Directory.htm	Postcode-Enumeration District directories (annual). A list of unit postcodes and the EDs in which they nest (England and Wales)
http://census.ac.uk/cdu/Datasets/Lookup_tables/Postal/Scottish_Postcode_Index.htm	Scottish postcode indexes (annual). A list of unit postcodes and the OAs in which they nest (Scotland)

Box 5.2 Additional CDS resources for Chapter 5.

Linking 1971–1981–1991 Censuses	Software for the linkage of 1971–1981–1991 Censuses, including data based on 1981 ward-level SAS
1981–1991 ward lookup tables	Lists of 1981 wards in terms of 1991 EDs and OAs and 1991 wards in terms of 1981 EDs
Census monitor data	Includes boundary maps and re-aggregated data published in census monitors
Functional and city regions lookup table	A list of 1991 wards and the localities in which they nest

Part II

Boundary Data and Visualization

6

Handling and Accessing Census Boundary Data

William Mackaness and Alistair Towers

SYNOPSIS

This chapter introduces the principles of digital boundary data (DBD) and provides an overview of the data available for census geographies. Issues of data size are considered and the need for map generalization is explained. An outline of the UKBORDERS system for access to UK census boundary data is given.

6.1 INTRODUCTION

The subject of this chapter is the digital representation of census area boundaries, generically termed *DBD*. Census data describe characteristics or 'attributes' of groups of persons and households at a particular point in time—census day—in the form of 'counts' or aggregates. Because it is aggregates that are published, census data are inherently spatial.

Comparison between one or more groups in different areas is one of the principal uses to which census data are put. The results of such spatial and (where data from more than one census is used) temporal comparisons provide the basis for key demographic, socio-economic, and cultural insights. As a result, census data have application in all sectors—academic, local, and national government, public utilities, health service, and commerce—as a source of basic information for research and teaching, as well as decision making in resource allocation.

Because they are inherently spatial, the mapping of census data has a long history. Computer mapping of census data has been going on since the 1970s (Smith, 1985). But it is only in the last few years that the mapping of census data by computer has been possible for most users. The combination of falling cost, increasing power and 'user-friendliness' of desktop computers, mapping, and geographical information systems (GIS) software, has meant that it has become a fairly straightforward exercise for most users with little previous experience to map census and related data (Rhind, 1997), so long, that is, as the data, which they wish to map, can be linked to the digital representation of points or areas.

6.2 CENSUS GEOGRAPHIES

The aggregate areas for which census data are published are known collectively as 'census geography'. Because the censuses for England and Wales and Scotland are conducted on a different geographic basis, there are, in fact, two separate census geographies for Great Britain. The basic levels of aggregation for these geographies for 1991 are described in Table 6.1. It should be noted that the areas for England and Wales on the one hand and Scotland on the other, 'nest' into each other. In other words, all of the areas for England and Wales above enumeration district (ED) can be constructed from EDs; similarly, for Scotland the basic 'building blocks' of the higher aggregate areas for Scotland are the output areas (OAs). It is notable that, because the basic 'building blocks' of census reporting areas are different for England and Wales (EDs) and Scotland (OAs), it is only from the level of the 'district' up that the same types of area are available for Great Britain as a whole. In other words, there is no common 1991 Census geography for Great Britain below the level of the local government districts that existed between 1974 and 1996 to 1999.

There are, however, some compensations for this lack of common geography. In Scotland, the 1991 OAs are based on postcode units 'frozen' by General Register Office Scotland (GROS) for census purposes on 31 December, 1990. From this, all postcode sectors in Scotland were digitized by GROS to provide the geographical basis for the construction of OAs. Thus, while census counts are not available at the unit postcode level, the postcode provides a powerful link between the increasing amount of non-census data that is available to researchers and teachers and census data itself. This means that data from an increasingly wide variety of other sources that are postcode based can be used in conjunction with census data. Similarly, in England and Wales, because EDs nest directly into wards, the census can be used in a more direct manner by local government, for example, in calculating resource allocations.

Census boundary data and boundaries derived from them also have uses outside the mapping of census data. In England and Wales, for example, local government district ward boundaries can be used as the basis of chloropleth maps of electoral results, whereas in Scotland OAs could be used to display market research results collected on a postcode basis. In addition, the large number and small size of the census building blocks available means that users can construct new, 'designer', aggregates for their own specific purposes,

Table 6.1 1991 Census—principal reporting areas for which digitized boundaries are available.

Area	England and Wales	Scotland	Great Britain
Enumeration districts	113 196	N/A	113 196
Output areas	N/A	38 255	38 255
Electoral wards	9930	N/A	9930
Postcode sectors	N/A	1003	1003
Districts	403	56	459
Counties/regions	54	12	66

which may or may not be census related. Five geographic categories have been identified as containing all the geography types that can be derived from the basic census building blocks: census, administrative, electoral, postal, and 'other' (Burnhill and Morse, 1993). Often, the same boundary set can be fitted into more than one category. For example, district boundaries can be placed under census, administrative and electoral (as the aggregate of their wards); for Scotland, sectors can be placed in both census and postal. The multiplicity of uses to which boundary data can be put makes them a key product of the census that is of great utility outside its primary purpose.

6.2.1 What are Digitized Boundary Data?

Digitized boundaries can be best understood as a series of lines connecting points related to a geographical projection, which locates the points to places in the real world. The elements used in the construction of digitized boundaries are illustrated in Figure 6.1. The basic item of information required to construct digitized boundaries is the geospatial reference for each point in the form of an x, y coordinate. Connecting two points having the coordinates x_1, y_1 and x_2, y_2 with a line, forms an arc. A digitized boundary represents an enclosed or 'bounded' region (i.e., a region enclosed by a series of three or more arcs). A polygon may be a recognizable topographic unit, for example, a property boundary delineated in the real world by a fence; more frequently, digitized boundaries represent cultural areas that are not delineated by topographic features, for example, postcode units or census EDs. It should be noted that while expressions such as 'arc', 'polygon', 'region', and so on, are commonly understood in relation to elements of GIS, each software package uses its own terminology. In part, this is a symptom of the fact that computer mapping and GIS are developing technologies, allied with the desire of each software company to establish its product as a 'standard'.

Each polygon, which represents a digitized boundary, normally has one other type of point associated with it, which is not joined to any of the other points by an arc, but is of critical importance for the mapping of attribute data, the centroid. A centroid, as its name suggests, is usually in the centre of a polygon. It is critical because it is

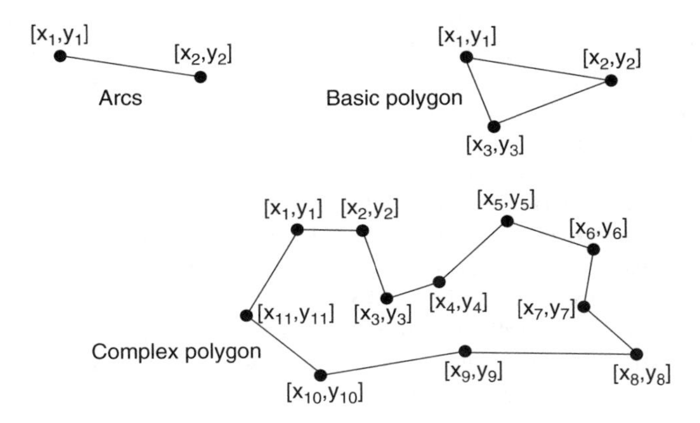

Figure 6.1 The graphic primitives used to define boundary data.

with the centroid that data for a polygon is associated. When producing a choropleth map, the data attached to a centroid are either spread out within the area defined by its boundary or represented at the centroid by, for example, a pie chart, bar graph, a symbol, or text.

6.2.2 A Brief History of Digitized Boundary Data

Census-related DBD first became available widely to researchers and teachers at UK Higher Education Institutions (HEIs) in the late 1980s, when DBD down to the level of the electoral ward were purchased for England and Wales by a consortium of HEIs, and for Scotland down to the postcode sector level by Edinburgh University Data Library. But it was not until the digitization of boundaries down to the level of EDs had been completed by the ED-Line consortium for England and Wales and OAs by the General Register Office for Scotland GROS, as part of the preparatory work for the conduct of the 1991 Census, that a central purchase of boundary data for Great Britain was completed on behalf of all UK HEIs by the Economic and Social Research Council (ESRC), under their 1991 Census Initiative. The 1981 Census was the first census for which a suite of boundary products was purchased by a consortium of higher education institutions. Because of the then high cost of line boundary capture, data down to the aggregation of the ward was captured, but not the EDs. The wards nest within the travel to work areas and the parliamentary constituencies as well as both the districts and counties. With these key boundary types captured in 1981 it made sense to capture these boundaries in the following census to ensure the ability to look at social and economic changes over the decade. Thus, the boundaries used for research post-1981 had a large influence on those captured in the following census.

6.3 METHODS OF ACCESS—THE UKBORDERS SYSTEM

UKBORDERS was set up under the ESRC 1991 Census Initiative to provide an on-line extraction system for the UK Higher Education Community, allowing access to a wide range of census and other digitized boundary data. UKBORDERS is now jointly funded by the ESRC and JISC as one of the range of services provided by EDINA National Data Centre (see *http://edina.ed.ac.uk*). UKBORDERS also provides technical support for users and continuing quality control of the digitized boundary data to which it provides access. The main aim of the UKBORDERS system is to ensure that users obtain the key spatial data they require from a data warehouse in as timely a manner as possible. To this end, a considerable amount of effort has been put into the design, construction, implementation, and maintenance of the extraction system.

Figure 6.2 provides an outline of the UKBORDERS system. A request (step 1) is processed by the 'Perl Gateway' (Morse *et al.* 1998), which then (step 2) sets the parameters required by querying the Ingres meta database in which information describing the DBD is held in the GIS (ARC/INFO) (Morse *et al.*, 1998). This step is repeated until the user has defined their search to identify the correct time period, boundary type, and area that they require. The Perl Gateway subsequently uses the information acquired (step 3) from the meta database to extract the required DBD from ARC/INFO and return the output to the user.

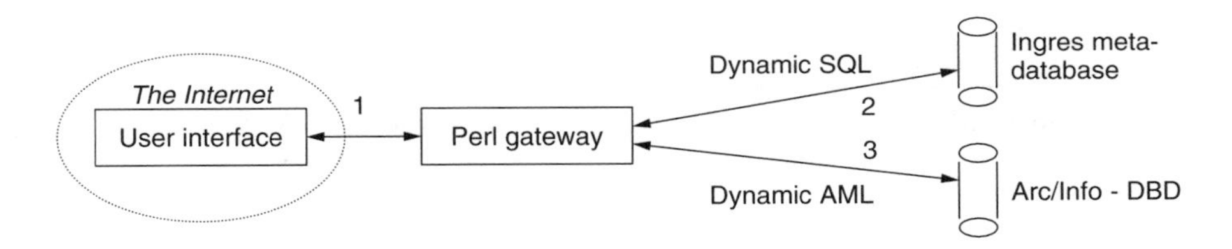

Figure 6.2 Outline of the UKBORDERS system.

6.3.1 Uses and Users

Since its inception in August 1994, more than 2000 users, representing 117 HEIs, have registered in order to gain access to digitized boundary data via the UKBORDERS system. To access the system, users must first be eligible to access the data and secondly be registered at a recognized institution. Eligibility can be broadly stated as being open to resident UK higher education staff and students, and eligibility status may be checked by contacting EDINA at *edina@ed.ac.uk*. Registration is a two-step process, involving first the registration of the institution followed by the registration of the individual. The registration status of an institution can be checked via the on-line database of registered institutions at the EDINA Web site. This gives contact details of the site representatives at each institution responsible for the distribution of end user registration forms. On completion of the registration process, a username and password is issued that is required to authenticate users each time the system is accessed.

The boundary data for the 1991 Census have been digitized to a high degree of precision. In Scotland, postcode unit boundaries for urban, mixed urban and rural, and rural areas were produced using source maps with scales ranging from 1:10 000, 1:25 000, and 1:50 000, respectively, with some inner urban areas using scales as high as 1:1250. Similarly, for England and Wales, these four different scales of mapping were involved.

6.4 WHAT DATA ARE AVAILABLE?

Table 6.2 lists the boundary data available through the UKBORDERS delivery system. In order to extract one or more files (which may contain one or more coverages) users must make decisions in four basic stages.

6.4.1 Set Constraints

By default, all of the data held in UKBORDERS is selected. Users may choose to modify this blanket selection by constraining one or more of

- A time period of interest—any data that is time-stamped for any part of the period set is selected
- One or more of the five geography 'types' held in UKBORDERS—census, administrative, electoral, postal, or other
- One or more of the three countries currently represented in UKBORDERS—England, Wales, and Scotland. Supply of DBD for Northern Ireland has not yet been agreed with

Table 6.2 Available boundary types.

Scotland	England and Wales
Regions	Standard regions
Districts	Counties
Pseudo sectors	Districts
Council areas	Wards
Postcode areas	Enumeration districts
Postcode districts	Regional health authorities
Postcode sectors	District health authorities
Postcode units	European constituencies
Output areas	Parliamentary constituencies
Health board areas	Urban/rural areas
European constituencies	Rural areas
Parliamentary constituencies	National parks
New towns	Poor law unions
Civil parishes	Poor law union counties
Localities	Urban/rural districts
Police force areas	Registration districts
Sample of anonymized records	National parks
Regional electoral divisions	Police force areas
	Sample of anonymized records

Ordnance Survey Northern Ireland (OSNI) (this constraint may also be set by clicking on a map consisting of the boundary outlines of the three countries).

6.4.2 Select File(s)

- Depending on the constraints set in Stage 1, Stage 2 immediately provides a file list or a list of the level(s) of aggregation available (for example, ED, Ward, District) selecting from which then produces a list of available files.

6.4.3 Select Data Transformation(s)

- Generalization (default 'off')
- Precision—where appropriate in relation to ...
- Format—ARC/INFO ungenerate, polys; ARC/INFO ungenerate, lines; ARC/INFO ungenerate, points; ARC/INFO export; AUTOCAD DXF; MAPINFO; and MAPINFO Native Table (for use in Microsoft Excel).

6.4.4 Preview and/or Download Output

- Provision of a 'preview' option to allow the user to confirm that they have specified, and that the system has delivered, the required coverage is regarded by the UKBORDERS team as a prerequisite.

6.5 ISSUES OF ACCESS

The 1991 Census digitized boundary data are copyright of the data producers—the DBD for England and Wales is owned by the ED-Line consortium (consisting of the Office for National Statistics (ONS), London Research Centre, MVA Systematica and Ordnance Survey) whereas that for Scotland belongs to GROS and the Post Office. Crown copyright is applicable to both. The fact that it is copyright has an important effect on who can access it, from whom, and for what purposes. In the case of UK higher education, the 1991 Census digitized boundaries were purchased under licence from the vendors by the ESRC under its 1991 Census Initiative on the strict understanding that the data would be used for academic teaching and research. Commercial use of the data purchased by the ESRC is prohibited because the digitized boundary data is available as a copyright commercial product for purchase by others—the private sector, local government, the National Health Service (NHS), and so on, under separate conditions of licence.

6.5.1 Managing the Volume of Data Using Map Generalization Techniques

With the provision of such a unique product as the DBD, there was always going to be a need to tailor both the data and the delivery mechanism to best suit the needs of the user community. In addition to issues of interface design and database handling, there is a special need for rapid delivery of data over the internet, that is consistent and meets strict quality criteria (Edwardes *et al.* 1998). The existence of libraries cannot address the problem that a user's request may generate large volumes of data, with consequent problems in the transfer, application data loading, analysis, and display of the data. Automated methods, termed *map generalization techniques*, exist that can be applied to geographic data to reduce the size of such files. Map generalization is a science that has developed alongside GIS and is concerned with the optimum display of geographic information commensurate with resolution (choice of scale) and intended use (thematic form) (Muller, 1991). It encompasses a range of tasks such as data abstraction, selection, classification, and symbolization of data as well as more subjective processes such as selectively omitting data, merging or applying small displacements (Mackaness, 1994), simplifying networks (Mackaness and Beard, 1993), and simplifying polygonal data (Bader and Weibel, 1997). These are applied to improve the legibility and ease of interpretation of geographical map data. In the context of the UKBORDERS service, it is unreasonable to expect users to have the knowledge to manage and manipulate large volumes of data, particularly because, by definition, users of this internet service are remote from the resources and skills required to manage the data. The challenge in developing any generalization method is twofold: first the development of techniques that work independently of the user and secondly the need to generalize data within clearly defined quality constraints.

6.5.2 The Genaboundary Project

The complete UKBORDERS 'library' of digitized boundaries is approximately 3 GB in size. This is due, in part, to the fine geographic precision with which the base boundary data has been defined: the number of digitized coordinates (or vertices) used to define a given segment or polygon. This high precision in the data, required for large-scale, small area work, is not required for small-scale, large area use of the boundaries. This high precision has resulted in file sizes that are currently too large for practical use. The Genaboundary Project (Edwardes *et al.*, 1998) was undertaken to develop algorithms that reduced the size of files by reducing the number of points used to record the census boundary but without degrading its visual quality or compromising its use in further analysis and integration with census attributes, and to do this without requiring the user to be an expert in GIS! The boundary of a region can be a simple rectangle (as might be found in an urban area) or a complex shape reflecting its demarcation by natural physical features such as ragged coastlines or meandering streams. The amount of generalization achievable will therefore vary with theme, scale, complexity of line, and the data capture technique used to record the line. One enduring algorithm used to reduce the number of vertices used to record the boundary works by retaining points that demarcate significant changes in direction, and removing points in between (Douglas and Peucker, 1973). The important points of change are identified by taking offset measurements from the chord of the line, and a threshold or filter value is used to determine which points are retained. A series of experiments were undertaken to determine how changes in the data set might be approximated by mathematical functions, and what relationships might exist between these functions and properties of the data sets. The Douglas Peucker algorithm was applied in varying degrees to different data sets. The first observation made was that as the filter value is increased (1) fewer points (vertices) are retained, (2) the file size becomes smaller, and (3) the line becomes simpler, though more angular in form. It was noted that quite low tolerances (between 1 and 5 m) achieved very high reductions compared with higher tolerance values. As the number of vertices become fewer and boundaries more angular, so the sum of the length of all boundaries is reduced. Excessive generalization results in changes in topology of the data (a well-known problem). It is time consuming to expect a user to correct such errors, so the project concerned itself with (1) development of methods for analysing the form of a line (in order to set appropriate filter values and create a new form) and (2) develop methods of assessment, which are required in order to provide metadata (data about data) on the quality of the derived product.

Through empirical analysis of a range of data sets, it was clear that urban regions of a map could be distinguished from rural regions by examining a number of the properties of the boundary of a land parcel. Some of these characteristics were used to separate the boundary data into rural and urban data. Differentiating between categories allowed for the selection of a more appropriate parameter to be used, making the process more accurate than if it were to select a single parameter to apply to the entire data set. Tolerance values were set using an equation derived by empirical observations. After generalization the coverages were reconstituted into a single coverage. A polygon count revealed if self intersection had occurred during generalization. Those polygons were identified and offending boundaries removed and replaced with boundary sections from the original coverage. Cartometric evaluation was again undertaken and the values compared. If the changes were within prescribed tolerances, then the derived coverage was passed to the user, together with a metadata quality report (Figure 6.4), otherwise the original coverage (copied at the outset) was offered to the user. Figure 6.3 summarizes the key operations in generating the generalized form.

The purpose of the metadata is to provide the user with an indication of the changes to the data, thus helping them to decide whether the end product is suited to their needs.

Commencement of session	Cartometric analysis	Partition coverage	Generalize and evaluate	Quality control and report
Choice of geographical extent and boundary type made	- duplicate coverage - add meta variables - cartometrically analyse coverage	- assess purity of coverage and divide coverage into 'rural', 'urban', 'small polygons'	- set tolerance for each existing partition and generalize - reconstitute coverage - evaluate coverage	- replace self intersecting segments - generate meta data report

Figure 6.3 Schematic of the generalization algorithm.

INFORMATION AND RESULTS OF GENERALIZATION.
The data was generalized on the 14 of January 1998.
The generalization information was created at 13:56:37.

THE ORIGINAL DATA:
Number of polygons : 911.
Total area of polygons : 5784005832 square metres.
Number of lines : 3079.
Total length of lines : 6019473 metres.
Number of points in lines : 507474.

THE GENERALIZED DATA:
A value of 0.74 was used to remove excess points on urban lines (15 or fewer points).
With a value of 3.2 used on rural lines (more than 15 points).

Number of polygons : 911.
Total area of polygons : 5783998584 square metres.
Number of lines : 3079.
Total length of lines : 5980592 metres.
Number of points in lines : 126949.

CHANGES TO DATA DUE TO GENERALIZATION.
There are 0 more polygons in the generalized data.
There has been a 0.000% decrease in the total area of the polygons.

The generalized data contains 0 fewer arcs.
There has been a decrease of 0.646% in total line length.
The generalized data contains 25.02% of the points of the original.

Figure 6.4 An example of the information appended to the metafile by the simplification routine, in this case for EDs in the Welsh County of Dyfed.

The information present for both the unsimplified and simplified data is the number of polygons and their total area, the number of arcs with their total length, and the total number of points in the lines. In addition, a percentage change in area and length is given. Because a major reason for simplification is reduction of file sizes, the number of points in the end product is given as a percentage of the original. In the example given in Figure 6.4, a 75% reduction in the number of points was achieved.

6.5.3 Evaluation

The reduction in file size was as high as 80% for some data sets but more typically between 20 and 40%. Importantly, the differences between data sets before and after simplification were not visually discernible. Clearly this approach has very significant improvement over manual approaches as it achieves very significant reductions in file size, guarantees maintenance of shape, guarantees maintenance of topology, and achieves this without any intervention, visual inspection, or interaction with the user. Currently, more than 40% of users request the generalized form of the coverage. The fact that it was almost impossible to visually discern the differences between the original file and the generalized version is perhaps testimony to its success. This, together with its speed, means that the algorithm is effectively transparent from the user's point of view.

6.5.4 Findings from the GenaBoundary Project

The final algorithm was a combination of a rule-based system for preserving polygons, separating the data set into homogenous subsets and maintaining topological integrity, and a component for mathematically determining appropriate tolerances based on an evaluation of the statistical properties of the subsetted data. The approach satisfied its objectives by conserving the integrity of the data sets both in terms of topological consistency and linkage to third-party data sets. This was ensured by the application of rules to protect small polygons from simplification and to replace intersecting or self-intersecting simplified boundary lines with their original geometry. The objectives of maintaining data quality while maximizing data reduction were met by exploiting the empirical observation that a significant volume of data points could be removed at a minimal cost to loss of detail. Minimizing the loss of detail also ensured that the completeness of the data set was as far as possible retained and thus flexibility in the potential end uses of the data set was conserved. In addition, quantitative measures recording the changes in the statistical properties of the data set were supplied to the user as metadata. The objective of robustness was handled by the algorithm through a series of preliminary and post-processing checks and ameliorations. It should be stressed that the research focused on the automatic derivation of parameters for a line simplification algorithm without human intervention, together with evaluation techniques and provision of quality information. This autonomous simplification of data sets is illustrating the way in which more fully automated environments can support a broader community of users.

6.6 CONCLUSION: THE FUTURE

In this chapter, we have described the various forms of boundary data that can be used in conjunction with census data. Description of the UKBORDERS service has included a discussion of issues relating to data volume and fitness for use.

In anticipation of significant growth in the use of the UKBORDERS service, there are future developments that will address a number of key issues: the development of database technology capable of managing the delivery of larger volumes of data to a broadening and diverse user community (an example being the provision of more historical data, as described by Gregory *et al.*, 1998). The use of such systems from disciplines outside the geographic domain requires development of an interface that addresses issues relating to ease of use, of data compression, and the provision of metadata relating to the fitness for use of UKBORDERS data.

The diversity of users is mirrored in the diversity of applications now being utilized in the analysis of census information. It is therefore planned to extend the data format options considerably, providing output for use in a wider range of software formats.

With the development of geospatial data interfaces within various Internet protocols, issues of data standards and interoperability must also be addressed. For example, the settings used to select data within one system should be transferable to another system in order to avoid users having to answer similar questions in multiple interfaces. As a member of the Open GIS Consortium, EDINA hopes to stay at the forefront of these developments and thus provide boundary data of the census efficiently, accurately, and with the minimum of effort on the part of the user.

Box 6.1 Key web link for Chapter 6.

http://edina.ed.ac.uk/ukborders/	The UKBORDERS interface to census boundary data

7

Visualizing Census Data

Jason Dykes, Jackie Carter, and Danny Dorling

SYNOPSIS

This chapter considers the variety of ways in which census data may be visualized, particularly focusing on cartographic approaches. A range of static, dynamic, and hypermaps is presented, together with a description of the visualization tool Census Data Visualizer (CDV). Static maps are appropriate for conventional paper reproduction but include novel representations such as the cartogram. Dynamic maps offer the user the ability to interact with the data using techniques such as brushing, whereas hypermaps make use of Internet technology. Full colour figures and the CDV software are provided in the accompanying CDS Resources CD.

7.1 INTRODUCTION

The term *visualization* has become increasingly prevalent in the 1990s (Orford *et al.*, 1998) and is used in a variety of ways in social and physical sciences. It can refer to the production of graphical material from collected data, 'a method for seeing the unseen' (McCormick *et al.*, 1987), graphical software that can be controlled in some way by the viewer, a graphical interface to programming, the activity of exploratory data analysis (EDA) that involves repeated viewing of graphics, and even the hardware and software combinations that permit these processes. Whichever of these definitions is used, maps have ancient roots. Graphics of social activity were drawn on cave walls some 30 000 years ago, showing people dancing in a field (Dorling and Fairbairn, 1997). They have played an integral part in scientific enquiry for a long time, with Descartes and Faraday being notable early visualizers (Orford *et al.*, 1998). Recent changes in the technology that is used to visualize are affecting the ways in which social science is being researched. Hence, this chapter is concerned with a concept that has an extremely long history.

This chapter and associated materials provide an introduction to some of these interpretations of the term *visualization*, with specific examples of good and novel practice relating to the authors' areas of research. Developments in mapping and map use that have occurred since the 1991 Census are outlined and assessed. Specifically, this contribution introduces a series of additional digital resources that are provided in the accompanying *CDS Resources*. This chapter contains sections focusing on the production, use, and potential of three types of maps. These are static maps suitable for paper production, dynamic maps that rely upon a software medium, and hypermaps that use the communications technologies of the Internet. Some familiar issues are outlined and recent developments described. This chapter relies upon the supplementary text and has been provided with detailed, dynamic colour figures, and so should be used in conjunction with the CD. Figures that appear only in the CD are referenced here as Figure '7.*n* (CD)'.

'Visualization' is an immense, expanding field. Orford *et al.* (1998) include more than 3500 references in their review of visualization in the social sciences, and navigating through this mass of information is a difficult task. Here, we focus on enumerated data sets such as the small area statistics (SAS).

7.2 VISUALIZING ENUMERATED DATA: STATIC MAPS

Envisioning spatial information would have been referred to as *cartography* before the advent of powerful and affordable desktop computing. It is important to recognize the antecedents of visualization in cartography and not to forget some of the basic rules of cartography. Robinson *et al.* (1995) have a broad view of what cartography entails, identifying it as a series of three transformations from geographical environment to geographical information, from geographical information to a map representation, and from the map to the user's mental map image. Users of a survey, such as the census, have no retrospective control over the first of these transformations. Their concern is with the second and third transformations, turning geographical information into a suitable map representation that will allow the user to read, analyse, and interpret the information into a useful map image. The processes that the cartographer can use to create a representation include selection, classification, simplification, and symbolization.

Many of the techniques developed by cartographers for creating traditional static products are applicable for mapping enumerated data such as the SAS, for which the choropleth map is the standard means of presentation. Choropleths consist of a set of zone boundaries containing colour symbolism to represent numerical values. They are an unsophisticated way of 'visualizing' representations of the world created by the division of space into geographical areas and presentation of numerical information within them. Choropleths have been popularized in the last couple of decades by the fact that they are very easy to produce. However, a series of issues, relating to the numbers mapped, the areas used, and the symbolism chosen, make them a far-from-perfect device from which to create a map image. Users of the technique require an awareness of the issues involved and should assess some suggested solutions and alternatives that have developed as computer cartography has progressed. What is most important is to avoid producing maps that distort the underlying distributions. These issues are addressed in the following sections. Suitable resources containing additional information are referenced throughout and a more detailed account is available on-line (Dykes and Unwin, 1998).

7.2.1 Issues Concerning the Numbers

Although choropleth maps are a faithful reflection of the head count and boundary geometry used to record information about the population, they can be very poor representations of the underlying distribution that these data represent. The zones used to enumerate people often vary considerably in size, particularly when produced to equalize population for electoral purposes. If the population is distributed relatively uniformly, any counted

Figure 7.1 Choropleth maps of the wards of Greater London, 1999. A shading scheme using a linear intensity scale depicts data that represent, from left to right, a count, a density, and a ratio.

data will provide information about the size of the zones used in the sampling and that of the sample. Effectively, in a choropleth map symbolizing a count, the symbol used to represent the zone and the colour used to shade the zone are showing related information, obscuring the underlying distribution. Mapping a ratio better reveals the distribution of population. Examples are maps in which counts of people are expressed as a density or as a ratio to some other population total. In Figure 7.1, three example maps of Greater London are provided. The first shows the number of people, the second, population density, and the third, the ratio of cars owned per head of population. A more detailed digital version and visualization software are provided in the *CDS Resources* CD.

Other methods can be applied to refine the numbers for mapping. These can involve standardization of the denominator in ratios, for example, standardized mortality ratios show the ratio of the number of deaths in each area to those expected on the basis of national age- and sex-specific rates. Where the numerator or denominator is very small, ratios become unstable to small changes in the data. It is sensible to map more robust quantities in such circumstances, such as the probability of a more extreme value occurring than that which is observed (Bailey and Gatrell, 1995).

7.2.2 Issues Concerning the Areas

Whatever pattern the map displays, the result is just as dependent on the zones imposed upon a population when collecting the data as the phenomenon under investigation. An example is provided in Figure 7.1 (count) in which the dark colours displayed in wards in the District of Barnet (to the North) show higher population totals. In the instance of Greater London used here, the zones are designed to ensure that district populations are comparable, meaning that ward populations are related to the number of wards per district. Hence, the dark shading simply indicates that Barnet has fewer wards than other districts rather than anything about the population. Where the zones are devised to equate the number of inhabitants, their sizes will vary as the population density varies. In zones of equal land area, such as many of the counties used for the US Census, the populations represented by the zones can vary enormously, whereas zone size remains relatively constant. Indeed, any numbers collected with relatively arbitrary areas, including ratios, depend greatly on the definition of those areas. This effect has been termed the *modifiable areal unit problem* (Openshaw, 1984).

The problem of the definition of areal units has a direct effect on one of the most interesting aspects of the human geography of Britain, which will be revealed by the 2001 Census, namely, the way in which the country has changed in terms of the distribution and the nature of its population over time. The reason for having a census every 10 years is that the geographical distribution of population and its social spatial structure change sufficiently such that that the government is willing to expend hundreds of millions of pounds to record them (Dorling and Simpson, 1999). To capture these changes, the large majority of the questions that are asked in each census were asked in the previous survey. Between each census, a few new techniques are usually developed for visualizing data, which can aid our understanding of the social structure of the country. However, the large majority of maps that are produced after each census provide very little new information. This is because although the social structure of the country changes, it does not often change fast enough to produce differences that are perceptible when the population distributions are mapped as a whole. For example, a map showing where people in high social classes lived in the country in 1991 looks remarkably like the same map for 1981. There are a few exceptions to this generalization. For instance, maps of the geography of mining look very different between these two dates. However, what is usually most interesting is to draw a map of the changes in the distributions. In the case of social class, areas can be identified that are undergoing gentrification, as can those that are being deserted by the upper classes. Mapping changes allows us to focus on the process and the trends rather than on the overall distribution. To draw a map of change, data have to be available for the same areas at successive census dates and for a number of reasons, they almost never are. Some of the most interesting maps that can be drawn from the census, those depicting how life has changed since the previous census, are thus some of the most difficult to draw. The data that would enable us to produce them are not available, usually because the units used to aggregate the population have changed. A number of techniques have been developed to overcome this problem. In the simplest case, areas are made up of different combinations of identifiable old areas and statistics aggregated for comparison. More sophisticated approaches involve redistributing the numbers on the basis of statistical models. Examples include the work of Bracken and Martin (1989) to produce continuous population estimates (discussed in Chapter 10) and Langford and Unwin's (1994) dasymetric maps that allocate populations to residential areas identified from satellite imagery. Additional details are provided in the supplement to Section 7.2.2 in the CD.

7.2.3 Issues Concerning the Symbolism

Methods for classifying the statistical values shown on choropleth maps received a lot of attention when map production was an expensive and lengthy procedure with a single static outcome. An excellent review of the possibilities is provided by Evans (1977). Advances in technology have been used to compute 'optimal' classifications according to some criteria (Jenks and Caspall, 1971) and to produce classless choropleths (Tobler, 1973; Muller and Honsaker, 1978). The huge range of colours available when using modern computers has made it easy to specify colour schemes and trivializes the production of classless grey-shaded choropleths. However, interpretation of colour requires considerable thought. The grey shades produced by a computer, although linear in terms of their intensity, are not perceived in a linear way by viewers who see more variation among lighter shades than among darker ones (see Figure 7.1). Different colour combinations also have a variety of effects and Monmonier (1991: 147) regards the use of colour in choropleths as a 'cartographic quagmire'. Mersey's study (Mersey, 1990) suggests that it can aid map interpretation. Brewer (1994) provides effective guidance by matching colour scales to data types and data scales, and her guidelines form an excellent resource.

7.2.4 The Cartogram Solution

Computing technology has made a series of novel static views of data increasingly feasible. The population cartogram is a response to the problem of land area, rather than population, dominating statistical maps that use geographic projections and providing a solution to the problem of mapping human geography. If the large majority of the population live at high population densities, on a conventional map, very little space exists to illustrate the variations among different groups of the population. In the population cartogram, every person surveyed is given equal representation on paper or on screen. Additional details are provided in the supplement to Section 7.2.4 in the CD.

While 'A New Social Atlas of Britain' (Dorling, 1995) uses a number of population cartograms as the basis for mapping spatial information about society, there are two main reasons why census users do not routinely employ cartograms. Firstly, they have been relatively difficult to draw. Secondly, for people used to conventional maps, cartograms are very difficult to read. Cartograms are difficult to draw because there are an infinite number of possible area cartograms that can be drawn, given any single population distribution, and so it is difficult to decide which to use. (The same is true of conventional equal area maps, but almost everyone is happy to use the national grid projection and decide upon a satisfactory combination of techniques for representing the information.) Cartograms are also difficult to draw because there are only a few generally available computer algorithms to produce them and none, yet, in any geographical information system (GIS). The algorithm used here is one of the simplest available. It turns all zones to be mapped into circles, with each circle centred on its zone's original centre and sized in proportion to its population. Each circle is then allowed to move in such a way that it is attracted to areas that it neighbours in conventional space, whereas being repelled by areas with which it is overlapping. Zones at the edges of the areas of interest can be given a higher degree of inertia to help preserve some of the shape. The process iterates a few hundred or thousand times and results in an equal population area cartogram. The result of running the process on the wards of Greater London is shown in Figure 7.2. Production then is not too difficult but the result remains unfamiliar and so difficult to interpret.

Cartograms are hard to read because it takes some effort to learn the new technique, and in doing so to become fully aware of the flaws with the old one. This is particularly true because of the familiarity of the traditional representations and our belief that we can interpret them successfully. People tend to think that they can read conventional choropleth maps with ease. If you show a geographer a map of unemployment in England by district, he will attempt to explain the distribution. The fact that the map does not show that the highest concentration of unemployment is in central London and that it makes the Grampian mountain chain look like the labour exchanges of New York after the Wall Street crash will not detract the reader from pointing out 'the north/south divide' or 'the urban/rural gradient'. But how much are we learning from these maps and to what extent are we confirming our initial beliefs? Cartograms hardly ever convey the same message as

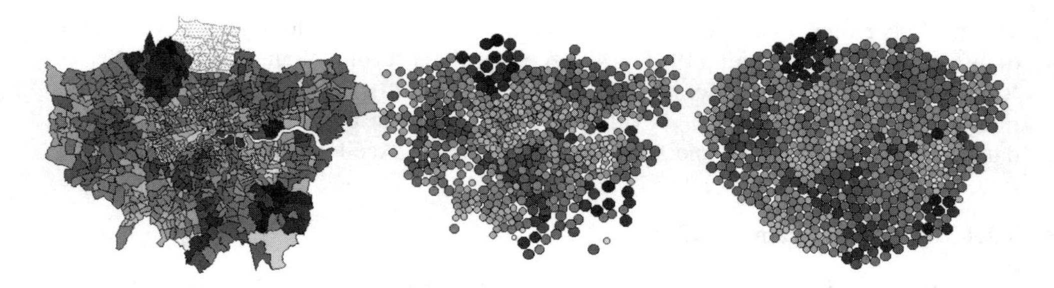

Figure 7.2 A population cartogram for the wards of Greater London using Dorling's algorithm. The map on the left shows total population as recorded by the 1991 Census. The central map shows overlapping symbols located at ward centroids with symbol size representing ward population. This is the starting point of Dorling's cartogram algorithm. The map on the right shows the result of the algorithm, a non-continuous cartogram of the population of Greater London in 1991.

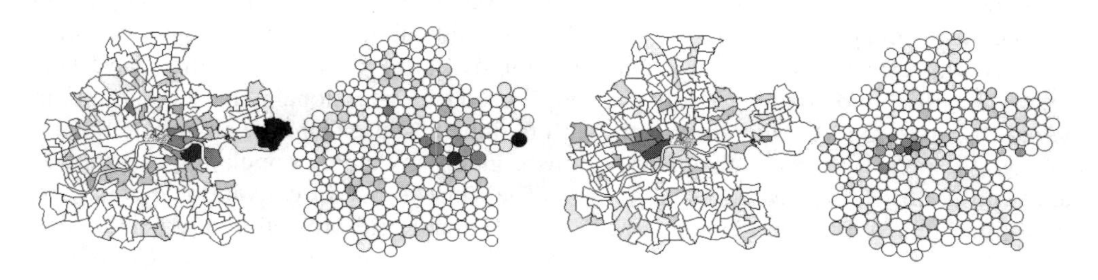

Figure 7.3 Cartograms showing changes in population in Inner London between 1981 and 1991. These pairs of images demonstrate how cartograms can be used to show changes in population. A regular choropleth and cartogram show ward shapes and locations and the 1991 population totals for each ward, respectively. Each is shaded in a continuous scheme that represents percentage change in population between 1981 and 1991. The pair of maps on the left show wards in which population totals increased, with a maximum of 75% change, the maps on the right show wards that experienced a decrease in total population, with a maximum of 52% change. Increases in population along the Thames and areas with a reduction in population to the west of Inner London can be identified. Colour versions of the figures provided in the CD use a divergent scheme to show areas that have lost and gained population on a single map.

equal area maps. This is even true of the ones shown here of the relatively evenly densely populated area of Greater London. If conventional map and area cartograms of the same distributions convey such different impressions, then the new mapping technique is helping to convey ideas that the old one could not. New representations such as the cartogram can be difficult to interpret initially. One solution for the cartographer is to provide a population cartogram in parallel with a conventional map. The reader can look between the map and the cartogram to familiarize themselves with the locations of places that they know, while seeing a picture undistorted by land area. This precludes the need for insets of the areas of high population density spread around the map. Figure 7.3 uses pairs of maps of the wards of Inner London to show the changes in population that occurred between the censuses of 1981 and 1991. Significant changes in the distribution of population within Greater London are demonstrated in Figure 7.4 (CD), which concentrates on the changes among age cohorts using population cartograms. These pairs of static maps provide useful insights into enumerated data sets. Dynamic software maps give us more flexibility in the ways in which we can interpret cartograms and make this type of view even more suitable, as we shall discover.

7.3 VISUALIZING ENUMERATED DATA: DYNAMIC MAPS

Advances in computing mean that the cartographic process is no longer restricted to using manual production techniques and the paper medium. While significant emphasis has been placed upon automating traditional cartography, which has restricted the development of novel techniques that take advantage of the digital medium (Fisher, 1998), an entire host of electronic media and methods now exist through which dynamic representations of spatial phenomena can be constructed. Maps that move, make noise, and provide successive alternative interactive views of data are becoming increasingly easy to produce and are increasingly expected by those familiar with the graphic and interactive nature of the internet.

7.3.1 Observer-Related Map Behaviour

A useful typology of the kinds of dynamism that dynamic computer maps exhibit was produced by Shepherd (1995). It encompasses a wide range of dynamic map 'behaviours', each of which can originate from a variety of sources. 'Observer related behaviour' originates from actions on the map made by the user. Map responses include observer motion, object rotation, dynamic comparison, dynamic re-expression, and brushing. In two-dimensional dynamic maps, observer motion and object rotation are analogous to changing the viewpoint by rotating, zooming, and panning, just as when navigating, we often fold a street map, turn it around, and shift our view to an inset showing greater detail. The remaining types of behaviour, dynamic comparison, dynamic re-expression, and brushing involve the user changing the way in which maps symbolize information and can be used by the map reader to modify the map to their exact specification. Dynamic maps can use these properties to overcome some of the deficiencies of the single static map. Some of these relate to familiar map roles, such as navigating. Others are able to address issues concerning interpretation, while additional properties have been applied to use maps for new types of task that relate to the kind of analysis for which much of the information collected in the census is used.

Brushing refers to a variety of data selection procedures that allow symbols in dynamic maps and graphics to be identified with labels, removed from the analysis, and highlighted in different views. An example of brushing is shown in Figure 7.5 (CD). 'Dynamic re-expression' involves alternating more than one graphical version of a data set in the same coordinate space by changing the cartography used. This technique can overcome some of the limitations inherent in static maps and use some of the 'problems with the symbolism' to assess the significance of identified distributions. For example, users can classify choropleth attribute distributions interactively allowing the spatial extents of values above and below certain local figures or values of particular significance to be compared. The varying patterns that can be achieved are demonstrated by the three classifications of the same data set shown in Figure 7.6 (CD). 'Dynamic comparison' involves displaying more than one data set in the same coordinate space at different times to show variation. By rapidly switching between these maps, comparisons between the distributions can be made. A dynamic example is provided in Figure 7.7 (CD). These figures and additional details can be found in the supplement to Section 7.3.1 in the CD.

The development of tools that encourage transient mapping using these techniques has resulted in a change of emphasis regarding map use. While maps were used primarily as presentation tools, DiBiase *et al.* (1994) identified an increasingly popular and evolving form of map use termed *visual thinking*, in which a map responds to the user's train of thought to confirm or contradict ideas developed by viewing successive graphics early on in the research process. This use of highly interactive maps for visual thinking at the exploratory stage of research has been termed *visualization*. For example, DiBiase *et al.* (1994) define cartographic visualization software as comprising dynamic displays tailored to a specific set of users, which encourage experimentation with different combinations of data and graphic symbols for use in fostering discovery rather than presenting conclusions.

7.3.2 Cartography for Visualization

Little evidence exists to support rules or guidelines for the most appropriate use of dynamic map behaviours for visualization. Indeed, there is little documentary proof that dynamic maps are genuinely advantageous for analytical use. MacEachren (1995) notes that we have not yet begun to explore the potential for using multiple cartographic representations of data sets to allow researchers to see their data from different perspectives. The International Cartographic Association (ICA) identified the changes in map use that were occurring and responded by commissioning a working group on visualization in response to the need for information about the impacts of these developments upon cartography. The need to perform some evaluations of exploratory visual methods has been stressed at recent meetings of the commission despite the difficulties in performing such a task. How do we test whether our techniques reveal things that we might have missed otherwise? Extremely flexible software environments that encourage experimentation with cartographic representation are essential if we are to make a start.

Many currently available computer maps do not provide the kinds of interactivity required to enable an analyst to instantly change their perspective of a data set through selection and transformation to reveal meaningful relationships within those data, and yet we are familiar with instant response and transient graphics when using the buttons and symbols of our graphical user interface (GUI) to interact with computer programs. Most commercial products require extensive customization to produce flexible analytical maps of this type. Here, a cartographic tool is introduced that has been developed since the 1991 Census to demonstrate the potential utility of dynamic mapping when applied to enumerated data and to propose suitable functionality and representations for the process of visualizing the 2001 Census data. It enables a range of users to access and assess the techniques and for good practice and utility to be evaluated and discussed. The software has been used in research and for demonstration and teaching across the country. This implementation is provided as part of the *CDS Resources* to demonstrate that the types of map and map use that are appropriate are developing along with information technology. Hopefully, it will result in a refreshing and modern perspective on the use and expectations of dynamic census maps.

7.3.3 CDV—A Graphical User Interface to Enumerated Data for Visual Thinking

The CDV software for cartographic data visualization is provided in the *CDS Resources* CD accompanying this chapter. Documentation supporting the software is also available and includes a series of 'Release Notes' and a 'Quick Start' page to enable the inexperienced user to get the CDV running immediately. This section briefly introduces the functionality of the software, the rationale behind its design, and the relationship between the software and the developing techniques identified earlier. Additional details and figures that relate to this section are provided in the supplement to Section 7.3.3 in the accompanying CD. The software, guide, and accompanying data should also be used to gain familiarity with the techniques detailed.

CDV maps the types of data addressed in this chapter, namely, a series of enumerated zones with associated numerical information such as the census SAS variables. It provides a series of dynamic graphics specifically to encourage visual thinking. When the CDV is first run, a map is produced with polygons representing enumeration units, each shaded in a grey intensity dependent on the value of a recorded attribute for the zone. The shades vary from white to black in a linear scale representing the lowest and highest values, respectively. This initial view is the 'base map' and while it would appear to be extremely similar to Tobler's static classless choropleth, it has dynamic properties that aid interpretation. The symbols of enumeration units can be touched to reveal their names and data values. The user can zoom in to areas to reveal detail and pan across the mapped region (see Figure 7.8 (CD)). Brushing is also used in the CDV to overcome the inherent problems associated with non-perceptual colour scales and non-familiar spaces in static maps. The CDV enables the user to change their perspective on a data set by displaying data in a variety of user-selected views. A dot plot view of a distribution uses distance along a single axis to represent data values. By linking the dot plot to the map so that brushed cases are highlighted in each view, the user can take advantage of the linear statistical representation in conjunction with the Gestalt provided by the spatial distribution (Figure 7.9 (CD)).

Pre-computed cartograms of the type described in the previous section can be loaded into the CDV, displayed, and dynamically linked to the more familiar geographic representation as shown in Figure 7.10. These features aid interpretation and make the representation more usable. The software also contains functionality for determining the topology of enumerated data sets and performing the iterative cartogram calculation process.

Dynamic comparison can be performed between successive variables by using the sliding scale option on the control panel (see Figure 7.11 (CD)). Dynamic re-expression can be useful in a number of ways. The statistical range shown on any dot plot can be classified using a number of pre-defined schemes or user-selected intervals, and the classification applied to any other view. This overcomes the age-old problem of determining class intervals and allows the stability of a pattern to be ascertained as its persistence can be checked under a variety of cartographic conditions. A whole variety of cartographic parameters can be varied in circle map views, which are appropriate for displaying counts. Some of these are shown in Figure 7.12 (CD). The CDV contains techniques for creating focused data subsets to be used in subsequent views enabling the full range of symbolism to be used to reveal local variation that is not detected at the global level as demonstrated in Figure 7.13 (CD).

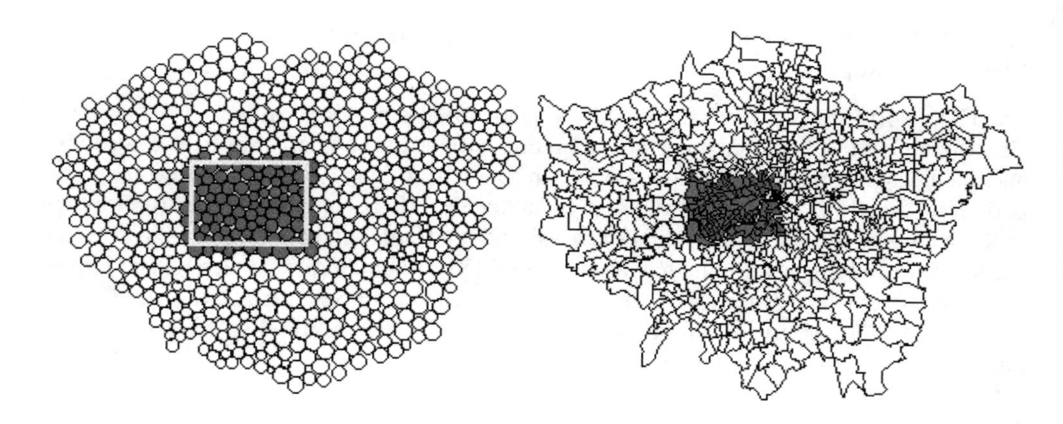

Figure 7.10 Dynamic links between multiple views of a data set. Symbols can be selected and highlighted in multiple views with the brushing functionality of CDV. The dynamic version of these figures provided in the CD demonstrates a number of examples and means of linking between views.

A number of novel ways of viewing multivariate spatial data are achievable in dynamic form using the CDV. These include specific implementations of colour composite maps (Ware and Beaty, 1998) and parallel-coordinates plots (Inselberg, 1985). The former are shown in Figure 7.14 (CD) in which the proportions of the population in three age groups are displayed by combining red, green, and blue components that represent individual cohorts. In the CDV, the ability to switch colour components on and off, combine them as desired, and link them to other views aids their interpretation significantly.

Parallel-coordinates plots have emerged as a viable means of representing information about a large number of variables concurrently. The technique was originally developed by Inselberg (1985) and has been widely used in the field of graphical statistics. They are similar to scatter plots in that data values are displayed by locating points along axes representing different attributes. However, whereas scatter plots represent a pair of values by a point at which the two values intersect orthogonal axes, parallel plots represent a pair of values by a line that joins the locations relating to these values on parallel-scaled axes. Whereas scatter plots use one dimension per variable, parallel plots have the advantage of being able to display any number of variables in two dimensions by extending the line to join additional parallel axes representing additional attributes. Outliers, clusters, multiple modes, and other qualities of multidimensional data sets can be determined with a little experience of such representations (Wegman, 1990). By adding dynamic links to a spatial data set, the multivariate nature of geographic distributions can be assessed with parallel plots. The CDV provides an extremely dynamic parallel representation allowing users to brush the symbols, produce scatter plots, and change the ordering of axes interactively. Figure 7.15 (CD) uses a parallel plot to show the distribution of four age cohorts in London and the Barnet district.

While the CDV was partly developed as an experiment in designing for visualization for which no agreed practices exist, particular attention was paid to keep views clear and simple as advocated by Tufte (1990). This seems particularly appropriate in a medium in which the user can always interrogate a view for more information. The figures presented here and in the related Web pages demonstrate some of the techniques used in the CDV and the rationale behind the software. An appreciation of the types of map and map use that are possible and suitable for visualization in the digital medium can only be gained through the use of software. The reader is encouraged to try the CDV from the CD, use the associated support documents, and provide feedback to the authors and the wider community.

7.4 VISUALIZING ENUMERATED DATA: COMMUNICATIONS TECHNOLOGY AND HYPERMAPS

The development of Internet technologies and programming languages such as Java and Tcl/Tk that facilitate communications means that dynamic maps can connect to remote machines. This enables them to take advantage of rapid update from large remote databases containing data such as those collected in the census, as well as additional less formal, contextual, or exogenous information that might aid analyses.

7.4.1 Hypermapping Enumerated Information

Many examples of dynamic hypermaps are currently available over the network, either using communications protocols to transfer data between applications or through a Web browser interface. Dynamic maps can be used as a geographical interface to georeferenced information, for example, information about attractions, hotels, and restaurants in the United Kingdom is available at *http://www.multimap.com*, or for downloading census information, as is the case with CASWEB (*http://census.ac.uk/cdu/casweb/* and Chapter 8) and SURPOP (*http://census.ac.uk/cdu/surpop/* and Chapter 10). The US Tiger Service provides an on-line choropleth mapping system that enables data to be viewed in a geographical context at *http://www.tiger.gov*. An implementation of Dorling's cartogram algorithm that visualizes the iterative process is available in Adrian Herzog's Web pages (1998–2000), and Dykes (1996a; 1996b) provides examples of dynamic maps with the code. The software for implementing interactive techniques for visual thinking can also be run over the Internet using Java or other Internet-equipped languages. Andrienko's Descartes (Andrienko and Andrienko, 1999) is an excellent example. One resource that has a collection of such software along with data and support materials is the Janus Visualization Gateway Project.

7.4.2 The Janus Visualization Gateway Project (*http://www.kinds.ac.uk/kinds/janus.htm*)

The Janus Visualization Gateway Project was initiated in 1996 to integrate software tools and resources for visualization, such as those developed by project ARGUS, and Web interfaces to the national data sets, such as those developed by the Knowledge-based Interfaces to the National Data Sets project (KINDS). The result is an interactive, on-line visualization gateway to the 1991 UK Census and associated digitized boundary data for the academic community.

The gateway has incorporated the key deliverables from the ARGUS and KINDS projects, which were transferred into an integrated service hosted on Manchester Information and Associated Services (MIMAS). As a result, both the CDV software and project ARGUS Web pages are now available together with the KINDS Web pages. By promoting, using, and supporting these products through a series of workshops and seminars, the academic community has been made aware of visualization techniques suitable for exploring the census data. In addition, software is being developed that will enable users to perform exploratory spatial data analysis of the census data over the Internet.

The CDV software can either be downloaded for local use or, for registered users, accessed on-line through the MIMAS server. Documentation supporting the use of CDV is also available on the Internet (Carter, 1998; Dykes, 1998). When downloaded from MIMAS or run from *CDS Resources*, CDV includes a demonstrator data set containing SAS counts and generalized polygons for the county of Leicestershire. Additionally, subsets of the 1991 Local Base Statistics (LBS) have been produced for each county in Britain at ward level together with an associated set of generalized digitized boundary data, so that other counties can be visualized. The production of a set of cartograms for each of these data sets is planned. Documentation is provided to assist users who want to use the CDV with these data, with additional instructions available for those who plan to use the CDV with data sets accessed using UKBORDERS and CASWEB Web interfaces (see Chapters 6 and 8, respectively).

The final stage of the project is concentrating on enhancing the KINDS service by incorporating software that contains the environment required for the remote visual exploration of the census data. Through the resulting visualization gateway, users will be able to use dynamic hypermaps of the census in ways comparable to those demonstrated by the CDV. Exploratory analysis of the census data will be achievable via the Internet, before more formal analysis is carried out locally with suitably selected data. The software being used is Descartes (Andrienko and Andrienko, 1999), which is Java based and has been developed at the German National Research Institute for Information Technology. The Leicestershire sample data set can be examined with the Descartes software on-line (MIMAS, 1998). Only authorized users of the census and boundary data will be enabled to access the full array of data provided by the service, which in the first instance will comprise the county data sets at ward level.

The fully realized gateway will instruct a user on how to access the CDV (on-line or for downloading), support them in the use of pre-generated data sets, checking that they are registered for these data, and point out to them information on how they can use the software with their own data. In addition, it is planned that the gateway will enable academic users to visually explore additional census data sets of their choice using the Descartes system within a Web browser. Links to pages that document and demonstrate these resources are provided in the list of references contained in the digital supplement to this chapter. By following these readers can keep up to date with developments. An outline of the steps that will be required to use the gateway is also provided in the CD.

7.5 CONCLUSION

In this chapter, we have attempted to introduce many new methods for mapping the census and performing visualization that were either unavailable or not easily accessible when the 1991 Census was collected. The uptake of these emerging techniques, their assessment, standardization, and incorporation into commercial software and mapping will of course lag behind their development. Most of the maps and diagrams that have been produced from the last census use far more traditional methods than those presented here, many by professionally trained cartographers on behalf of colleagues and clients. As it is now possible for anybody with a computer and Internet connection to produce highly detailed and visually stunning maps and perform visual analyses with them, this chapter and the associated resources have aimed to introduce the reader to mapping issues, techniques, and developments associated with visualization that will be pertinent to users of the 2001 Census. The popularization of mapping and visualization is welcome, and the widespread availability of the tools required to achieve both will ensure that the trend continues. It is useful, however, to remember that there were reasons for the production of maps to be the sole preserve of a small group of skilled technicians. Most simply, the issues outlined here mean that it is very easy to produce maps badly and equally easy to make erroneous inferences about the populations represented by data such as those available in the SAS from bad maps. Here, we have aimed to provide some context, examples, and a few warnings about visualizing enumerated census information as a result of these trends and in anticipation of their consequences. We have provided some insight into developments since the 1991 Census and have pointed to some of the resources and assistance that are available to take advantage of them. The software and data from the 1991 Census provided in the *CDS Resources* CD that accompanies this chapter enable the reader to explore and assess some of the material covered here, in the interactive and user-centred fashion typical of visualization. With the software and the knowledge gained from using the materials presented here, the reader will be in a position to acquire data from the 2001 Census and produce maps and graphics from which the changes in population distributions that we identify as being such an important aspect of visualizing the census can be assessed.

Acknowledgements

This chapter uses data that are provided with the support of the Economic and Social Research Council (ESRC) and Joint Information Systems Committee (JISC) and contains images of material that is copyright of the Crown and Post Office. Much of the work was undertaken by Project Argus and the Janus Visualization Gateway Project, both of which acknowledge JISC funding. We are grateful to the Office for National Statistics and to the ED-LINE Consortium for permission to distribute the census variables and the boundary data that are provided with the CDV software.

Box 7.1 Key web links for Chapter 7.

http://www.multimap.com	General-purpose multi-scale interactive mapping with address and postal geographies
http://census.ac.uk/casweb/	CASWEB interface to census SAS/LBS with digital map boundaries
http://census.ac.uk/cdu/surpop/	SURPOP interface to population surface models based on 1981 and 1991 Census SAS
http://www.tiger.census.gov	Interactive US Census mapping
http://www.kinds.ac.uk/kinds/janus.htm	The Janus gateway to visualization and software tools for 1991 Census mapping

Box 7.2 Additional CDS resources for Chapter 7.

CDV installation guide	Installation instructions for CDV software
CDV	CDV software help files, detailed user guide and related resources
Figures	Links to all figures from this chapter including the dynamic and interactive maps not able to be reproduced on paper

Part III

Area Statistics

8

Disseminating Census Area Statistics Over the Web

James Harris, Justin Hayes, and Keith Cole

SYNOPSIS

This chapter reviews the development of electronic dissemination of census area statistics. The role of the Web in widening access and the need for intelligent interface tools are discussed. The census area statistics on the WEB (CASWEB) system is introduced and illustrated as a tool by which relatively inexpert users may obtain the area statistics and associated data that they require.

8.1 INTRODUCTION

Area-based statistics of the type produced by the UK Census Offices from the 1981 and 1991 Censuses pose a number of challenges with respect to delivering the data to the end user. Firstly, the data are voluminous, precluding the provision of printed outputs for all but the most generalized reports. Secondly, the structure of the data is not straightforward and close linkage with information about the data, such as the cross tabulations contained within a given census table, is crucial if access is to be user-friendly. It is not the purpose of this chapter to provide a usage guide to the CASWEB interface because any such documentation would be rendered obsolete by ongoing redevelopment of the system. CASWEB is fully documented at *http://census.ac.uk/casweb/*. Rather, this chapter examines how recent developments have improved access to census data for academic users via the World Wide Web.

8.2 THE DEVELOPMENT OF ON-LINE ACCESS TO CENSUS DATA

The introduction of computer processing for the 1961 Census vastly increased the potential for the production of large and complex data sets (Dewdney, 1983). Improvements in the automation of the data collection and tabulation processes have only been matched recently, however, by improved arrangements for accessing census data outputs. It was not until the development of dedicated software for the 1981 Census that local authorities and other core users gained widespread access to digital census outputs. This software, called SASPAC—*small area statistics package*, made sense of the thousands of data items that were available from the small area statistics (SAS) (Davies, 1995). SASPAC enabled users to define and extract subsets of data by means of a simple scripting language-driven batch program that was able to run on the many different mainframe systems in use by local authorities at the time. During the same period, the growth of computer networks and, in particular, the establishment of the joint academic network (JANET) provided a means for universities across the country to access similar mainframe computing facilities based at regional computing centres. The establishment in 1984 of a Manchester University–based census support unit led to the establishment of the census dissemination unit (CDU), providing an Internet-based data service allowing university staff and students to access census data from any networked computer in the country.

These developments were at the forefront of the available technology and represented enormous advances in data accessibility for social scientists. The system was redeveloped for 1991, and under the aegis of the ESRC/JISC 1991 Census Programme has expanded to incorporate an increasingly diverse resource of census outputs and derived data sets. As the community of census users has broadened to embrace less proficient computer users, including the growing numbers of undergraduates, certain limitations of the dissemination system have become apparent. These centre on the difficulties novice users experience in gaining access to the data. Firstly, there is the problem of finding out which of the varied and complex census-related data sets contains the required information. The next hurdle is registering to use the data, a condition imposed by the licencing arrangements under which the data were purchased by the academic community (Rees, 1995a). The user will need to become familiar with the content and structure of the data set and learn of any limitations or special considerations pertaining to it. Finally, the user must get to grips with the particular access arrangements for their chosen data set, which will often involve becoming proficient in the use of bespoke extraction software, such as SASPAC in the case of area statistics.

Efforts have been made to address each of these concerns with the aim of bringing the census to a wider audience. The Web has been vital to this process and has been increasingly relied upon for the provision of all aspects of census metadata and information and higher level resource discovery materials (*http://census.ac.uk/cdu/*). It has been the issue of data access software, however, that has presented the greatest challenge, mainly because of the unwieldy size of the data sets in question. While the various systems for on-line access to the census data sets have by and large proved extremely robust and reliable and have met the basic requirements of most traditional users, there has long been a perceived need for a simpler and more intuitive alternative means of accessing census outputs for less complex enquiries and for teaching (Martin *et al.* 1998). The natural focus of developments in this area is the SAS/LBS data sets, as they attract, by far, the greatest number of users (CDU, 1996–1999) and provide the basic demographic data that underpin all population-based studies. One of the principal deterrents to a more widespread use of these data is the esoteric nature of the Telnet-based access solution offered by the CDU. It requires users to log in to a remote UNIX server and operate SASPAC in a text window-only environment. Not only does this arrangement require the user to have some familiarity with the operating system, but the software also demands familiarity with a proprietary command syntax and, perhaps most problematic of all, intimate knowledge of the structure of the data. Comparison with the offerings of other vendors suggests that this last issue, rather than being specific to any particular software package, stems from more fundamental limitations common to many census applications.

The most important of these is the lack of integration between the data extraction process and the census metadata, making operations such as the selection of variables for output overly complicated and necessitating reference to complex code books and manuals. This linkage is particularly problematic in the case of the 1991 SAS/Local Base Statistics (LBS) because much of the metadata, including the arrangement of the census tables themselves, were originally designed for hardcopy distribution and were consequently optimized for the printed page with no consideration given to on-screen viewing. Another drawback was the separation of the procedures for acquiring the statistical data and the digitized census boundary data. This meant that in order to define the geographical extent of a data extraction task, users had to work with lists of area codes, rather than selecting them on the screen from a gazetteer or map. Integrating the statistical and boundary data in a proprietary geographical information system (GIS) or mapping system may involve a surprisingly complex series of operations, as illustrated by Charlton *et al.* (1995).

However, none of these difficulties is insurmountable, and the various private and public sector systems that have emerged over recent years for accessing census statistics have addressed them with varying levels of success. There has been a move away from command-driven interfaces, which required of the user considerable knowledge of the complexities of census geographies and data structures. The most recent version of SASPAC, which continues to be the preferred software for most local authority users, comprises a self-contained 'point-and-click' Windows package, allowing users to interact with the data on their PC rather than executing command files on a mainframe server, as done previously.

By contrast, academic users have continued to access data remotely over the Internet. The principal reasons for this centre on the logistics of distributing data to a disparate community of users based at large number of institutions across the country. One of the challenges presented by such a diverse user base is the need for a data access solution that is simple enough for novices and yet sufficiently flexible to provide a rapid and efficient environment for experienced users to extract and manipulate the data. Minimizing prerequisites in terms of resources and equipment must also be a priority where computing services for students are concerned. In particular, it is felt that burdening large numbers of academic users with the task of installing software and loading large amounts of data locally on their PCs or fileservers, as is common in local authorities, would not be workable. The rationale behind networked access, therefore, has been to reduce the burden on the user with respect to the logistical and data management aspects of census data.

8.3 WEB-BASED ACCESS TO CENSUS DATA

The Web is now gaining rapid acceptance as a suitable medium for high-volume data dissemination, but the use of Internet technologies for the provision of census data has not yet gained widespread currency outside of North America. The Australian Bureau of Statistics' (ABS) primary product, CDATA96, comprises a 'canned' MapInfo-based CD-ROM software solution. The ABS reports a high volume of one-off requests but no provision has been made for a Web-based dissemination strategy (ABS, 1999), while in Japan, census metadata are accessible over the Web but not statistical information (*http://www.stat.go.jp*). In the United States where census outputs are available royalty free, both portable media and Internet and intranet solutions are widespread. Private sector companies such as Geolytics Inc. (*http://www.censuscd.com/*) and DecisionMark Inc. (*http://www.censuscounts.com/*) produce CD-ROM products offering census data packaged with other data and a lightweight extraction facility complete with mapping capabilities. The US Bureau of the Census, on the other hand, has been providing data via the Internet since the birth of the Web, the most widely known of which is the Tiger Map Server, which provides an open access system for producing high-quality thematic maps of census variables for a variety of geographical units (*http://tiger.census.gov*). Similar systems exist in Canada but are the preserve of registered users only because of similar copyright restrictions to those that apply in the United Kingdom. Independent initiatives in the Canadian academic sector also offer data on-line, although they are limited to statistical profiles, such as E-STAT at Concordia University, Montreal (*http://juno.concordia.ca/collections/camsim.html*) and Simon Fraser University, British Colombia (*http://www.sfu.ca/~rdl/dlib/data/survey/census/96census/96intro.html*). In the United Kingdom, the government statistical service (GSS) has recently launched an on-line catalogue and data delivery system called *Statbase*, which may be adapted to incorporate some of the outputs from the 2001 Census. (*http://www.statistics.gov.uk/statbase/mainmenu.asp*). Each of the services mentioned here is accessible via the *CDS Resources* CD.

8.4 CASWEB

During 1997, the CDU began to investigate Web-based delivery as a possible means of overcoming some of the difficulties of census access within the constraints of the networked dissemination model. This work became embodied in an ESRC/JISC-funded research project that set out to address the need for a more intuitive means of accessing the census area statistics (CAS) by developing an interface to the data that was easy to use, fast, flexible, secure, and platform-independent. It was felt that the Web could provide some important advantages: a generic and user-friendly interface environment, for example. Furthermore, the data should be held in an open format, preferably in an industry standard database management system (DBMS), to allow the development of alternative and complementary access tools in the future. The resulting system, CASWEB, was released into service in October 1998 and has become a mainstream service of the CDU, available at *http://census.ac.uk/casweb/*.

CASWEB consists of a large, relationally structured database of census counts and metadata from the 1991 SAS with which the user interacts through a series of menus and maps accessed entirely via the user's Web browser. The system is built around a client-server architecture, summarized in Figure 8.1, with the user logging in from their desktop computer and formulating and executing queries remotely on the server.

CASWEB is a secure system, with users having to authenticate themselves by user name and password to gain access to the interface. With the exception of an optional map component, the interface is not specific to any particular client platform and will run on any reasonably up-to-date Web browser on Windows 3.1, 95, 98, and NT, Apple Macintosh, and UNIX systems.

The CASWEB interface guides the user through a three-step procedure to select and extract specific variables from the 1991 SAS for user-defined study areas:

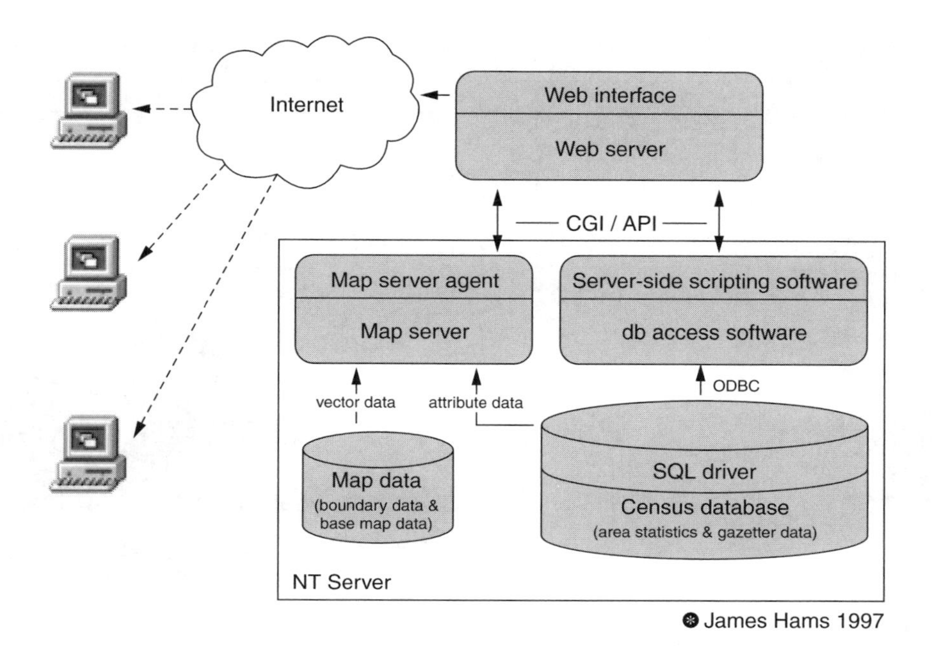

Figure 8.1 CASWEB system architecture.

8.4.1 Selection of Study Areas

CASWEB provides users with the choice of either textual or map-based methods of selecting areas (such as counties or wards) for which census data is required. One of the most advanced features of the interface is the option of using a dynamic map to define regions for which data is required, as illustrated in Figure 8.2. This has been developed using third-party software to embed desktop mapping functionality in the user's Web browser. The map consists of census digitized boundaries and digital base map data providing human and topographic features to enable users to relate arbitrary census geography to recognizable locations. The user is thus able to identify their area of interest with reference to basic topographic information that does not necessarily form part of the census geography itself.

8.4.2 Selection of Census Data

Figure 8.3 illustrates the way in which a user selects census data using CASWEB.

8.4.3 Creation and Download of Output File

The Census data selected in the previous two steps are output in the form of an ASCII data file or as the result of recent developments, in the form of textual data integrated with appropriate digital boundary data in one of a variety of formats native to industry

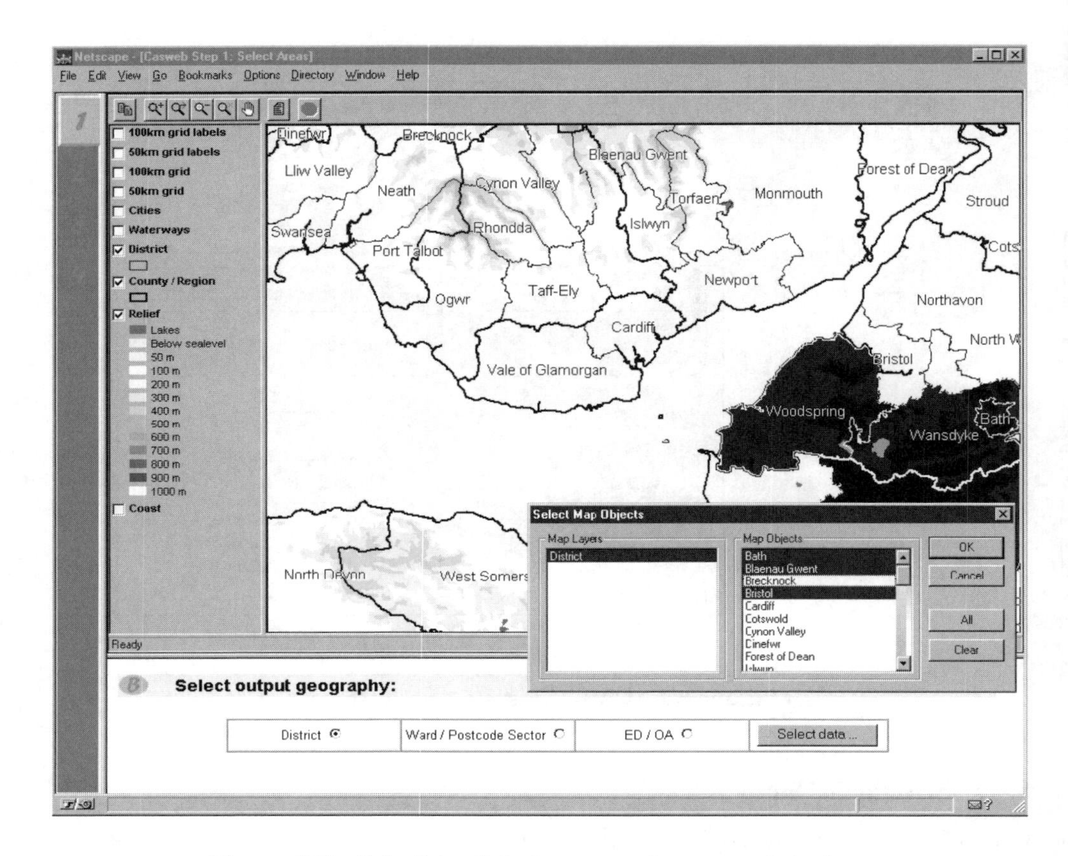

Figure 8.2 Selection of study areas using a map interface.

standard GIS mapping software packages, as shown in Figure 8.4. The GIS format outputs allow users to rapidly produce thematic maps of selected census variables, without the requirement of sourcing boundary data elsewhere, and subsequently combining them with the census data.

Output is temporarily stored on the server until it is retrieved by the user and saved on the local computer. Alternatively, the user can request whole census tables to be output in HTML format for printing or on-screen viewing. Further manipulation and analysis of the data takes place locally, although limited analytical functionality is envisaged as CASWEB is developed further.

The database has been constructed using an enterprise level DBMS on a moderately powerful NT server. The arrangement of the SAS and LBS data sets into tables has been retained to maintain compatibility with the existing coding schema used by other census software. However, an entirely new relational data structure has been devised to optimize the performance of the data engine for large queries. A central component of this structure is that tables of descriptive metadata reference the many thousands of data items in such a way that the complexities of census geography, the census table coding schema, and the special rules governing the data items available for each geographical area are entirely

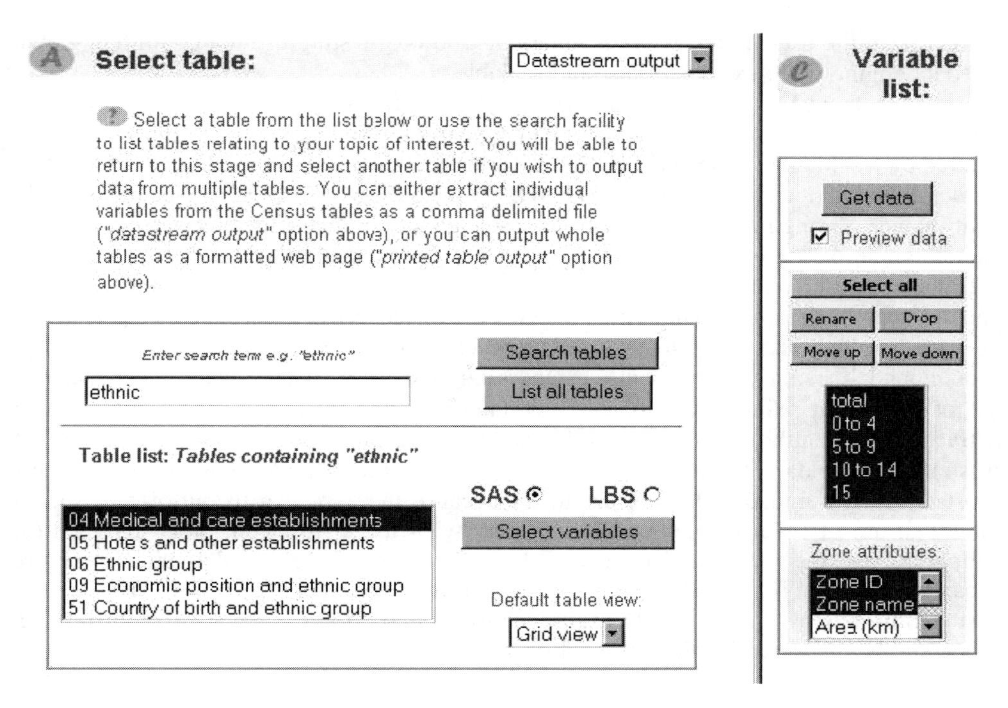

Figure 8.3 Selection of census variables using CASWEB.

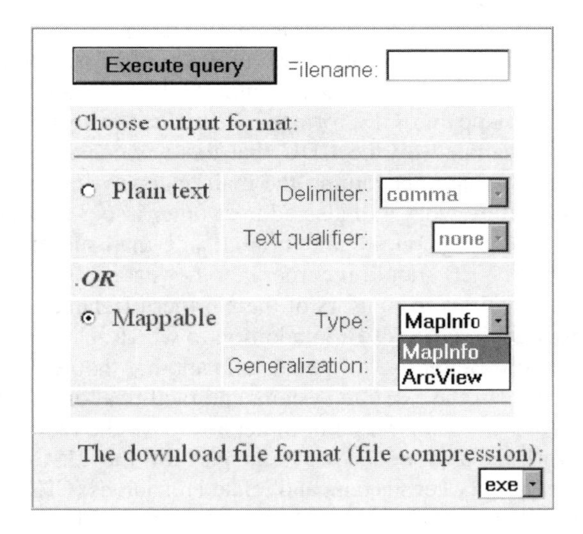

Figure 8.4 Selection of map boundary formats using CASWEB.

hidden from the user. Extensive benchmarking and beta-testing of the system has shown the data engine to be extremely fast and scalable.

Since its launch as a supported service, CASWEB has proved popular with users, logging over 8000 user sessions in its first year of service. Much of this activity is centred on teaching, as CASWEB is used for undergraduate and postgraduate courses at 14 academic institutions during this period. While the success of CASWEB has undoubtedly contributed to bringing new users with lower levels of computer literacy into the census-user community, there is evidence to suggest that a great deal more can be done to make the data that users extract from CASWEB more useful in the context of their research and teaching applications. The initial CASWEB project recognized that existing methods of accessing the data failed to meet less proficient users' needs on two levels. Firstly, a command line interface presents an enormous barrier to users—whose expectations are of a familiar Windows graphical user interface—wishing to perform even simple extractions. Secondly, users who have learned to operate the extraction software are often frustrated by the data it outputs. This is either because the data are too detailed and must be statistically summarized to be useful, or because they are tied to out-of-date spatial units that are of little relevance to the geography of users' research, or in many cases, simply because the statistical data needs to be manually joined to digital boundaries before it can be mapped. The CASWEB data-extraction interface, therefore, represents only the first step in making the CAS truly usable for users who do not possess the requisite data processing skills to realize the potential of the data.

8.5 CASWEB DEVELOPMENTS

Clearly, there is scope for the development of an enhanced second-generation product that goes some way to meeting the aforementioned needs. CASWEB2 is currently under development with further ESRC/JISC funding, and the first results of this work in the form of the GIS format outputs described previously are already beginning to be made available to the academic community as enhancements to the existing CASWEB service. The main focus of this work is on providing functionality to increase the utility of the data via a suite of tools that allows users to re-aggregate census counts to other geographies (e.g., postcodes), to extract summary statistics and profiles, and to output the results in readily mappable GIS formats. It is clear from the volume and nature of user-support queries handled by the CDU that there is considerable demand for these features to be added to what is becoming the main gateway into the CAS for increasing the number of users. Future work will also address other issues that have emerged from consultation with users attending courses and through an e-mail survey. First among these is the expectation that CASWEB should incorporate other data sets in addition to the GB SAS/LBS. Strong representations from users of the Northern Ireland SAS make a clear case for inclusion of those data in CASWEB, in addition to which the 1981 GB SAS would also seem a logical extension. Better use of metadata is another theme that the project team acknowledge as a weakness of the existing system, and facilities for searching the Census tables by topic and keyword are a natural point of departure for the redevelopment of the CASWEB interface.

Work is also under way, as part of the JISC-funded collection of historical and contemporary census data and related resources (CHCC) project, to develop a link between CASWEB and DESCARTES (see Chapter 7). The result of this will be an on-line interface to the CAS that will allow users to select data using CASWEB and then proceed directly to visually explore the data using DESCARTES.

8.6 CONCLUSION

Although Web-based access to census data by no means replaces the requirement for desktop applications and hard copy distribution routes, CASWEB has demonstrated that there is a great demand for such a system. In some respects, the need for any type of third-party system such as CASWEB and SASPAC is an anachronism in the modern computing environment. Openshaw observes that'the Office of Population Censuses and Surveys (OPCS) should have been distributing census information systems not raw data. These systems should have provided access to census data without any need for SASPAC, GIS and separate mapping packages.' (Openshaw, 1995b: 410)

It is also worth remembering that even by the standards of today's hard disks and portable media such as DVD, the Census is still moderately large. As a consequence of this fact alone, there will be a role for network-based dissemination systems well beyond 2001. It seems likely, however, that by whatever route the data are served up to the user, advances in the scalability and sophistication of off-the-shelf statistics and GIS packages mean that there will eventually no longer be a requirement for dedicated extraction software at all. For the time being, however, the Web provides a very useful opportunity for getting census users closer to the data.

Box 8.1 Key web links for Chapter 8.

http://census.ac.uk/casweb/	Web site providing full documentation on CASWEB
http://census.ac.uk/cdu/	CDU Web site
http://www.stat.go.jp/english/1.htm	Japanese census metadata Web site
http://www.censuscd.com/	Geolytics Inc. Web site (information on CD-ROM of US Census data with mapping)
http://www.censuscounts.com/	DecisionMark Inc. Web site (information on CD-ROM of US Census data with mapping)
http://tiger.census.gov/	US Bureau of the Census Tiger Map Server
http://juno.concordia.ca/collections/camsim.html	Concordia University, Montreal E-STAT statistical profiles for Canadian census
http://www.statistics.gov.uk/statbase/mainmenu.asp	UK GSS on-line catalogue and Statbase data delivery system

9

Deprivation Indicators

Martyn Senior

SYNOPSIS

The chapter provides a guide to the numerous indices developed over the past two decades to capture the deprivation faced by the residential populations of small areas. Deprivation indices are used for a wide variety of purposes. Censuses have traditionally provided input variables for composite indices, though the most recent indices developed by academic researchers (Noble and colleagues) contain mainly administrative rather than census data. In the first part of the chapter, theoretical constructs and data supporting deprivation indices are described along with a review of the units for which deprivation can be measured. The chapter then describes, with a critical commentary, the ways in which indices are built. The final section references both academic and government sites where information on the indices can be found.

9.1 WHAT IS THE STUDY'S PURPOSE?

Prompted by concerns over increasing social and economic inequalities in the context of tightening public expenditure constraints, the quarter century since Holterman's study (1975) has witnessed a remarkable, and somewhat bewildering, growth in the use of deprivation measures. Many of these depend wholly (e.g., Jarman, 1983) or partly (e.g., Bradford *et al.*, 1995) on census data. To make intelligent use of these census (and other) data resources, a set of questions needs addressing.

Purposes can range from basic research, through myriad policy applications, to the simple mapping and description of deprivation (Forrest and Gordon, 1993; Gordon and Forrest, 1995) for various information purposes. The purposes for which planners might use 1991 Census data are usefully discussed by Hirschfield (1994).

A substantial body of more basic research investigates deprivation's influence on welfare and behaviour, especially in the health field. The Townsend Index (Townsend, Phillimore and Beatty, 1986; 1988) was developed to operationalize the concept of material deprivation emphasized in the Black Report as the major cause of inequalities in health (Townsend, Davidson and Whitehead, 1988). This has spawned a stream of research on deprivation and mortality (Carstairs and Morris, 1991; Morris and Carstairs, 1991; Phillimore and Reading, 1992; Phillimore *et al.* 1994a; McLoone and Boddy, 1994; Higgs *et al.*, 1998) and increasingly, with the question on long-term illness in the 1991 Census, on deprivation and morbidity (Bentham *et al.*, 1995; Gould and Jones, 1996; Shouls *et al.*, 1996; Senior, 1998). There is also a substantial literature on deprivation and educational achievement, given the need to contextualize school league tables (Higgs *et al.*, 1997; Gibson and Asthana, 1998), and some studies of crime and disadvantage (Hirschfield and Bowers, 1997).

Such deprivation-welfare analyses point to more policy-oriented questions on the need for, and resourcing of, public services. The incorporation of deprivation measures into health care funding has been researched (Royston *et al.*, 1992; Smith *et al.* 1994) and a range of deprivation-related variables has been used as need indicators in regression formulas for Standard Spending Assessments (Flowerdew, 1994; Thomas and Warren, 1997). Deprivation measures may be used not only in policy and resource targeting but also in policy assessment. Cole and Gatrell (1986) use census data to identify mismatches in the locations of needy groups and the supply of library services and suggest how mobile library facilities could be reorganized to improve service delivery.

Often, however, policy-oriented deprivation indicators are less well informed by basic analytical research. For example, the validity of Jarman's (1983; 1984) UPA index for allocating deprivation payments to GPs has been fiercely criticized (Carr-Hill and Sheldon, 1991). Even more problematic is the construction of deprivation indicators prior to their precise purposes being specified. Thus, because of central government's requirements, the Index of Local Conditions (ILC)/Deprivation (DoE, 1995a; 1995b; DETR, 1998) was developed as a general-purpose index, but was capable of disaggregation for more specific purposes and applicable at varying geographical scales (Bradford *et al.*, 1995). In a similar vein, the recent Indices of Multiple Deprivation (IMD), produced by Mike Noble's team at Oxford, are aggregates of six sectoral deprivation domains (DETR, 2000a; NAfW, 2000).

As a general rule, the more precisely the purpose of the study can be specified, the more readily can an appropriate conceptual framework be chosen or devised to inform the selection of indicators.

9.2 WHAT THEORETICAL AND CONCEPTUAL FRAMEWORKS CAN INFORM THE STUDY OF DEPRIVATION?

Deprivation is a controversial concept and defined with difficulty (Coombes *et al.*, 1995). Common usage of the term embraces anything from poverty to inequality. Greater clarity can be achieved by considering the breadth of meaning of deprivation and related concepts (Hirschfield 1994; Lee *et al.*, 1995). Thus, poverty, defined as low income or material resources in absolute or relative terms, can be regarded as narrower in scope than deprivation, which Townsend (1987, p125) defines '... as a state of observable and demonstrable disadvantage relative to the local community or the wider society or nation to which an individual, family or group belongs'. In turn, deprivation is narrower than both social exclusion (implying inadequate social participation and integration and poor access to social institutions affecting life chances) and inequality, which may refer to differences across the entire social spectrum. In addition, there are subdivisions of the aforementioned concepts. Thus, income and expenditure poverty are not necessarily coincident. Similarly, Townsend (1987) distinguishes material from social deprivation and suggests more specific subcategories of each.

9.3 WHICH UNITS OF ANALYSIS ARE TO BE USED?

As Hirschfield (1994) stresses, deprivation can affect individuals, households, families, groups, institutions (such as schools), and areas. Census data are collected from individuals and households, with limited information on persons in institutions (e.g., hospitals) rather than in private residences. Additionally, individuals or households are classified into various groups (e.g., social classes, socio-economic and ethnic groups, households by housing tenure). Before the 1991 Census, information was provided only for individuals or households aggregated by census areas, such as wards and local authorities. With the 1991 Samples of Anonymised Records (SARs), data are now available for individuals and households, although their locational identification is by relatively large areas. With area data, there is always the risk of ecological fallacies, namely, attributing area deprivation to all the persons and households resident therein. Fieldhouse and Tye (1996) have used the SARs data to show that many deprived people are resident outside those local authorities consistently identified as deprived.

9.4 WHAT DATA ARE AVAILABLE FROM THE CENSUS (AND ELSEWHERE)?

Although the census has limitations in its coverage of deprivation (e.g., no information on income, welfare benefits, service provision, or environmental conditions), it provides comparable data for small areas throughout Britain on unemployment, lack of a car, housing amenities and ownership, ill health, and lack of educational qualifications. Table 9.1 identifies the main 'deprivation' tables in the 1991 Local Base Statistics (LBS), indicating any cross-classifications with other person or household attributes. Although used quite often (see Table 9.2), groups, like lone parents, at the risk of deprivation, have been criticized as indicators because some members of such groups will not be deprived (Townsend, Phillimore and Beatty, 1986). Significantly, the DoE's deprivation index for 1991 deliberately excludes vulnerable groups (Bradford *et al.*, 1995); this represents a change from its 1981 index, which included ethnic minorities, lone parents, and the lone elderly (DoE, 1983). However, even the 'direct' measures of deprivation may include non-deprived persons (e.g., wealthy non-car owners living in central London). Excluding 'at risk' groups as deprivation indicators should not prevent deprivation measures being constructed for sub-groups of the population where appropriate cross-tabulated data are available. For example, Townsend deprivation scores for ethnic groups could be constructed using census data on ethnicity by unemployment, overcrowding, non-owner occupation, and no car, respectively (see 1991 SAS/LBS Tables 9 and 49). Interestingly, Bush (1996) develops and critiques measures of deprivation affecting women.

Whereas the LBS provide limited cross-tabulations of census variables, the availability of the SARs permits virtually any cross-classification of two or more variables for individuals or households. Thus, Fieldhouse and Tye (1996) could measure for individuals the prevalence of multiple deprivation across nine SAR variables simultaneously.

Of course, the census is not the only source of deprivation data. The NOMIS (National Online Manpower Information System) has for some time provided, for example, far more recent and updatable data on monthly unemployment than a dicennial census can offer. The use of increasingly available non-census data for small areas is a key feature

Table 9.1 Deprivation indicators in the 1991 local base statistics.

Deprivation indicator	Table number	Cross-classified variables
Unemployment	8, 34	Age, gender, marital status
	9	Ethnic group
	14	Limiting long-term illness
	36	No earners by dependent children
	37	Young adults' gender and age
	40	Lone parents not employed
	45	Migrants by age and tenure
	84	By educational qualifications
	91	Social class and gender
	92	Socio-economic group and gender
	94	Former industry
	95	Former occupation
Housing:	20, 23	Tenure
lack of bath/shower, WC, central heating, and/or overcrowding	41	Shared accommodation & household size
	42	Household composition
	46	Households with dependent children
	47	Households with pensioners
	48	Households with dependants Ethnic group
	49	Type of dwelling
	57, 58	
No car	20	Tenure
	21	Persons aged 17 and over in household
	41	Shared accommodation
	42	Household composition
	46	Households with dependent children
	47	Households with pensioners
	48	Households with dependants
	49	Ethnic group
	82	Means of transport to work
	83	Employed persons per household
	86	Socio-economic group
	87	Family type
Limiting long-term illness (Permanent sickness in Tables 8, 9, 14, and 34)	6, 49	Ethnic group
	12, 13	Age and gender
	14	Economic position
	29	Dependants per household
	44	Household composition; economic activity
	47	Households with pensioners
	75	Gender and hours worked
	86	Socio-economic group
Lacking higher educational	84	Age, gender, & economic position
qualifications	85	Ethnic group

Note: Also, see the keyword index in Dale and Marsh (1993: 236–247).

Table 9.2 Components, weights, and characteristics of some deprivation indices.

Deprivation indicator	DoE81	Jarman UPA	Townsend d	Carstairs	ILC (DoE91)	ILD (DETR)	Matdep	Socdep	Bread-line Britain	Oxford	Welsh Office SEC	Welsh Jarman	Scottish Office
Unemployment	2	3.34	1		1	1*		1	0.0943		1	3.34	XX
Male unemployed				1						1.3904			
Youth unemployed								1					XX
Non-earners in households													XX
Children in no or low earner households					1	1*							X
Economically (in)active										0.1088	1		
No car			1	1	1	1*	1		0.2174				
> 1 person per room	1	2.88	1	1	1	1	1			0.7250	1	2.88	
Households below occupancy norm													XX
Vacant dwellings													X
Lacks basic amenities	1				1	1	1				1		X
No central heating							1						
(Not) owner occupier			1						0.2025	−0.069			
Rent from local authority										0.3101			
Children in flats					1								
Limiting long-term illness								1	0.1079				X
Permanent sickness											1		XX
17-year olds not in full-time education					1W	1W							
Households with dependants only								1					
Single pensioners	1	6.62						1				6.62	

(*continued overleaf*)

Table 9.2 *(continued)*.

Deprivation indicator	DoE81	Jarman UPA	Townsend d	Carstairs	ILC (DoE91)	ILD (DETR)	Matdep	Socdep	Bread-line Britain	Oxford	Welsh Office SEC	Welsh Jarman	Scottish Office
Pensioner households													X
Lone parents	1	3.01						1	0.1597	2.0262		3.01	XX
4 or more dependent children													X
Unskilled manual		3.74										3.74	
Social classes IV or V				1					0.1585		1?		
Ethnicity	1	2.50								0.3717		2.50	
Below 5 years		4.64										4.64	
Population loss, ages 20–59											1		
Residential mobility		2.68										2.68	
! House condition												+D	
! Long-term male unemployment					1D	1D							
! Income support					1D	1D							
! Low GCSE grades					1D	1D							
! SMR					1D	2D					1	+D	
! Derelict land					1D	1D							
! House insurance weights					1D	2D							
Standardization	Z	Z	Z	Z	chi	chi	%/ range	%/ range	% poor	Z	Z	Z	?
Transformation	log of 5	arc	log of 2	no	log chi	log chi	no	no	no	no	no	arc of 8	no
Weighting method	arb	surv	arb	arb	arb	arb	arb	arb	regr	regr	arb	surv + arb	pca >

Note: W—used at ward and district levels; !—not a census variable; D—used at district level only; +D—measured at district level, but added uniformly to all wards within a district; X—weighting of Scottish office indicators not given in Duguid (1995); XX—indicator used in the 'general deprivation' measure suggested by principal components analysis (PCA); ?—details unclear; Z—z-score; Chi—signed chi-square; Log—logarithmic; Arc—arcsine square root; Arb—arbitrary; Surv—survey; Regr—estimated by regression; Pca—estimated by principal components analysis.
*Updated using non-census data at local authority district level only.

of the IMD (IMD2000) recently produced for central governments in England and Wales (DETR, 2000a; NAfW, 2000). These IMD make considerable use of data on persons or households in receipt of various government benefits, although this does pose the problem of measuring, and compensating for, lack of uptake by some of those eligible. The IMD2000 is constructed for wards from 33 (29 in Wales) indicators grouped into the following 6 deprivation domains (with weights in brackets): Income (25%); Employment (25%); Health and Disability (15%); Education, Skills, and Training (15%); Housing (10%); Geographical Access to Services (10%).

9.5 HOW ARE DEPRIVATION INDICES CONSTRUCTED?

9.5.1 To Combine or not to Combine?

Although the combination of deprivation indicators into a composite index is commonly undertaken, this may be neither necessary nor desirable. Indeed, the more theoretically informed and analytically sophisticated studies often use multiple indicators simultaneously (e.g., the regression analyses of Smith *et al.*, 1994). Before committing oneself to a composite indicator, it must be decided whether the research should use multiple, but not combined, indicators, or even whether a single indicator might suffice (Campbell *et al.*, 1991).

However, composite deprivation indices are attractive because they capture multiple dimensions in a single, manageable scale. Essentially, this deprivation construct can be defined as the sum of a set of weighted component indicators:

$$\text{Deprivation} = \text{weight1} * \text{indicator1} + \text{weight2} * \text{indicator2} + \text{weight3} * \text{indicator3} \ldots \text{etc.}$$

In constructing such composite indices, one is faced with a number of thorny technical questions. These technical choices can have significant influences on the deprivation ranking of areas and the allocation of public resources to them (see Martin *et al.*, 1994 for an example).

9.5.2 To Standardize or not to Standardize?

Standardization is the process of converting indicators to a common metric so that they can be combined. For example, income could not be added to percentage unemployment as their units of measurement differ. It is debatable whether indicators expressible as proportions or percentages (as all census variables are) need to be standardized as they have a common 0 to 1 or 0 to 100% scale (Martin *et al.*, 1994). The arguments in favour of standardizing indicators essentially boil down to eliminating hidden weights. For example, if an overcrowding indicator, with a range of 0 to 10%, is to be combined with non-car ownership, varying from 5 to 82%, then the car ownership values will dominate the overcrowding ones (Bradford *et al.*, 1993). Whether this 'problem' (if indeed it is one) should be corrected by weighting or standardization (or both) is, however, an open question. This author believes that resort is often made to standardization because of the difficulty of explicit weighting by other means (see Section 9.5.5).

Preferred standardization methods have involved: (i) z-scores in the 1980s, used by Jarman (1983), Townsend, Phillimore and Beatty, (1988), Carstairs and Morris (1991), and by the DoE (1983); (ii) signed chi-square in the 1990s as used in the DoE's 1991 and

DETR's 1998 indices (Bradford *et al.*, 1995; DETR, 1998); and (iii) ranking now used in the IMD2000 (DETR, 2000a; NAfW, 2000). The first two methods require values of the deprivation variables for a reference area, often the nation (e.g., Jarman, 1983) or region (e.g., Townsend *et al.*, 1986).

Standardization issues have become intimately bound up with the reliability of data based on small numbers. Because deprivation indicators are usually measured for zones, varying zone size means that counts will not accurately reflect relative concentrations of deprivation (Simpson, 1996). Percentages or proportions will eliminate zone-size effects, but they, and z-scores based on them, are insensitive to the size of the denominators used in their definition. Thus, three zones with 3, 300 and 3000 unemployed, out of, respectively, 10, 1000 and 10 000 economically active persons, will all have 30% unemployment. Similarly, assuming a reference (national) unemployment rate of 20% and a standard deviation of 5%, all three zones have identical z-scores of 2 ({30 − 20}/5). Percentages, proportions, and z-scores ignore the low robustness of values based on small denominators, which often occurs at enumeration district (ED) level, especially if the 10% sample census data are used. This problem is compounded by the pseudo-random modification (by ±1, once in the SAS, twice in the LBS) of most ward and ED non-zero census counts (but not the 10% sample census counts). Such modifications have most impact where denominators are small, as 3/10 (30%) unemployed might signify unmodified data of 2/11 (18.18%) or 4/9 (44.44%); even larger errors may be involved if numerators and/or denominators are totals of modified counts (see Cole, 1993, pp. 220–225).

The signed chi-square addresses this small denominator problem by incorporating relative and absolute values and scaling the local area and reference percentages or rates by the local denominator, LD (economically active population):

$$\text{Signed chi-square} = \text{LD} * (\text{local\%} - \text{reference\%})^2/\{\text{reference\%} * (100\% - \text{reference\%})\}$$

with the sign determined by the sign of (local% − reference%). For the aforementioned 3 areas with 30% unemployment:

$$\text{Signed chi-square} = \text{LD} * (30\text{–}20\%)^2/\{20\% * (100\text{–}20\%)\} = \text{LD} * 0.0625$$

The three LDs of 10, 1000 and 10 000 produce signed chi-squares of 0.625, 62.5, and 625, respectively. However, the signed chi-square is now discredited because it causes, as we shall see later, more problems than it solves.

A more transparent and statistically respectable way of solving the small denominator problem is to use 'shrinkage estimation' (DETR, 2000a,b). This takes 'unreliable' zone values on an indicator and adjusts them towards the more reliable mean score on that indicator for a larger set of neighbouring zones, for example, the score for the local authority in which a ward is located.

Ranking is the latest method used to standardize indicators (DETR, 2000a; NAfW, 2000). Problematically, ranked values are symmetrical and involve equal distances between adjacent indicator values. This implies that, when combined, a high deprivation rank on one indicator could fully cancel out a low deprivation rank on another. Such full cancellation is undesirable, especially with zonal data, as say the 20% of households deprived through lack of car ownership are not likely to be the same 20% not deprived on a housing indicator. In other words, the various components of deprivation should be regarded as largely non-compensatory. The response adopted to this problem in the IMD

(DETR, 2000a,b) is to transform the ranks to an exponential distribution that emphasizes the 'tail' of most deprived zones.

9.5.3 To Transform or not to Transform?

Three arguments are advanced for transforming deprivation variables. First, transformation can reduce or eliminate the domination of a composite index by highly skewed indicators (Thunhurst, 1985; Senior, 1991). Second, the indicators may be used with statistical techniques that assume Normal distributions (Bradford *et al.*, 1993). This latter argument is only valid if such techniques as Principal Components or Factor Analysis are used to identify the underlying dimensions of multiple deprivation (e.g., Duguid, 1995). Moreover, Normally distributed indicators are not essential for certain types of factor analysis. Thus, the authors of the IMD claim that normal-theory maximum likelihood, as an estimation method in factor analysis, 'is not invalidated by non-normality ... It is therefore possible to use untransformed variables in the analysis' (NAfW, 2000, p51). Many popular deprivation indices (Jarman, Townsend, Carstairs) have been constructed without recourse to such statistical techniques, although they leave themselves open to criticism of double counting aspects of deprivation measured by highly intercorrelated indicators (e.g., see Carr-Hill and Sheldon, 1991 on the Jarman index). Thirdly, as noted at the end of Section 9.5.2, transformation has been used to prevent cancellation of high deprivation on some indicators by low deprivation on others.

In practice, various transformations have been used (Table 9.2). Jarman (1988, see page 103 for an example calculation) takes the arcsine of the square root of each indicator measured as a proportion. The logarithmic transformation is more commonly used (logarithmic, square root, and reciprocal transformations are special cases of the Box-Cox family of power transformations). Thus, the DoE's (1983) 1981 index logs all variables except pensioners living alone and the Townsend index incorporates logged values of the unemployment and overcrowding indicators (Phillimore *et al.*, 1994b). However, a logarithmic transformation of an indicator containing one or more zero values is impossible, necessitating the addition of an arbitrary constant (usually 1). Typically, indicators are transformed prior to being standardized, but the Indices of Local Conditions/Deprivation are constructed by computing the chi-square standardization of all indicators first, then logging the standardized values, and finally attaching a positive or negative sign as appropriate.

One major disadvantage of transformations is that they make the scores less intelligible to users. For example, the readily understandable 0 to 1 range of a proportion becomes a range of 0 to 1.5708 if Jarman's (1983) arcsine square root transformation is used. The natural 0 to 100% range must have an arbitrary constant added (say 1) before natural logarithms (as in the Townsend index) are taken to give a transformed range of 0 to 4.615. It is perhaps not surprising, therefore, that some deprivation indices avoid transformations (Table 9.2), although the motives for doing so are often not explicit.

9.5.4 Are there Problems with Standardized and/or Transformed Indices?

Yes! Apart from their insensitivity to unreliable data for small areas, z-scores based on percentages are criticized for concealing concentrations of deprivation within large areas, which are usually more heterogeneous than small areas (see Simpson, 1996, p541, for an example comparing Leeds and Grimsby). While scale-independent percentages or z-scores thus work 'in favour' of smaller areas, positive values of the scale-dependent signed chi-square 'favour' larger areas. Even the authors of the ILC imply that this scale-dependence might overstate deprivation for large areas when they justify their use of logarithmic transformations: '...transformation also means that the contribution through the signed chi-square of the absolute as against the relative value of deprivation is dampened down, especially at district level. Thus, ...the varying size of districts does not dominate the deprivation values' (Bradford *et al.*, 1995, p522). Connolly and Chisholm (1999) provide examples that contest this claim. Moreover, for local areas with percentages below the (national) reference percentage, the resulting statistics can be counter-intuitive and may imply perverse resource allocation decisions. In response to this latter problem, DETR's (1998) revised Index of Local Deprivation (ILD) sets negative statistics to zero. Again, Connolly and Chisholm (1999) show that this amendment modifies, but does not overcome, the problems with this index. Another possible problem is that the choice of reference area can affect the ranking of local areas by the chi-square statistic. Further examples of these problems are available in Simpson (1996).

9.5.5 To Weight or not to Weight?

This is a tricky question. The combination of deprivation indicators into a composite index means that issues surrounding the weighting of the component variables are unavoidable. Yet, weighting often does not attract the attention it deserves, probably because justifiable weightings are not easily accomplished. There are three main methods of weighting.

First, and often, it is assumed, more or less arbitrarily and sometimes implicitly, that the indicators should be equally weighted (see Table 9.2). Arguments have been advanced for differential weightings (e.g., DoE, 1983; DETR, 1998; DETR, 2000a; NAfW, 2000), but again these are essentially arbitrary rather than rigorously validated by empirical evidence. Of course, empirical evidence may be difficult, costly, or even impossible to obtain, but two alternatives have been tried.

The Jarman index is a good example of the second method, the use of surveys that seek the subjective weightings of a relevant population. Jarman (1983, p1706) asked for a 10% sample of British GPs to weight each of the eight variables in his UPA index on a scale of zero to nine 'according to the degree to which it increases workload or contributes to the pressure of work when it is present'. As the index was later used to allocate deprivation payments to those GPs, such a survey seems a reasonable procedure, but weights may be distorted by respondents' vested interests and personal values (Coombes *et al.*, 1995). Carr-Hill and Sheldon (1991) further note that GPs were not asked to weight the transformed variables actually used, that standardization involves an unjustified weighting by the inverse of each variable's standard deviation, and that failure to address the intercorrelation of variables leads to double weighting.

A third and less subjective method is to estimate weights on deprivation indicators statistically. Because deprivation, as commonly conceived, is a latent construct (i.e., it cannot be directly measured), Bartholomew (1988) advocates the use of factor analysis and regards an associated goodness-of-fit test as essential. Measurable indicators reflecting deprivation have their common variance extracted and combined into a single deprivation index or factor. An area's factor (deprivation) score can be derived as:

$$\text{estimated factor score} = \text{weight1} * \text{indicator1} + \text{weight2} * \text{indicator2} + \cdots$$

where the weights are factor score coefficients. Indeed, maximum likelihood factor analysis is used to weight and combine the indicators, which do not have a common metric, in the health, education, housing, and geographic access domains of the IMD (DETR, 2000a). However, the possible validation of the six domains in the IMD by factor analyzing the original variables has not, apparently, been attempted. Weights can also be derived from principal components analysis (PCA), although Bradford *et al.* (1993) reject the use of PCA scores because of their sensitivity to the particular variables used. Instead, they used PCA to explore the interrelationships of deprivation variables and to suggest the indicators that should be included or excluded.

Alternatively, some form of regression analysis (Aitkin *et al.*, 1989) of the general form:

$$\text{response variable} = \text{weight1} * \text{indicator1} + \text{weight2} * \text{indicator2} + \cdots + \text{random error}$$

may be possible. Here, the weights are regression coefficients. Crucially, this requires the definition of, and observations on, a response variable. This should be possible in two contexts: first, where the response variable measures behaviour or outcomes hypothesized to be influenced by deprivation; and second, where more sensitive, but less easily obtained, measures of deprivation are available perhaps from specialist sample surveys or for specific study areas. Examples of the first situation abound in health research. Thus, weights on the eight Jarman indicators could be estimated using sample data on GP workloads, operationalized as consultations with patients (e.g., frequency or time spent per patient), as the response variable. If necessary, additional variables could be used to control the influences of supply on demand (see Carr-Hill *et al.*, 1994, for an example on secondary health care). Similarly, weights on the component variables of the Townsend and Carstairs indices, both devised to examine hypothesized relationships between health and material deprivation, have been estimated using premature mortality or morbidity as response variables (e.g., Phillimore and Reading, 1992; Senior *et al.*, 1998). Moreover, such statistical analyses provide significance tests of hypotheses, so certain indicators of a suggested index might be invalidated (i.e., zero weighted) by the empirical evidence. This author would claim that this is the acid test of the worth of a deprivation index. The second situation is illustrated by the Breadline Britain and Oxford indices (Lee *et al.*, 1995; Gordon and Forrest, 1995). Indicator variables are weighted by regressing less readily available (at least for small areas) poverty and benefit data on census indicators. Statistical assessments of which census variables can predict poverty, and how well they do so, can be made.

9.6 HELP! HOW DO I CHOOSE A DEPRIVATION INDEX?

Many composite deprivation scores are now available for downloading (see next section). The primary choice criterion is fitness for purpose. If you are researching in the fields of health or urban regeneration, you have the most choice. The identification of deprivation domains in the IMD2000 widens the range of choices. Given the ease of selecting indices 'off the shelf', a word of caution is appropriate. You will still need to evaluate carefully the content, construction, and context of any existing index to be aware of its limitations, its undesirable attributes, and its idiosyncrasies. For example, many health applications automatically use the Jarman index as a deprivation measure but fail to appreciate that its main purpose is to identify excess workloads on doctors, whether deprivation-induced or not. Given its origins, there are also complaints that the Jarman index is biased towards primary health care in London.

There may well be circumstances in which existing indices are deemed unsatisfactory. Thus, urban bias (e.g., the non-car ownership indicator in the Townsend and Carstairs indices) is a common complaint of those interested in rural deprivation (Higgs and White, 2000; Martin *et al.*, 2000). Building your own indices can be off-putting, especially given the technical considerations detailed earlier. However, Martin *et al.* (1994) provide a justification, on the basis of aggregating from deprived individuals or households to geographic areas, for index construction in general to be simply the sum of (weighted) proportions or percentages *without any requirement to standardize or transform the indicators*. Notably, the income and employment deprivation domain scores of the IMD2000 are simply proportions obtained by summing the component indicators, which have a common metric, namely, counts of persons on various welfare benefits (DETR, 2000a). You can also examine the Matdep and Socdep indices formulated by Forrest and Gordon (1993; see also Lee *et al.*, 1995) to see how relatively easy to construct and transparent indices can be, especially if weights on indicators are set arbitrarily. One, however, has to watch out for indicator scores based on small denominators and be prepared to invest time to understand and apply 'shrinkage estimation' to deal with this problem (and avoid signed chi-square like the plague!). If one must combine deprivation indicators lacking a common metric (e.g., age-gender standardized morbidity or mortality ratios, proportion of low weight births, and access to health facilities) one will typically require statistical procedures such as factor analysis or regression. Additionally, if indicators and their weights are to be validated (and preferably they should be), use of the same statistical methods will be needed.

So what is the best available composite deprivation index, and why? My choice is the Breadline Britain index (Gordon, 1995; Saunders, 1998). It has a *clear focus* on purpose. It is a simple weighted sum of untransformed census indicators that have been *validated* against poverty data. And the outcome scores are estimates of the percentages of households in poverty, which are *transparent* and *easily interpretable* because of the 0 to 100% scale.

9.7 WHERE DO I OBTAIN MORE INFORMATION? WHAT READY-MADE DEPRIVATION SCORES ARE AVAILABLE AND WHERE?

For those seeking to replicate commonly used deprivation indices, identification of the 'official' (i.e., chosen by the author of the index) census cells used in their construction can be found in Townsend *et al.* (1986), Gordon and Forrest (1995), and Phillimore *et al.* (1994b); Carstairs and Morris (1991) for their 1981 index only; and DoE (1995a; 1995b) and DETR (1998) for the Index of Local Conditions/Deprivation.

On-line information on how various deprivation scores are calculated, the particular census cells used, and often downloadable files of the scores themselves, are available at the following Web sites. Note that some of the details provided here may have changed by the time you read them, but some intelligent searching of the relevant Web site should yield the information you require.

9.7.1 The Census Dissemination Unit (CDU) Web Site

The Manchester Information and Associated Services (MIMAS) service of Manchester Computing (University of Manchester) maintains the Web site: *http://census.ac.uk/cdu/*, which has collected a wealth of information on deprivation indices based on 1991 and 1981 Census data. The information includes the methods and formulae used to construct each index, various SASPAC and SPSS programs that implement the formulae with 1991 Census data, and for some geographical coverages and spatial scales, pre-computed values of the indices in ASCII (column delimited or csv) format) or SPSS format. These files have been contributed by a variety of census users.

To get to the relevant Web pages, select the following sequence of pages on the CDU Web site:

Datasets ⇒ 1991 Census datasets ⇒ Area stats ⇒ Derived data ⇒ Deprivation scores based on 1991 area statistics.

Information is then available on the following topics:

- Pre-calculated deprivation scores
- Calculating Townsend, Jarman, and Carstairs scores for 1991 and 1981
- Townsend index
- Carstairs index
- DOE ILC
- Underprivileged Areas
- Breadline Britain weighted index
- Noble *et al.* index
- SOCDEP and MATDEP
- Bradford District Report
- 1981 deprivation scores

Note that to access the pre-calculated scores, census users need to be registered for access to the 1991 Census SAS and LBS data and to have user names and passwords for the server on which the information is held (currently *irwell.mimas.ac.uk*). It is likely in the future that most of these data can be made directly accessible from the Web to all users.

9.7.2 The Department of the Environment, Transport, and the Regions (DETR) and the National Assembly for Wales Web Sites

The University of Oxford's Department of Social Policy and Social Work has produced IMD (IMD2000) for the DETR in England and for the National Assembly for Wales and a deprivation index for Northern Ireland for NISRA. The Oxford Web site (*http://www.apsoc.ox.ac.uk*) provides links or information on all three indices. The DETR (now DLTR) Web site, accessed via *http://www.regeneration.detr.gov.uk/* or *http://www.housing.detr.gov.uk/*, contains information on this latest in a sequence of DETR indices for England. While the previous 1998 ILD still included many census variables and non-census variables, the IMD2000 incorporates more recent information, mostly referring to 1998 and 1999, derived from central government records in six 'domains' identified earlier. Only 1 of the 33 variables used to construct this English IMD2000, limiting long-term illness, was derived from the 1991 Census.

Currently, the DETR (now DTLR) Web site makes available two reports on the IMD2000: (1) *Indices of Deprivation 2000* (DETR, 2000a) and (2) *Response to the Formal Consultations on the Indices of Deprivation 2000 (ID2000)* (DETR, 2000b). Users can also download spreadsheets containing the scores for wards in England and district level summaries of the ward information. So, users can track deprivation using DETR sponsored indices from a 1991 version comprising only 1991 Census information (see Section 9.7.1), through mid-decade versions incorporating administrative data along with census data (Index of Local Conditions 1995, Index of Local Deprivation 1998) to end-decade versions in which administrative data have replaced census indicators. The 2001 Census will provide an opportunity to validate the IMD2000 methodology to measure change in deprivation between 1991 and 2001 and to provide deprivation indices for areas smaller than wards.

The Web site of the National Assembly for Wales:

http://www.wales.gov.uk/keypubstatisticsforwalesfigures/

contains most of the contents of the published report on the Welsh IMD2000 (NAfW, 2000) and downloadable spreadsheets of ward deprivation scores. Deprivation scores and reports for Northern Ireland are available on *http://www.nisra.gov.uk*.

9.7.3 The Greater London Authority (Formerly London Research Centre)

Ward and ED scores and values of component variables for the 1998 ILD (and for the 1995 ILC) can be obtained by *bona fide* census users from the Data Management and Analysis Group, Greater London Authority (formerly the London Research Centre), 81 Black Prince Road, London SE1 7SZ, tel 020 7 787 5500, fax 020 7 787 5606, email *dsinfo@london.gov.uk*, and Web site *http://www.london-research.gov.uk/ds/dshome.htm*.

9.7.4 Information on 'Official' Jarman UPA Scores

These scores can be accessed at *http://www.med.ic.ac.uk/divisions/63/upa.info.asp*. This provides information on the nature and calculation of the scores and on their use for deprivation payments to GPs. 1991 UPA scores and census proportions for the constituent variables can be downloaded (in Excel format) for 1991 wards, 1991 local authorities, and 1996 health authorities. Downloadable scores for EDs are to be made available, and those interested are provided with a contact name.

Box 9.1 Key web links for Chapter 9.

http://census.ac.uk/cdu/	Deprivation indices deposited by users for a variety of geographies, all derived from 1991 Census data
http://www.apsoc.ox.ac.uk/	Department of Applied Sociology, University of Oxford, contractor to DETR/DLTR for the development of the Index of Multiple Deprivation
http://www.regeneration.detr.gov.uk/ or *http://www.housing.detr.gov.uk/*	DETR Web sites providing information on current deprivation indices. (Note that the DETR, the Department of Environment, Transport and the Regions, became the DTLR, the Department of Transport, Local Government and the Regions, in June 2001 and the cross-links from old Web site references may not be maintained.)

http://www.wales.gov.uk/keypubstatisticsforwalesfigures/	National Assembly for Wales Web site, which contains information on the Welsh IMD2000
http://www.nisra.gov.uk	The Northern Ireland Statistics Web site with links to ward deprivation scores and reports
http://www.london-research.gov.uk/ds/dshome.htm	Data Management and Analysis Group, Greater London Authority (formerly the London Research Centre) Web site with ED and ward deprivation scores
http://www.med.ic.ac.uk/divisions/63/upa.info.asp	Web site giving information on 'official' Jarman UPA scores

Box 9.2 Additional CDS resources for Chapter 9.

Deprivation scores overview	Guide to the wide range of ward and ED level deprivation indices available in the CD
(i) Carstairs (for wards)	Carstairs index for wards in Great Britain and Northern Ireland
(ii) DoE, Jarman, Carstairs, and Townsend (for wards)	DoE, Jarman, Carstairs, and Townsend indices for wards in England and Wales
(iii) Townsend (for wards and EDs)	Townsend scores for wards and EDs in England and Wales

10

Census Population Surfaces

David Martin

SYNOPSIS

Although the provision of digital boundaries for 1991 Census enumeration districts (EDs) has enabled many users to engage in thematic mapping and develop geographical information system (GIS) applications using census data, the very nature of the boundaries imposes a number of limitations on the uses to which the data can be put. One alternative is to remodel the geography of the small area statistics by redistributing the ED counts into the cells of a regular grid. The technique reviewed here has been applied to a selection of small area statistics from both 1981 and 1991, and the resulting census population surfaces are available from Manchester Information and Associated Services (MIMAS) through a Web interface known as SURPOP and can be downloaded in a variety of file formats. The chapter reviews the surface construction method and the models available to census users.

10.1 INTRODUCTION

The digital ED boundaries created following the 1991 Census have proved invaluable to many census users and have seen widespread use, both in simple thematic mapping and in more sophisticated GIS applications in which the census provides one of the data layers (Charlton *et al.* 1995). The systems available for accessing these digital boundary data are described in some detail by Mackaness and Towers in Chapter 6. The small size of EDs and the hierarchical nature of the census geography (described more fully by Denham, 1993) has allowed the investigation of detailed spatial pattern and re-aggregation to form meaningful spatial units that are incompatible with the larger areas in the census outputs.

Unfortunately, a number of problems stem from the fact that all census boundaries have an essentially arbitrary and unknown relationship with the underlying social phenomena that they are used to represent. The well-known modifiable areal unit problem (Openshaw, 1984) comprises both scale and aggregation components, in that any observed pattern in the zonal data will be to some extent an artefact of the size and configuration of the zonal system used. If the zone boundaries had fallen in different locations, this would have brought about differences in the observed patterns and statistical relationships between variables, although the underlying phenomena would have been the same. Further, Morphet (1993) demonstrates that even the most detailed census boundaries frequently fail to coincide with significant breaks in social phenomena such as housing tenure. The ED boundaries are primarily designed to facilitate efficient enumeration, but they must be made to fit within all higher level statutory boundaries. Where possible, EDs follow identifiable physical features on the ground. This combination of design constraints can lead to wide variations in ED shape, size, and social composition. The requirement that EDs should cover the entire land surface means that the zones with the lowest population frequently encompass extensive unpopulated areas, and choropleth (shaded area) mapping of this geography can produce results that are highly visually misleading, a problem addressed by Dorling (1993), who seeks to make the shaded areas on the map proportional to the population represented, a technique also demonstrated in Chapter 7. The use of population cartograms provides very valuable tools for the visualization of area-based census data, but it does not actually alter the underlying area-based data model, which frequently forms part of statistical analyses that do not involve any direct mapping.

A useful alternative to the use of census data referenced by ED boundaries is to remodel the data into a geographically referenced regular grids, in which each cell contains estimated values of the various census counts. Cells containing zero estimates will represent unpopulated areas. This is not simply a matter of rasterizing the ED boundaries because this would have merely produced a raster version of the shaded area map, nor is it a matter of interpolating the population values found at ED centroids, which would have generalized the data without reconstructing the unpopulated regions within it. Rather, it is necessary to devise an appropriate technique for the remodelling of the geographical distribution of the population captured by the census that recaptures the underlying population distribution to the maximum extent possible.

The following section describes an adaptive kernel estimation technique that has been used for the construction of national population surface models from the 1981 and 1991 Census small area statistics (SAS). These data sets have been assembled at MIMAS and may be accessed by registered census users through the SURPOP Web interface, and the central section of the chapter describes the SURPOP data sets in more detail. We then consider the applications of the surface models and seek to demonstrate the types of situations in which they provide census users with a useful complement to the traditional area-based statistics.

10.2 CONSTRUCTING POPULATION SURFACE MODELS

The conceptualization of population density as a surface is by no means new, and Tobler (1979) demonstrates a volume-preserving method for the interpolation of population values within the boundaries of the original data-collection zones. Langford and Unwin (1994) illustrate the construction of surface models using ancillary data derived from satellite remote sensing, and identify the important advantages that can be gained by a more realistic representation of settlement geography and a wider range of variations in population density than are possible with conventional choropleth mapping. The technique for population surface construction used in the construction of the data set described in this chapter was originally proposed in Martin (1989). It was developed specifically as a response to the inadequacies of mapping UK Census data by ED, but may be generalized for use with any area data set for which counts and weighted centroids are available.

Since 1971, population-weighted ED centroid locations have been available as part of the SAS data sets. These locations were determined by eye at the time when the ED boundaries were initially designed, and were recorded to 100 m (10 m in urban areas in

1991). These centroid locations thus provide local summary points for the population distribution: in relatively uniformly populated EDs, they fall close to the centre of each ED, whereas in rural areas, they fall in the largest populated area. The distribution of ED centroids is itself a fair indication of the overall population distribution, as illustrated by Gatrell (1994). The objective here is to redistribute the census counts associated with each ED centroid to regular geographical cells in such a way that the settlement geography is rebuilt and unpopulated regions are retained in the final model. This surface-type model can then form the basis for many forms of spatial analysis in which the zonal model is misleading. Goodchild *et al.* (1993) describe the key task when interpolating socio-economic data between one set of areal units and another as that of estimating the underlying density surface. Here, the underlying surface is made explicit and data are transformed from the source zones into surface form, on which the user may choose to undertake further aggregation or morphological analysis as appropriate.

Surface construction is illustrated in Figure 10.1. Figure 10.1(a) shows an ED with its centroid and the centroid of a neighbouring ED. In Figure 10.1(b), a regular grid

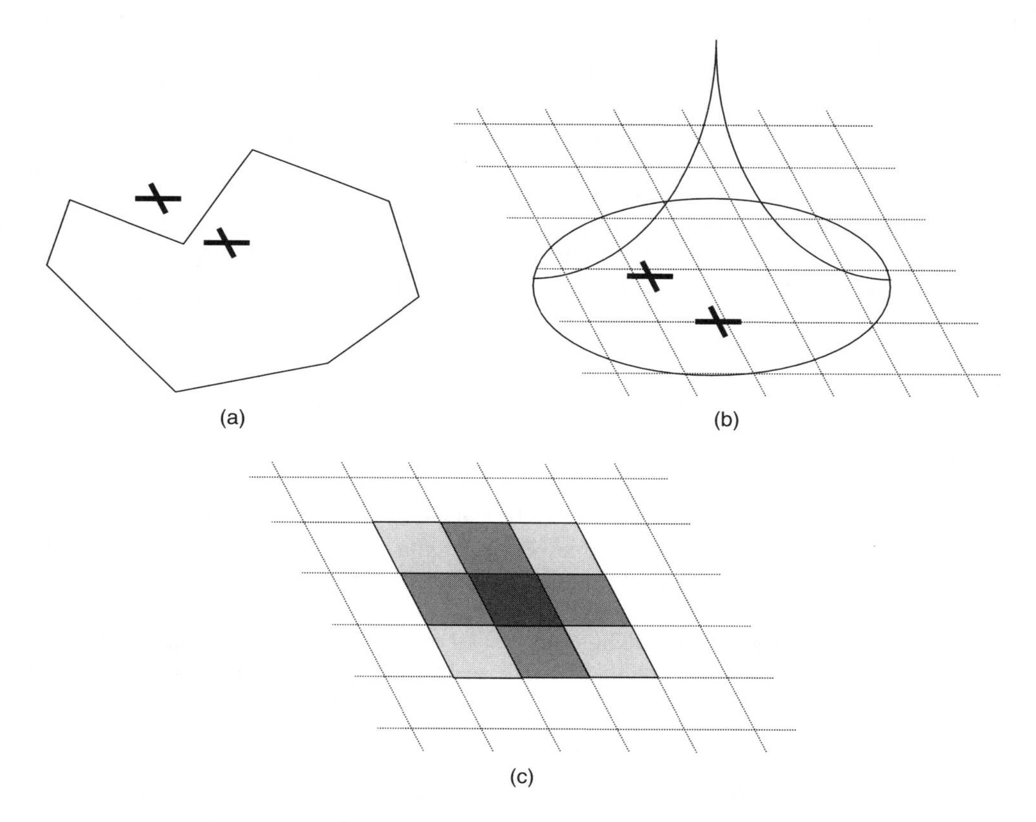

Figure 10.1 Population redistribution around zone centroids. (a) An enumeration district with its centroid and the centroid of neighbouring ED; (b) a regular grid is placed over the area to be modelled and a redistribution kernel is centred over the current centroid; (c) weights within the kernel are used to redistribute the total count from the current centroid into the surrounding cells.

is placed over the area to be modelled and a redistribution kernel is centred over the centroid, its initial width being adjusted in proportion to the local density of centroids. In sparsely populated regions where no other centroids are found, the kernel width remains unchanged, but in densely populated regions, the kernel width is reduced proportionally and may cover only part of one cell in the output grid. The kernel is thus adaptive to the local population structure. Weights are assigned to each cell falling within the kernel according to a distance decay function, and these weights are subsequently used in the redistribution of the total count (of population, households, unemployed persons, etc.) from that centroid location into the surrounding cells, as illustrated in Figure 10.1(c). In the application to 1981 and 1991 Census data described here, the weighting w_{ij} of cell i with respect to centroid j has been determined by the function:

$$w_{\mathrm{ij}} = \left(\frac{k^2 - d_{\mathrm{ij}}^2}{k^2 + d_{\mathrm{ij}}^2} \right)^{\alpha}$$

where k is the kernel width and d_{ij} is the distance between the centre of cell i and centroid j. α is a parameter that controls the shape of the distance decay function. In the creation of the SURPOP data sets, the initial kernel width has been set to 250 m (used in conjunction with a cell size of 200 m) and α has been set to one, which effectively gives a straight line decay function from the centroid to the edge of the kernel, where it falls to zero. The kernel width primarily influences the extent of the populated areas in the resulting model, whereas the form of the distance decay function affects the distribution of population within these areas (e.g., a steeply falling decay function results in high central population densities around each centroid, whereas a gently falling function produces more uniform population densities within settlements). When each centroid has been visited, many cells remain unpopulated, while others in densely populated regions receive population from more than one centroid.

10.3 1981 AND 1991 MODELS AVAILABLE USING SURPOP

In 1993, the technique described earlier was used to construct a series of population models covering England, Wales, and Scotland for selected counts derived from the 1981 and 1991 Census small area statistics, and the data set was made available to registered census users at MIMAS using a simple Fortran data retrieval program run from the MIMAS UNIX command line. This first phase of the work has been described by Bracken and Martin (1995). In 1997, data for Northern Ireland were processed to extend the 1991 models to cover the entire United Kingdom and an entirely new SURPOP interface using the World Wide Web was designed. Interface aspects of the Web version of SURPOP are considered in more detail by Martin *et al.* (1998), and the result may be viewed at *http://census.ac.uk/cdu/surpop/*.

The confidentiality restrictions imposed on 1981 and 1991 SAS led to counts being suppressed for some EDs that fall below the minimum thresholds for data publication, and these counts (from 'exporting EDs') are combined with those for nearby ('importing') EDs. In the suppressed cases, only the total numbers of persons and households are available in association with their true ED identifier and hence centroid location. In order to make maximum use of the locational information provided by the centroids, counts for exporting EDs have been allocated back to their original centroids *pro rata* the appropriate denominator variables. There are, thus, no centroids missing from the data set used for surface construction because of data suppression.

A further complexity when attempting to build comparable surface models from different censuses is that there is no continuity between the centroid locations, even in areas where there is no boundary change or significant substantive change in population distribution. It has, therefore, been necessary to construct a common centroid set for the construction of the two sets of census population surfaces. Four situations were identified: those in which 1981 and 1991 centroids occurred in exactly the same locations, those in which they occurred close together but not in exactly the same locations, and those cases in which either 1991 or 1981 centroids occurred in spatial isolation from any centroids in the other time period.

The 1981 data were allocated onto the 1991 centroids as follows. Where the ED centroids fell in exactly the same locations, or the difference was less than 100 m (i.e., the change in resolution between the most precise 1981 and 1991 centroid coordinates), the 1981 counts had simply been allocated directly to the 1991 locations. A weighted moving window approach similar to that used for surface construction was used for the situation in which 1981 and 1991 centroids fell in the same local area but with no direct one-to-one correspondence. The 1981 totals were assigned to the 1991 locations through a simple distance-weighted reallocation. In the two situations in which either 1981 or 1991 centroids occurred in isolation from any centroid at the time of the other census, these points had been retained unaltered, as there are no simple logical rules as to how their population counts should be treated. Areas of new residential development or demolition during the 1980s and redesign of sparsely populated rural EDs accounted for most of the unmatched centroids. This reallocation of 1981 data to 1991 centroids resulted in a point set (comprising all 1991 centroids and any unmatched 1981 centroids) to which modelled 1981 and actual 1991 counts were attached. This point set then formed the basis for surface modelling using the method described in the previous section.

In 1997, the same surface modelling technique was applied to 1991 centroid data for Northern Ireland, but unfortunately, the corresponding 1981 values were not available. Careful matching of SAS cell definitions was required to ensure that the Northern Ireland surfaces represented the same phenomena as those created for the rest of the United Kingdom. In the construction of each of these models, data for special EDs had been excluded, as these were frequently unrepresentative of local area populations, were not likely to fit the same distributional assumptions, and could not be easily identified by users in the final model. Each of the national surfaces was held in a custom-designed highly compact file format allowing rapid data retrieval for any specified geographical area. The full set of SAS variable definitions that had been used in the creation of the SURPOP data sets is given in Table 10.1, which also indicates the subject coverage of the surface models from the three source data sets (1981 and 1991 for England, Wales, and Scotland and 1991 only for Northern Ireland).

The surface data sets are available via the SURPOP Web interface, which seeks to provide a complete resource for users by integrating the models with appropriate metadata and background information. Any user can access the basic description of the modelling technique, details of the variables available, and lists of citations and existing applications, which are periodically updated. A casual user can also specify variables, areas, and file formats for download, using a simple html form, and retrieve header information relating to the specified download, including the full SAS definitions and estimated output file

Table 10.1 Cell definitions of surface variables for GB and NI.

Variable	1981 GB SAS definition	1991 GB SAS definition	1991 NI SAS definition
Population count (1991 base)		S010064	SN010064
Household count (1991 base)		S200001 or SS200001	SN200001
Population count (1981 base)	C50	S010057	SN010057
Household count (1981base)	C929	S270020	SN270020
Population aged 0–4	C57	S020008	SN020008
Population aged 5–15	C64 + C71 + C78	S020015 + S020022 + S020029	SN020015 + SN020022 + SN020029
Males aged 65+ and females aged 60+	C155 + C162 + C169 + C176 + C183	S020110 + S020113 + S020120 + S020127 + S020134 + S020141 + S020148	SN020110 + SN020113 + SN020120 + SN020127 + SN020134 + SN020141 + SN020148
1-year migrants	C642	S150001 or SS150001	SN150001
Residents permanently sick	C422	S140061	SN140061
Residents unemployed	C422	S140037	SN140037
Households owning and buying	C967	S200002 + S200003 or SS200002 + SS200003	SN200002 + SN200003
Households renting from a public authority	C983	S200008 or SS20008 + SS200009	SN200006
Households with no car	C949	S200010 or SS200011	SN200010
Households overcrowded at more than 1 ppr	C929 + C946	S230003 + S230004 or SS230003 + SS230004	SN480002
Households lacking inside wc	C929—C930	S200001 – S200011 or SS200001 – SS200012	SN200001 – SN200011
Households lacking central heating		S200031 + S200061 or SS200034 + SS200067	SN200031 + SN200061
Roman Catholic			SN510002
Presbyterian, CofI, Methodist, or			SN510003
Other			

size. Only registered users of the census data sets at MIMAS can actually download the data, and are prompted at this point for their MIMAS user name and password. The three source data sets are each covered by separate purchase agreements and require separate registration. Registration forms are downloadable from within SURPOP.

Robinson and Zubrow (1997) note the difficulties inherent in attempting to evaluate the accuracy of surface models and attempt an approach through simulation of a true surface, although this is very difficult to achieve in the case of census population distributions. The results from this surface modelling approach are compared with conventional zone-based representations by Martin (1996a), taking data directly from the national surface models described earlier. In this study, a series of areal units with known population counts were reconstructed from the cells of the surface model by aggregating the counts for the cells falling within each area. At the level of the ED, the correlation was poor ($r = 0.6486$), with a mean absolute error of 107 on a mean population size of 469. At the ward level, the correspondence was much closer, with $r = 0.9968$ and a mean absolute error of 190 on a mean population size of 5552. In both cases, the mean figures are strongly influenced by a few outliers, particularly reflecting very small urban EDs with high populations whose boundaries are poorly represented by the 200 m resolution cells. As part of the construction of SURPOP surfaces for Northern Ireland, the ED centroid locations were checked by computing a new population-weighted mean location for each ED from their associated postcode locations, on the basis of the pseudo-ED codes shown in the postcode/ED directory (with the addition of 50 m in X and Y directions to overcome the systematic bias inherent in grid references for the SW corner of each 100 m cell). Analysis of the grid references reveals that there is, in fact, a very strong correspondence between the two centroid sets, with a mean difference of 90.31 m over the 3729 EDs, with very few large differences. In other words, the postcode-derived locations (of which there are several available for each ED) confirm the plausibility of the ED locations. The SURPOP surfaces were constructed from the best available set of centroids held by the Census Offices in each case, and these were further screened for gross locational errors, some of which were corrected directly by the addition or subtraction of 10 or 100 km to the given coordinates. Accuracy measurement is particularly difficult, but it is a feature of the technique that no population counts can be allocated further than the initial kernel radius (250 m in the case of SURPOP) from their own centroid locations. These various tests confirm that the model provides reliable population estimates at the level of the ward and above, and that the counts for the cells should be treated as building blocks rather than individual estimates.

10.4 APPLICATIONS

A 1991 household count surface downloaded from SURPOP and mapped in ARC/INFO for a 25 × 25 km area covering Southampton is illustrated in Figure 10.2. The figure includes the coastline for reference. Cell shading is proportional to population count, with the darkest shading representing the highest counts and the majority of the image being unpopulated. Discrete settlements are clearly identifiable, with the main urban area in the centre and a number of distinct outlying urban areas. Urban form is reproduced in considerable detail, including the presence of population density gradients and of internal unpopulated areas such as parks and commercial land uses. The surface models are thus particularly useful in applications in which a more realistic representation of settlement pattern is required, but should not be used if the requirement is to obtain precise counts for any area that can be created by direct aggregation of existing census areas. However, if the spatial characteristics of the settlement pattern such as density, distances between populations, or the spatial extents of settlement and neighbourhoods are required, then

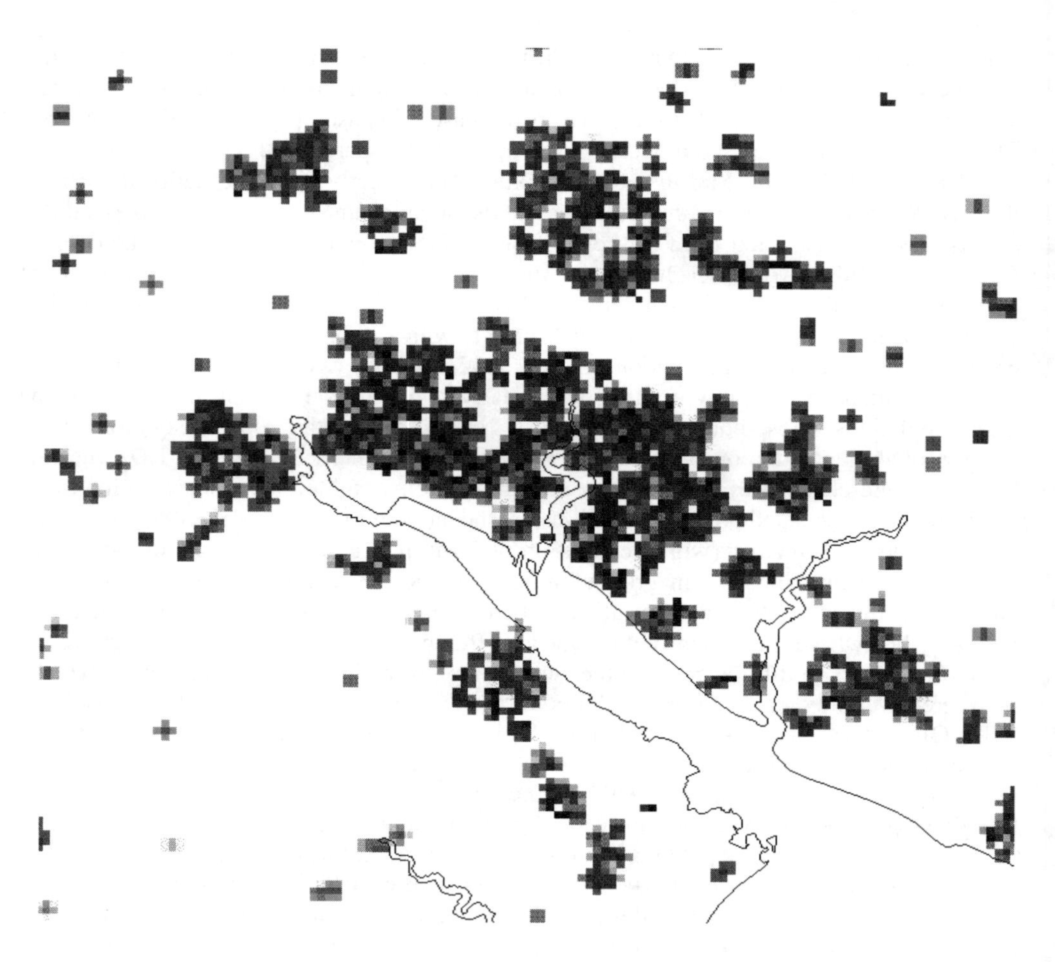

Figure 10.2 1991 household density surface of Southampton, downloaded from SURPOP. Darker cells represent highest densities. Cell size is 200 × 200 m.

the surface database may offer considerable advantages. Cartographic representation of the surfaces can provide valuable visualizations of the population distribution at regional scales, but the cellular structure tends to make interpretation at local scales more difficult. The SURPOP does not provide cartographic visualization tools, although some interesting three-dimensional surface visualizations are presented by Wood *et al.* (1999). Rather, SURPOP has been designed to allow the users to download the data for manipulation in their own GIS or analysis software. A variety of grid formats is provided that can be read by the majority of standard raster GIS packages, including Arc/Info, IDRISI, and generic grid file formats.

Martin (1996b) illustrates the use of standard GIS routines for the manipulation of the 1981 and 1991 SURPOP surface data sets to extract information about population change in discrete settlements, by an analysis of surface form. The surface offers the opportunity to extract information on a settlement basis, although settlements do not form a part of the input census information; they are, rather, features of the actual population distribution that is recaptured by the surface modelling process. Similar procedures can be used to identify neighbourhoods within the populated cells having distinct social characteristics, allowing the subdivision of an urban area with reference to both built form and neighbourhood type (Martin, 1998b). Availability of the models for 1981 and 1991 in a standardized format also permits region- and settlement-based analysis of intercensal population changes, although cell-by-cell change analysis is subject to small number problems.

Mesev *et al.* (1995) make use of a series of surface models of different tenure types to aid in the classification of remotely sensed imagery for an urban area, while Brainard *et al.* (1996) use the model in the evaluation of appropriate transportation routes for hazardous waste, both examples of applications in which a conventional zone-based model does not adequately capture the cross-boundary and within-zone clustering of population. Other existing applications based on the 1991 surface models include the modelling of populations exposed to noise around airports, populations accessible to local radio broadcasting, and as a base population for disease incidence mapping. A full listing of known citations and applications of the method is available from the SURPOP Web pages.

In situations in which postcoded data are available, a surface model constructed from postcode locations would provide a more appropriate basis for incidence mapping than the standard ED centroid-based models. The directory of EDs and postcodes (OPCS/GROS, 1992) provides information on the degree of overlap between each ED and postcode, associated with a postcode grid reference (to 100 m resolution). The census counts associated with each ED may be reallocated onto its associated postcode locations *pro rata* the household count, and the postcode locations subsequently used as the basis for surface modelling. Other refinements to the modelling technique are also possible, such as the constraint of the redistribution process from each centroid to the boundaries of the corresponding ED, as illustrated in Martin (1996a). A series of these possibilities, including use of postcode-based centroids, is demonstrated using the Northern Ireland data by Martin *et al.* (2000). The inherently geographical nature and simple structure of the surface models, without the need for associated boundary data, makes them an attractive basis for Web-based population mapping of extensive areas, using interfaces similar to CASWEB (described in Chapter 9 of this volume). These refinements do not currently form part of the SURPOP data sets.

10.5 CONCLUSION

A technique has been described for the remodelling of area-based census counts into surface models, represented as population estimates for the cells of a georeferenced grid. This approach has been applied to national 1981 and 1991 Census data for selected variables, and is available for download via the Web to registered users of the census data. The data are primarily intended for subsequent manipulation and modelling within raster GIS software. These models are well suited to applications that require a more realistic representation of the geographical distribution of census characteristics, particularly those involving concepts of settlement structure, distance, and adjacency. They incorporate estimated values for restricted Eds and are thus partially freed from the problem of unshaded areas in conventional area-based census maps resulting from suppressed data for certain EDs. A range of experimental and operational applications has been noted.

The use of a full address base for census geography design in 2001 (as discussed in Chapter 3) will permit the direct calculation of address-weighted centroids for output areas (OAs) as part of the 2001 data sets. Greater standardization of population sizes will result in a larger number of 2001 OAs than 1991 EDs. The closer identification of census OAs with unit postcodes may also make the direct use of groups of postcode locations in association with output area population counts a simpler operation. Either of these centroid sets would provide an ideal basis for the construction of new surface models for 2001, with more data points and greater locational precision than has been possible for 1991 and 1991 data. These data developments would provide increased opportunities for the use of census population surfaces to tackle appropriate classes of geographical analytical problems based on 2001 Census data.

Acknowledgements

The surface data sets described here were created with the support of ESRC awards A507265014 and R000222045, and the assistance at various stages of Richard Gascoigne, Jason Sadler, Nick Tate, and the late Ian Bracken are gratefully acknowledged.

Box 10.1 Key web link for Chapter 10.

http://census.ac.uk/cdu/surpop/	SURPOP interface to 1981 and 1991 Census population surface models

11

ONS Classifications and GB Profiles: Census Typologies for Researchers

Philip Rees, Chris Denham, John Charlton, Stan Openshaw, Marcus Blake, and Linda See

SYNOPSIS

Classifications are means of summarizing a large body of information into a small number of categorical variables. In the United Kingdom, this process began with the classification of towns by Moser and Scott (1961) using statistics from the 1951 Census. The recent history of census classifications is reviewed in the first section of the chapter. The second section provides an account of the ONS Classification of local government and health authorities, which has been widely used. An Office for National Statistics (ONS) classification for wards, linked with the local government areas (LGA) /Health Authorities classification but separate from it, is referenced in the third section. The fourth section of the chapter describes a widely used academic classification—GB Profiles (Blake and Openshaw 1995), which employs census data for enumeration district (EDs) or output area (OAs) in Great Britain. It also describes a particularly useful utility in which a researcher can enter a postcode and will receive in return a succinct but highly diagnostic profile of the type of residential area in which the postcode is located. The chapter concludes with a brief discussion of the updated ONS classification of Local Authorities/Health Authorities, based on 1991 Census information but re-tabulated for the new LGAs defined from April 1, 1998.

11.1 A LITTLE HISTORY

Census information has been extensively used to create area classifications. The aim of this chapter is to introduce readers to two widely used schemes, the results of which are available in the public domain, together with accounts of the methods used in their construction. The ONS has produced classifications of local authorities in Great Britain and of electoral wards for three decades. The ONS classifications based on 1991 Census data and boundaries are described in Section 11.2 for local and health authorities, and in Section 11.3 for wards. Openshaw *et al.* (1995) have produced a classification based on neural net methods called *GB Profiles*, which has been applied to EDs (England and Wales) and OAs (Scotland) in Great Britain. The associated software system enables researchers to link postcode records to the ED/OA types. GB Profiles are described in Section 11.4 of the chapter. The final Section 11.5 discusses future developments.

The Census of Population provides a rich set of variables describing the residential populations of the country. To make sense of this richness, the researcher uses classifications to reduce the multivariate complexities to a simple set of classes. If constructed properly, these enable the researcher to keep the variety of the original data, without being overwhelmed by information.

Multivariate classifications of small area census data, or geodemographic classification systems, are a well-established means of providing a simple-to-use and easy-to-understand and therefore useful descriptive summary of the characteristics of residential areas (Webber and Craig, 1976; Openshaw, 1983; Openshaw and Wymer, 1995). The general concept underpinning geodemographic classifications is that people who live near to each other tend to have similar characteristics and behave in similar ways.

Early work was carried out in the United States by sociologists (Shevky and Williams, 1949; Shevky and Bell, 1955) who used a small number of census variables to define social dimension indicators for census tracts in cities. These indexes were then used to place the tracts into 'social areas', which were divisions of the multi-indicator space into arbitrary subspaces. Tyron (1955) computed the correlations between census tracts and used these to group tracts into clusters. Numerous studies refining these techniques using principal component analysis to simplify the dimensions of variation and various cluster algorithms were carried out in the 1960s (Rees, 1972 provides a review).

Application of these techniques to British Census data followed. The first one was a study by Moser and Scott (1961) of British towns with populations over 50 000 using 57 indicators covering population size and structure, population change, households and housing, economic activity, social class, voting, health, and education (the latter three sets of variables from non-census sources). Principal components analysis was used to reduce 57 variables to 4 components. Towns were grouped into clusters, largely by inspection of their position in Component I versus II space, with distances in Components III and IV space used to assign outliers. Some 13 groups in 3 families were defined, the families being 'Mainly resorts, administrative and commercial towns', 'Mainly industrial towns', and 'Suburbs and suburban type towns'.

In the 1970s, Webber, working at the Centre for Environmental Studies, London, developed a typology of Liverpool's EDs using 1966 Sample Census data (Webber, 1975). In collaboration with Office of Population Censuses and Surveys (OPCS), Webber developed a 36-cluster ward or parish level classification and a classification of enumeration districts, which became known as ACORN (A Classification of Residential Neighbourhoods) when used commercially (Webber, 1989). Once 1981 Census data became available, many classifications were developed by market analysis firms (Brown, 1990). A selection is listed in Table 11.1, while Box 11.1 provides relevant Web links to the classifications. Most classifications define a 2- or 3-layer hierarchy of classes, so that clients can select the appropriate level of detail for their problem. Some geodemographic classifications include non-census data, such as credit agency data, unemployment statistics, county court judgements, and lifestyle data, which are available for post-census years. However, two features of these geodemographic products reduce their utility for researchers: they cost considerable sums to license and documentation on methods and utility is not available in the public domain.

Table 11.1 A comparison of selected current geodemographic systems.

Classification	Supplier	Data sources	Var.	Areas	Classes
ACORN	CACI	1981 Census	41	EDs	12, 39
ACORN91		1991 Census	79	EDs	6, 17, 54
CAMEOUK	EuroDirect	1991 Census		EDs	9, 44
DEFINE	Infolink	1981 Census, credit data, electoral roll, PAF data	67	Postcodes	10, 47, 423
DEFINE 91		1991 Census, credit activity data, electoral roll, unemployment statistics		Postcodes	
IMAGES	Equifax Europe (UK)	1991 Census, NDL data		Postcodes	
MOSAIC	CCN	1981 Census, credit activity data, electoral roll, PAF data, CCJ data	38, 16	Postcodes	12, 38, 57
MOSAIC 91	Experían	1991 Census, credit activity data, electoral roll, PAF data, CCJ data, retail access data	87	Postcodes	11, 62
PiN	PinPoint	1981 Census	104	EDs	12, 25, 60
PiN 91	CACI	1991 Census	49	EDs	6, 17, 42
SUPER PROFILES	CDMS	1981 Census, trading data	55, 10	EDs	11, 37, 150

Source: Brown (1990), Sleight (1993), current company web pages or brochures (see Table 11.5).
Note: Var. = number of variables used in classification. Wherein two numbers are given, the first refers to census variables, the second to non-census variables; Classes = hierarchical sequence of categories produced in the classification; EDs = enumeration districts; CCJ = county court judgements.

OPCS and its successor, The ONS, have continued to produce classifications at higher spatial scales—local government districts, health authority areas, and electoral wards. The local and health authority classifications have been published and made available in the public domain, while the ward classifications have been licenced at reasonable costs to user communities.

The OPCS produced area classifications using 1971 Census data, working with the Centre for Environmental Studies, and published reports on the district classification (Webber and Craig, 1976, 1978). The exercise was repeated after the 1981 Census and extended to cover health authorities (Craig, 1985). In preparation for the 1991 Census, the Census Offices (OPCS and the General Register Office Scotland (GROS)) proposed to develop general-purpose area classifications for local and health authorities and for wards. The Department of Health saw a need for such standard typologies to be used by the

National Health Service, and arranged for OPCS to provide the new 1991 classifications (Wallace *et al.*, 1995: 2). In the next section of the chapter, an account of the ONS classifications based on 1991 Census data is provided, beginning with the local authority classification.

11.2 THE ONS CLASSIFICATION OF LOCAL AND HEALTH AUTHORITIES

11.2.1 Objectives and Uses

The objectives of the ONS classification were (1) to produce a general-purpose classification incorporating the demographic, employment, socio-economic, household, and housing characteristics of Britain's population, (2) to produce area clusters that were internally homogeneous and differentiated from one another, (3) to structure the clusters in a nested hierarchy so that users could select an appropriate stratum, and (4) to retain broad comparability with previous OPCS 1971 and 1981 classifications.

The classifications could be used in three broad ways: (1) to organize the reporting of other phenomena that vary across social-economic categories and geographical space, (2) to help in the analysis of other data as an intermediate step, and (3) to help identify specific types of area for public policy or commercial purposes. Typical uses of classifications are, for example, to convey geographic patterns in the population, to code geographically sparse data to an area typology for further analysis, to monitor public expenditure by type of area, or to help stratify sample selection where it is known there are strong links between the subject being investigated and demographic and socio-economic attributes of the population being sampled. A general-purpose classification saves time for the researcher by summarizing a battery of census indicators.

11.2.2 Methods

A classification based on census data involves a series of decisions, some of which are based on objective criteria, and some of which must be based on informed judgement. Decisions must be taken on the following steps: (1) careful choice of variables, (2) standardization of the variables, (3) orthogonalization of the variables, (4) classification of the areas using cluster analysis, and (5) labelling and interpretation the clusters (Openshaw and Wymer, 1995).

Choice of Variables

A final set of 71 variables was chosen to represent the main 'dimensions' in the census data. The majority was the same as or equivalent to variables used in the 1971 and 1981 classifications. Several variables new to the 1991 Census, such as ethnic group or limiting long-term illness, were included. Through examination of the correlations of variables across 457 districts, the number of variables was reduced to avoid duplication and overweighting of particular dimensions. A final set of 37 variables was used. Full definitions in terms of underlying Local Base Statistics (LBS) counts are given in Appendix Table A in Wallace and Denham (1996).

Standardization

Before being entered in the cluster analysis, variables were expressed in a standard form in relation to the distribution of values. Although most variables are in a percentage form, a number of them are not (e.g., average number of household residents per household, directly standardized rate for limiting long-term illness). Range standardization was adopted in the 1991 classification rather than the more usual Z score standardization, which implies that the original variables are normally distributed. The range standardized value for variable X_k in district i, R_{ki}, is computed as

$$R_{ki} = (X_{ki} - X_{k\,\min})/(X_{k\,\max} - X_{k\,\min})$$

where $X_{k\,\min}$ and $X_{k\,\max}$ are the minimum and maximum local authority for variable X_k.

Orthogonalization

A principal component analysis was used to check the set of variables for redundancy and that no dimension of variation had been omitted.

Clustering

The starting point of the clustering process was the production of an interdistrict distance matrix. The squared distance between districts i and j, D_{ij}^2, in the space formed by the chosen 35 variables, X_{ki} and X_{kj}, used and is defined as

$$D_{ij}^2 = \sum_{k=1}^{35} (X_{ik} - X_{jk})^2$$

A two-stage clustering process was employed. Ward's hierarchical clustering algorithm was applied to the distance matrix using the single linkage technique. It was necessary to select stages in the clustering process, after which there was a jump in within-cluster variance. Inspection of the results suggested that the 34-, 12-, and 6-cluster solutions showed these features. A final clustering step was to re-sort districts using an interactive K-means algorithm so that they belonged to clusters, the centroids of which were closest.

11.2.3 Labels for the Classes and Class Characteristics

The final step was to invent labels for each class, through inspection of both the class variable averages and the geographical grouping of the class. Table 11.2 lists the labels chosen for the 6 Families (in capitals), the 12 Groups nesting inside the Families and the 34 Clusters. The table records the number of districts falling into each class in the first column of numbers. The classification ensures that every Cluster has at least 2 members (local authorities) and every Group and Family has at least 17 members. There are no isolated authorities.

Maps of the Families, Groups, and Clusters of ONS local authority classification are provided in Wallace and Denham (1996) or Rees *et al.* (1996). Full profiles of each Cluster, Group, and Family are given in Chapter 6 of Wallace and Denham (1996). Here, the main characteristics of Families and Groups are outlined.

Table 11.2 Demographic statistics for Families, Groups, and Clusters in the ONS classification of districts based on 1991 Census data.

Family (6) Group (12) Cluster (34)	No. of districts	Population distribution % of GB 1981	Population distribution % of GB 1991	% Population change Census 1981–91	% Population change Estimate 1981–91	Net internal migrants No. (1000s) 1990–91	Net internal migrants Rate per 1000 pop. 1990–91
RURAL AREAS	137	17.6	18.4	4.7	6.1	55.1	5.5
Scotland	25	1.8	1.8	1.4	1.7	7.9	8.0
Highlands and islands	15	1.3	1.4	1.3	1.6	6.3	8.7
Uplands and agriculture	10	0.5	0.5	1.5	2.2	1.6	6.0
Coast and country	68	8.4	8.9	6.7	7.8	31.7	6.6
Remoter England and Wales	22	2.1	2.3	7.3	8.3	8.7	7.1
Heritage coast	15	1.6	1.7	7.1	8.2	4.7	5.2
Accessible amenity	31	4.7	5.0	6.4	7.4	18.3	6.8
Mixed urban and rural	44	7.4	7.6	3.3	5.2	15.5	3.8
Towns in country	18	2.6	2.8	7.5	9.1	19.6	7.1
Industrial margins	26	4.8	4.9	1.1	3.1	4.9	1.9
PROSPERING AREAS	115	21.4	22.2	3.0	5.4	36.4	3.0
Growth areas	88	16.7	17.4	3.4	5.8	28.1	3.0
Satellite towns	24	5.4	5.5	1.4	3.7	3.4	1.1
Growth corridors	26	5.1	5.5	6.6	9.0	13.2	4.5
Areas with transient populations	7	0.8	0.9	8.1	11.6	5.8	11.8
Metropolitan overspill	10	2.4	2.4	−2.3	0.4	−2.2	−1.7
Market towns	21	3.0	3.1	5.0	7.0	7.8	4.6
Most prosperous	27	4.7	4.8	1.7	3.6	8.3	3.2
Concentrations of prosperity	6	0.8	0.9	7.2	10.6	1.8	3.8
Established high status	21	3.9	3.9	0.5	2.2	6.5	3.1
MATURER AREAS	44	11.3	11.3	0.4	3.1	−5.9	−1.0
Services and education	20	7.0	6.9	−2.0	1.0	−16.1	−4.3
University towns	9	2.6	2.5	−0.2	1.1	−8.4	−6.2
Suburbs	11	4.5	4.4	−3.0	1.0	−7.7	−3.2
Resort and retirement	24	4.3	4.4	4.2	6.6	10.2	4.3
Traditional seaside towns	18	3.4	3.4	3.1	5.7	4.7	2.5
Smaller seaside towns	6	0.9	1.0	8.2	9.7	5.5	10.2
URBAN CENTRES	70	20.6	20.4	−1.0	1.8	−20.5	−1.9
Mixed economies	47	10.4	10.6	1.5	3.8	−5.5	−1.0
Established service centres	15	4.8	4.7	−1.7	0.8	−8.4	−3.3
Scottish towns	14	2.1	2.1	−0.3	1.3	0.7	0.6
New and expanding towns	18	3.5	3.8	7.1	9.6	2.2	1.1

Table 11.2 (*continued*)

Family (6) Group (12) Cluster (34)	No. of districts	Population distribution % of GB		% Population change		Net internal migrants	
				Census	Estimate	No. (1000s)	Rate per 1000 pop.
		1981	1991	1981–91	1981–91	1990–91	1990–91
Manufacturing	23	10.1	9.8	−3.7	−0.3	−14.9	−2.8
Pennine towns	14	4.6	4.5	−2.6	0.1	−2.4	−1.0
Areas with ethnic minorities	9	5.6	5.3	−4.6	−0.6	−12.6	−4.4
MINING AND INDUSTRIAL AREAS	74	22.7	21.5	−5.1	−2.3	−23.5	−2.0
Ports and industry	31	11.9	11.0	−7.5	−4.4	−23.8	−4.0
Areas with inner city characteristics	8	3.8	3.5	−8.7	−4.4	−12.8	−6.9
Coastal industry	16	5.5	5.2	−4.7	−2.3	−4.8	−1.7
Glasgow and Dundee	4	2.0	1.8	−13.2	−10.2	−5.7	−6.0
Concentrations of public sector housing	3	0.6	0.5	−7.1	−5.1	−0.4	−1.4
Coalfields	43	10.8	10.5	−2.5	0.1	0.3	0.1
Mining and industry, England	17	5.1	5.0	−2.3	0.1	−0.2	−0.1
Mining and services, Wales	10	3.3	3.3	−1.7	1.4	−1.2	−0.7
Former mining areas, Wales and Durham	16	2.3	2.2	−4.0	−1.9	1.6	1.4
INNER LONDON	17	6.4	6.3	−6.3	2.0	−41.7	−12.2
Cosmopolitan outer boroughs	6	2.6	2.5	−6.8	0.1	−15.8	−11.6
Central London	5	1.7	1.6	−6.7	1.6	−8.8	−10.0
Inner city boroughs	4	1.5	1.5	−7.9	2.8	−12.1	−15.1
Newham and Tower Hamlets	2	0.6	0.7	0.5	8.7	−5.1	−13.9
Total: Great Britain	457	100.0	100.0	−0.2	2.5	0.0	0.0

Source: Wallace *et al*. (1995), Rees *et al*. (1996). The 1981 and 1991 Census and 1991 population estimates are Crown Copyright. The 1991 Census data were provided by the ESRC/JISC Census Programme and the Census Dissemination Unit, Manchester Computing.

RURAL AREAS have high values for farming employment and relatively low values for indicators of deprivation. Within this family, the *Coast and Country* Group has a mature and prosperous population, while the *Mixed Urban and Rural* Group has higher values for employment in primary production and manufacturing. The *Scotland* Group is characterized by relatively high values for local authority housing, low ethnic minority populations, and greatest concentrations of workers in agriculture and social classes 4 and 5.

The PROSPERING AREAS districts have well above average values in car availability, the highly qualified, social classes 1 and 2, owner occupation, and large dwellings, with low levels of unemployment and other deprivation indicators. Prospering areas are concentrated in central and southern England outside the metropolitan centres. The *Growth Areas* Group has a young population and is located in open countryside with expanding smaller towns beyond the *Most Prosperous* Group, which has the highest values on the affluence indicators and is concentrated to the south and west of London, in the well-off commuting belt.

MATURER AREAS form the least coherent Family in socio-economic and geographical terms. The population age structure is the oldest, and the areas experienced major development some decades in the past. The *Services and Education* Group comprises suburban Boroughs in Greater London and many university towns, which have concentrations of students, in-migrants, the highly qualified, users of public transport, social classes 1 and 2, and singles. The *Resort and Retirement* Group districts are coastal and have the second highest rate of population growth, reflecting in-migration of retired people, and have high levels of owner occupation but are not particularly prosperous.

URBAN CENTRES are scattered throughout GB, with modest rates of population growth and concentrations of employment in manufacturing. The *Mixed Economies* Group was close to the national average on most indicators, with slightly greater deprivation and slightly lower affluence. The *Manufacturing* Group districts were concentrated in the Pennine industrial belt and West Midlands, with the highest values for employment in manufacturing and the lowest Group value for employment in finance and services.

MINING AND INDUSTRIAL AREAS are linked to the coalfields of Great Britain and associated heavy industries, often in coastal locations. The districts have experienced major job losses together with population decline and contain some of the least well-off households in Britain. The *Ports and Industry* Group has the highest levels of local authority housing and the second lowest levels of car availability. The *Coalfields* Group has the highest group levels in primary production and also in long-term illness. They are areas showing ‘traditional working class’ characteristics.

INNER LONDON constitutes both a Family and a Group of areas, with just over half of Greater London’s population. *Inner London* has a relatively youthful population, with a concentration in the young adult ages. The most marked demographic characteristic is the extreme concentration of Black and Asian groups in this Family of districts. Values of the deprivation indicators are high in all districts. Employment is dominated by finance and services, and manufacturing is at the lowest level in any Family. High use of public transport and a lack of access to cars are features of this Family.

11.2.4 Classification of the Health Authorities

One aim of the classification exercise was to produce a health authority classification comparable to that of local authorities. The method involved preparing the same set of 37 variables for health authorities, range standardized to the local authority maxima and minima. The data were passed through the K-means procedure with output centroids from the local authority Groups. Health authorities were assigned to the nearest local authority Group in 37 variable space. Appendix Table G in Wallace and Denham (1996) provides a list of health authorities with their assigned Family and Group membership.

11.2.5 Illustrative Applications

Two examples serve to demonstrate the value of the ONS local authority classification.

In the right-hand columns of Table 11.2 are presented statistics for population change and net internal migration, assembled as part of a UK case study within a project to investigate internal migration and regional population change in Europe (Rees *et al.*, 1996). According to census statistics (fourth column), as we proceed through the classification, the balance of population change moves from positive to negative. The picture changes somewhat if we add back the population missed in the 1991 Census and use mid-1991 population estimates corrected for the census under-enumeration (the fifth column). The comparison of population estimates in 1981 and 1991 shows reduced losses in Mining and Industrial Areas and a switch from loss to gain in Inner London. Growth rates shift upwards in Rural Areas, Prospering Areas, and Maturer Areas, but not to the same extent. The final two columns of statistics show the pattern of net internal migration (moves from origins and destinations within the country). Rural Areas and Prospering Areas are gainers, whereas Maturer Areas, Urban Centres, Mining and Industrial Areas, and Inner London are net losers. Decadal population change and end-of-decade migration flows are in broad agreement across Families, except that Maturer Areas, Urban Centres, and Inner London post population gains but net internal migration losses, particularly in Inner London. This is due to the compensating net inflow of migrants from outside the United Kingdom. The ONS classification thus helps us understand the pattern of demographic change in 1981 to 1991 across 457 districts in GB.

A second illustration (Charlton, 1996) is provided in Table 11.3 for local authorities in England. The table tracks changes in life expectancy by Family and Group of local authorities

Table 11.3 Life expectancies at birth by ONS district groups, England and Wales, 1981 and 1992.

Family: Group	**Men**			**Women**		
	1981	**1992**	**1992/1981**	**1981**	**1992**	**1992/1981**
Prospering areas: most prosperous	72.6	75.8	103.2	78.3	80.4	102.1
Prospering areas: growth areas	72.2	75.1	102.9	77.7	79.8	102.1
Rural areas: coast and country	72.0	75.0	103.0	77.6	80.1	102.4
Maturer areas: services and education	72.2	74.8	102.6	78.1	80.4	102.3
Maturer areas: resort and retirement	72.1	74.4	102.3	77.9	80.1	102.2
Rural areas: mixed urban/rural	71.5	74.1	102.7	77.4	79.3	101.9
Urban centres: mixed economies	71.0	73.8	102.8	77.3	79.2	102.0
Mining and industrial areas: coalfields	69.9	72.8	102.9	76.0	78.1	102.1
Urban centres: manufacturing	70.0	72.4	102.4	76.0	77.9	101.9
Inner London: Inner London	70.2	72.2	102.0	76.9	78.7	101.9
Mining and industrial areas: Ports and industry	69.6	71.7	102.1	75.8	77.5	101.7
England and Wales	71.1	73.8	102.7	77.1	79.2	102.1
Range	3.0	4.1	1.2	2.5	2.9	0.7

Source: Adapted from Charlton (1996), Table 2.

Table 11.4 The structure of the ONS district lookup table.

Rows	Columns	Order
2–367	English districts	1991 Census order
368–404	Welsh districts	1991 Census order
405–460	Scottish districts	1991 Census order
Columns	**Content**	**Comment**
A	ONS/GROS Census code	Alphanumeric or numeric code
B	Long name	District official name
C	Family code	Sequence number
D	Family label	Text description
E	Group code	Sequence number
F	Group label	Text description
G	Cluster code	Sequence number
H	Cluster label	Text description

Source: file = */db/census91/area_class/district.xls* on *irwell.mimas.ac.uk*.

between the early 1980s and early 1990s. The classes are arranged in descending order of male life expectancy in 1992, which is related to a descending level of prosperity. For all types of area, both men and women experienced improvement over the 11 years, but differences between areas widened with greater gains in Prospering Areas and Rural Areas and poorer performance in Urban Centres, Inner London, and Mining and Industrial Areas.

11.2.6 Access to the ONS Classification of Local and Health Authorities

The ONS allows 'extracts from the study relating to the classifications of local authority or health districts to be reproduced without a licence provided these form part of a larger work not primarily designed to reproduce the extracts *and* provided that Crown Copyright and the source are prominently acknowledged . . .' (Wallace and Denham, 1996). Appendix Table F in Wallace and Denham (1996) provides a list of district names in alphabetical order, with Family codes as Roman numerals, Group codes as letters, and Cluster codes as numbers.

Two files, *district.cvs* and *district.xls*, have been prepared from the ONS description to provide the information in the form of a lookup table between local authority districts and the ONS Families, Groups, and Clusters. The left-hand column of Box 11.2 (top panel) gives the file path names on the Manchester Computing data server, which can be accessed by registered academic users. Copies of these files are provided on the CD-ROM accompanying the book. Table 11.4 sets out the structure of the lookup file. Note that the class codes have been converted into sequence numbers for easy use in further analysis.

11.3 THE ONS WARD CLASSIFICATION

ONS has prepared a classification of wards (postcode sectors in Scotland) from the results of the 1991 Census using similar methods to those described for local authorities in the

Table 11.5 The structure of the ONS ward lookup table.

Rows	**Content**	**Order**
2–823	Scottish wards	1991 Census order
824–9342	English wards	1991 Census order
9343–10186	Welsh wards	1991 Census order
Columns	**Content**	**Comment**
A	Ward code	Alphanumeric code
B	Long name	Official name of ward
C	Grid reference	OS Tile, Easting, Northing
D	Cluster ID	Cluster sequence number
E	Cluster name	Cluster label
F	Group ID	Group sequence number
G	Group name	Group name
H	Count	Indicator (=1)
I	Census resident population	LBS Cell L020001
J-AT	37 census variables	See Wallace and Denham (1996)

Source: file = */db/census91/area_class/ward.xls* on *irwell.mimas.ac.uk*.

previous section. Some details are given in Wallace and Denham (1996). The detailed, ward-by-ward results are available on electronic media from ONS (Census Marketing, ONS, Segensworth Road, Titchfield, Hants PO15 5RR, tel 01329 813800, email *census.customerservices@ons.gov.uk*). The ESRC/JISC Census Programme has licensed a copy of the classification for academic use, which has been added to the 1% Household Sample of Anonymized Records. The ward lookup file locations are given in the left-hand column of Box 11.2 (top panel), while Table 11.5 reports the information contained in the file. The file includes ward values of the 37 variables input to the classification in percentage or ratio form. Smaller wards or postcode sectors (with fewer than 1000 residents or 320 resident households) were merged with neighbours.

The classification resulted in 14 Groups containing 43 Clusters. The Group and Cluster labels and code numbers are given in Table 11.6. The Group and Cluster labels in this classification place greater emphasis on housing attributes, which differentiate wards within city regions. When using the classification, users may need to link address data for respondents, for customers, or for medical cases to the ward classification. The way to do this is to enter the postcode in the address into the PC2ED utility on *irwell.mimas.ac.uk* to obtain the ED or OA code, which can then be linked to the ward.

11.4 THE GB PROFILES CLASSIFICATION OF ENUMERATION DISTRICTS

Wards (postcode sectors in Scotland) are quite large geographical areas with populations between 1000 and 25 000. Both market analysis firms (see Table 11.1) and academics (Openshaw and Wymer, 1995) saw the need to develop classifications for the smallest areas for which census outputs were available, such as enumeration districts or output areas. Market analysis firms, by and large, combined conventional classification techniques

Table 11.6 The ONS ward classification: Groups and Clusters.

Group no.	Group name	Cluster numbers and names
1	Suburbia	1 Classic commuters, 16 Leafier suburbs
2	Rural areas	2 Agricultural heartland, 12 Remoter coast and country, 41 Accessible countryside
3	Rural fringes	5 Town and country, 17 Industrial margins, 36 Edge of town
4	Industrial areas	4 Traditional manufacturing, 6 Better-off manufacturing, 15 Scottish public housing, 34 Growth points, 37 Primary production
5	Middling Britain	7 Small towns, 11 West Midlands manufacturing, 20 Mixed economies, 40 Welsh coalfields, 43 Expanding towns
6	Prosperous areas	8 Established prosperity, 39 Affluent villages
7	Inner city estates	9 High rise housing, 24 London public housing, 30 Concentrations of affluence
8	Established owner-occupied	13 Green Belt, 28 Outer suburbs
9	Transient populations	14 Transient populations
10	Metropolitan professionals	18 Urban achievers, 26 Young singles
11	Deprived city areas	10 Inner London, 19 Scottish inner city, 25 Cosmopolitan London
12	Lower status owner-occupiers	21 Miner terraces, 27 Textile town terraces, 29 Industrial towns, 32 Margins of deprivation, 38 Declining resorts
13	Mature populations	3 Better-off retired, 22 Remoter retirement areas, 35 Retirement areas, 42 Coastal very elderly
14	Deprived industrial areas	23 Low amenity, 31 Ethnic groups in industry, 33 Heavy industry

Source: File *ward.xls* on directory */db/census91/area_class* on *irwell.mimas.ac.uk*. This contains a list of wards, ONS classes, and variable values in spreadsheet format.

with improved computer processing power to deal with the nearly 150 000 ED/OAs in GB. Openshaw *et al.* (1995) applied neural network classification methods and high performance computers to generate a set of classifications of EDs called GB Profiles. The research was supported in 1993 to 1994 by the ESRC/JISC 1991 Census of Population Programme (award R000234436).

11.4.1 Objectives and Uses

The research objective of this project was to provide a census data representative small area classification of Britain's residential areas. The aim was not to provide a specific classification targeted at a policy initiative (cf. the indexes of deprivation discussed in Chapter 9) but to provide social scientists with a means of summarizing the variation in residential populations across Britain. Many researchers do not want to tackle the complexities of multivariate census analysis themselves; they want to place their locationally referenced research observations in the context of a general class of area (defined in terms of population mix, housing type, occupations, jobs). Because most of these observations were georeferenced by postcode, the second objective was to enable users to link their postcode observations to EDs or OAs, and hence to one of the GB Profiles classifications. A software system called GB PROFILER was developed to achieve this goal.

11.4.2 Methods

The classification steps were (1) to choose and compute the variables, (2) to feed the data into two different classifiers—Openshaw's (1983) Census Classification Program (CCP), which is based on a k-means method and a neural net classifier that trains itself over millions of computer iterations to recognize patterns (clusters of areas), and (3) interpret the nature of the clusters by examining variable profiles for clusters. Openshaw *et al.* (1995) summarize the advantages of the neural net classifier approach, which allows for some uncertainty in the input data (owing to sampling, incorrect responses, or undercounts). However, the resulting classification is still dependent on variable selection and choice of the number of clusters from a range of solutions.

Choice of Variables

Blake and Openshaw (1995) reviewed and selected from the variables used in geodemographic systems, added new variables available from the 1991 Census, and assembled a set of 85 variables. The list of variables is given in Openshaw *et al.* (1995, Appendix 8.1) and the composition of variables in terms of Small Area Statistics (SAS) numerator and denominator counts is provided in Blake and Openshaw (1995, Appendix D). The set includes 15 *demographic* variables (age, marital status, pensioner, student indicators), 6 *ethnic* variables, 2 *migration* variables, 22 *housing* variables (tenure, housing type, amenity, density, and car availability indicators), 15 *household composition* variables (size, children, pensioner, and life stage indicators), 19 *socio-economic* variables (socio-economic group, industry, employment, unemployment, higher qualifications indicators), 2 *health* variables (limiting long-term illness), and 4 *travel-to-work* variables (mode of transport). No standardization or orthogonalization were implemented to retain the non-normal variation of the real world.

Clustering

The classification procedure used to create GB Profiles was based on an unsupervised artificial neural network called *Kohonen's self-organizing map* (SOM) (Kohonen, 1984). This is a very simple but powerful algorithm, which uses the concept of an array of competing neurons to create a process of self-organization. In its simplest form, a set of processing neurons competes to represent data using a measure of dissimilarity. The weights of the winning neuron, and those of the neighbouring neurons that are within a certain pre-defined distance of the winner, are updated to better represent the input data.

Over a number of iterations, the weights attached to each neuron provide a summary of any structure or patterns that exist in the input data set. The basic algorithm is as follows:

1. Define the geometry of the SOM to be used and its dimensions. Here, a grid with eight rows and eight columns is used.
2. Initialize a vector of M weights (one for each variable) for each of the neurons within the SOM.
3. Define the parameters that control the training process: block neighbourhood size, training rate, and number of training iterations.
4. Select a census ED at random, but with a probability proportional to its population size.
5. Add relevant noise to simulate data uncertainty in the input data.
6. Identify the winning neuron that is close to the input data.
7. Update the weights of the winning neuron and those of all the neurons within the neighbourhood.
8. Reduce both the training parameter and the neighbourhood size by a very small amount.
9. Repeat steps 4 to 8 a very large number of times.

A more detailed review of the application of Kohonen's SOM to census typologies can be found in Openshaw and Openshaw (1997). If step 4 is replaced by sequential selection and step 5 is dropped, then the algorithm is very similar to the K-means classifier used to fine-tune the ONS classifications. However, step 4 is very important because it enables spatial data uncertainty to be incorporated into the classification process.

11.4.3 Labels for the Classes

The GB Profile method was implemented many times with different target numbers of clusters. GB PROFILE makes available the 10, 49, 55, and 100 cluster solutions produced by the CPP algorithm (Openshaw 1983) and 10, 49, and 64 cluster solutions produced by the neural network method (see Section 11.4.5 for access details). Full labels for the different clusters are embedded within the system and are reported when the user supplies a set of postcodes. Table 11.7 sets out the labels used in the 64-cluster solution that are constructed hierarchically. The six major classes in this solution are mapped in Figure 11.1. The full cluster labels are provided by the GB PROFILER system. Labels were constructed by inspection of cluster variable profiles with sub-labels being associated with far-from-the-mean values of variables. A semi-automatic system of label generation could then be implemented as alternative cluster solutions were generated. Some geographic judgement was then exercised in selecting the best sequence of descriptive terms.

11.4.4 An Illustration of GB Profiles

Figure 11.1 maps the six major classes in the 64-cluster version of GB Profiles for the enumeration districts of Sheffield. These classes are arranged in a broad social order, reflecting the position and prospects of the residents. The unclassified category refers to special EDs, the data from which was suppressed and combined with neighbouring EDs. The unclassified EDs on the map pick out the Don Valley industrial and commercial belt at the heart of Sheffield. Around this core are neighbourhoods of the struggling class, where there are five sub-groups, all represented in Sheffield. West of these areas are the Aspiring districts of Sheffield, populated by students and single persons starting their

Table 11.7 The labelling system for the 64-cluster version of GB profiles.

Group	Sub-group	Name	Cluster no.
Struggling	Council tenants with multiple social problems	Multi-ethnic council tenants	1
		LA-rented semis	24
		Overcrowded council housing	33
		Council tenants in tower blocks	6, 7
		Single parents council tenants	29, 34
		Single parents in tower blocks	28, 30
		Unskilled council tenants	45
	Multi-ethnic, low income areas	Bangladeshi areas	4
		Indian areas	38
		Multi-ethnic bedsit areas	8, 27, 32
		Poor multi-ethnic singles	62
	Less well-off terraces	Terraces	2, 36
		LA-rented terraces	10, 35
	Fading industrial areas	Industrial terraces	43, 61
		Industrial council tenants	51
	Less well-off pensioners	Pensioners council tenants	17, 25, 31
		Pensioners in converted flats	18, 26
		Pensioners in HA-rented terraces	57
Aspiring	Young singles in flats	Poor young singles & students	3, 55, 60
		Singles in purpose-built flats	53
		Better-off singles	14, 54
	Better-off council tenants	Council semis	13
	Rural communities	Rural areas	44, 52
	Armed services	Young armed services families	12
Established	Semi-detached suburbia	Semis	56
		Mortgaged semis	63
		Owner-occupied semis	5
	Better-off pensioners	Pensioner migrants	15, 16, 23, 59
	Comfortable middle ages	Middle class suburbia	37
		Wholly owned semis	21
		The average	20
Climbing	Metro singles	Well-off singles in bedsits	14
		Well-off singles in purpose-built flats	19
		Well-off singles in converted flats	50
	Academic centres	Students in bedsits	41
Prospering	Wealthy achievers	Middle aged managers	46, 58
		Well-off middle aged managers	9, 47
		Self-employed managers	48
		Educated professionals	22
	Wealthy rural communities	Rich agriculturalists	11, 39, 49, 64
Unclassified			40, 42

Source: Openshaw *et al*. (1995), Appendix 8.2.

Figure 11.1 A GB Profiles 1991 Census classification of the EDs in Sheffield.

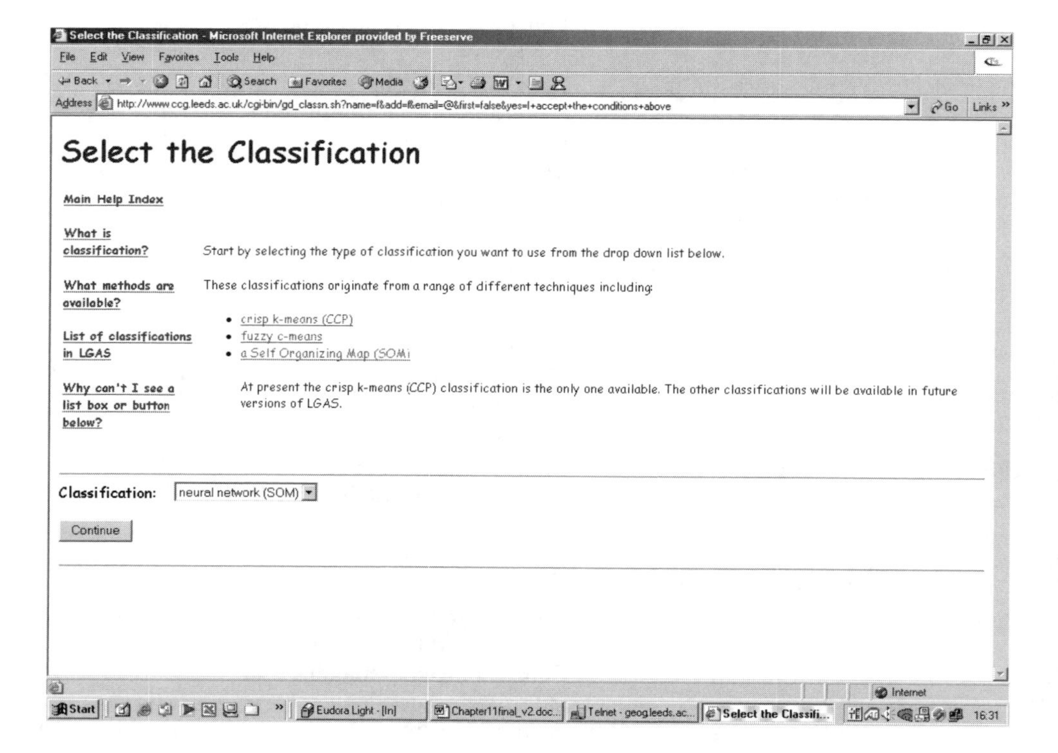

Figure 11.2 The GB PROFILES/LGAS facility for choosing the classification type.

careers, and better-off council tenants. To the South West of the city centre extending into the Peak District National Park are Prospering residential areas. The other suburban areas fall in the Established or Climbing categories.

11.4.5 Using GB Profiles

The GB Profile classifications have been made available in the public domain via a website located at the School of Geography, University of Leeds: *http://www.ccg.leeds.ac.uk/linda/geodem/*

The web-based version supercedes the PC-based version written by Marcus Blake in Visual Basic v.3, which does not run on 32-bit operating systems (Windows98 and later).

Users are required to abide by the terms and conditions set by the Census Offices for supply of the underlying SAS census data, to acknowledge Crown Copyright in that data and the associated 1991 Census Enumeration District/Postcode directory.

The website for GBProfiles (also referred to as LGAS or the Leeds Geodemographic Analysis System) provides a help facility and an FAQ with background information. To run GBProfiles, the user is taken through a series of steps. Figure 11.2 is an example of where the user can choose between a classification created using CCP (k-means) or a neural network. Once selected, the user will see the screen in Figure 11.3, where the

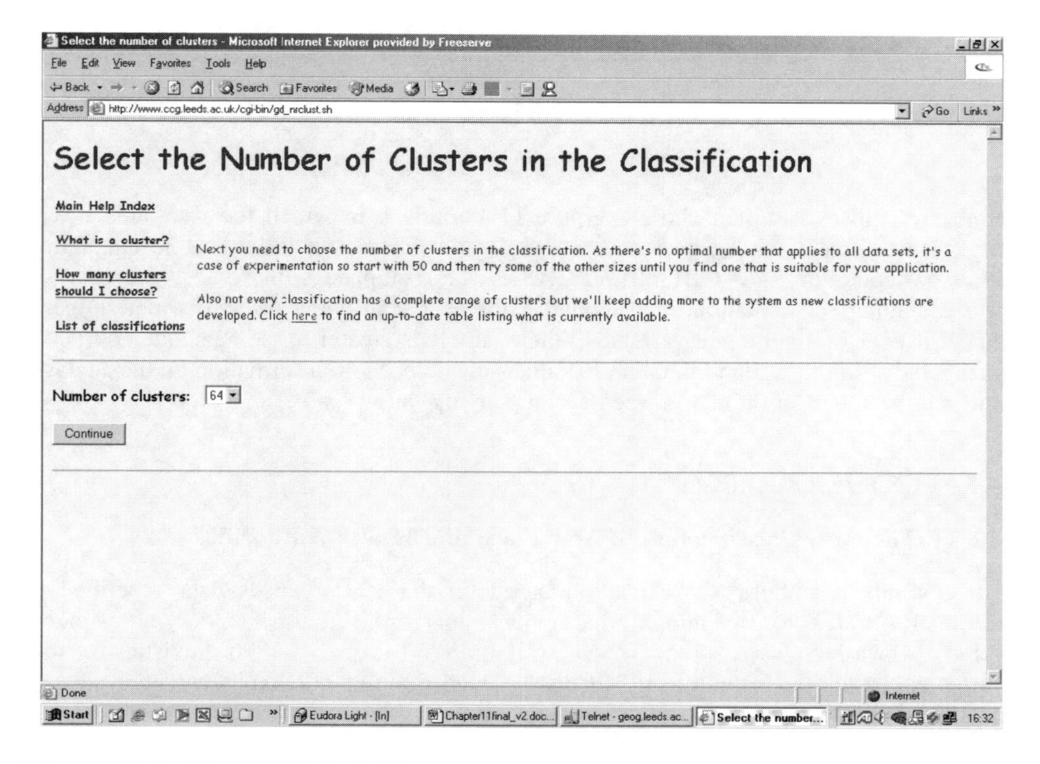

Figure 11.3 The GB PROFILES/LGAS facility for selecting the number of clusters in the classification.

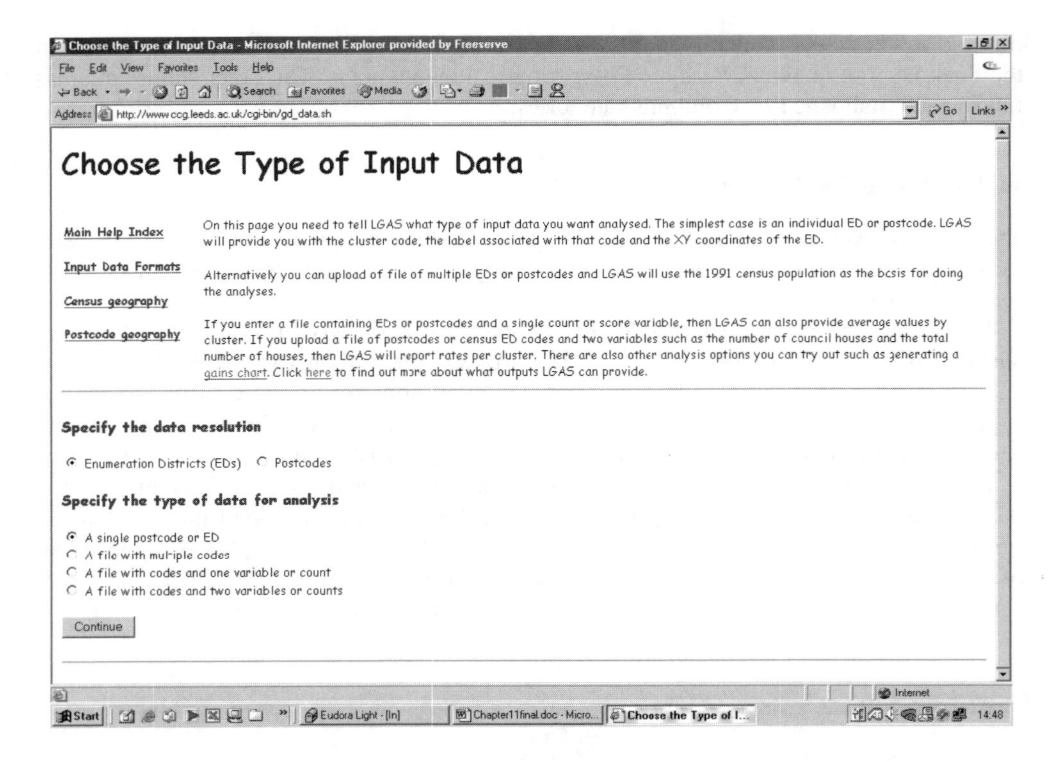

Figure 11.4 The GB PROFILES/LGAS postcode and enumeration search facility.

number of clusters is then chosen. Figure 11.4 follows, in which the user indicates if they wish to enter a single postcode or enumeration district or upload a file containing either postcodes or EDs. GBProfiles/LGAS is also equipped with some basic analysis tools including segmentation analysis and gains charts, in which users can determine if the occurrence of a particular variable in their sample is greater for some geodemographic types relative to those found in Great Britain as a whole. More information on the analysis tools can be found in the help screen accompanying the website.

11.5 FURTHER DEVELOPMENTS

11.5.1 The ONS Classification of New Local and Health Authorities

The ONS local authority classification based on the 1991 Census data described in Section 11.2 refers to a geography that applies, apart from minor boundary adjustments and shifts of a few wards, to the period April 1, 1975 to April 1, 1996. Statistics for that period and earlier data from the 1971 Census adjusted to that geography can be used with the 1991 classification.

However, the period April 1, 1996 to April 1, 1998 saw major reorganization of local and health authority boundaries (see ONS, 1999a and Wilson and Rees, 1998 for details). The ONS has carried out work to revise the classification, configuring it to relate to local authorities as they existed on April 1, 1998 and to health authorities as they existed on April 1, 1999. The same 37 variables were used and the same classification methods employed. Results are fully described in Bailey *et al.* (1999a, 1999b). The revised version of the classification was compiled using the same methods and approach as in the

Table 11.8 The Families, Groups, and Clusters of the 1999 classification of local authorities.

Family	Group	Cluster
Rural areas	Rural amenity	Rural amenity
	Remoter rural	Rural England and Wales
		Rural Scotland
Urban fringe	Established manufacturing fringe	Established manufacturing fringe
	New and developing areas	New and expanding towns
		Developing towns
	Mixed urban	Most typical towns and cities
		London and Glasgow periphery
Coast and services	Coast and country resorts	Seaside towns
		Traditional rural coast
	Established service centres	Established service centres
Prosperous England	Growth areas	Town and country growth
		Prosperous growth areas
	Most prosperous	Most prosperous
Mining, manufacturing, and industry	Coalfields	Mining and inner city
		Mining and industry
		Former mining areas
	Manufacturing centres	Manufacturing centres
	Ports and industry	Urban industry
		Liverpool and Manchester
		Clydeside and Dundee
Education centres and outer London	Education centres and outer London	Suburbs
		Cosmopolitan outer London
		Education centres
Inner London	West inner London	West inner London
	East inner London	Inner city boroughs
		Newham and Tower Hamlets

Source: Bailey *et al.* (1999b).

1996 version and the same 37 socio-economic and demographic variables from the 1991 Census, which were re-aggregated for the new local and health authority boundaries in Great Britain in April 1999. However, the revision produced a different hierarchical structure—that is, different numbers of Families, Groups, and Clusters, with slightly different socio-economic characteristics. Although the 1999 classification is not directly comparable with the 1996 scheme, where possible and appropriate, the same names were retained for Families, Groups, and Clusters. Table 11.8 lists the 1999 names for the groupings used in the classification.

11.5.2 Classifications Using 2001 Census Data

Publication of the Census Area Statistics and Standard Tables from the 2001 Census is expected in 2003. We can expect in 2003 and 2004 that ONS, commercial agencies, and academics will carry out a variety of classification exercises. What features should we be looking for from this activity?

Geographical coverage and scale will be improved. It should be possible to develop classifications for the whole of the United Kingdom because the Northern Ireland Census processing will be closely integrated with that of Great Britain. Output areas in 2001 to be used to publish the Census Area Statistics will be smaller than EDs in 1991, and so the small area classifications will be more geographically precise. However, the greater number of them will pose challenges to the software used. There will be a wider *choice of variables* available in 2001 (see Chapter 20), and so the classifications will have a richer content, particularly if religion and income questions are included. Both traditional and new *methods of clustering* will be used. We should expect to see some development of fuzzy classifications where areas can be partially assigned to different classes, to reflect internal heterogeneity where this is present. To date, *Labelling and Describing* clusters has been largely a judgemental process, but we should expect more automatic methods to be developed that link statistical profiles to set phrases. It will be important to provide census users with official classifications cheap or free, as has been achieved with the 1991 ONS classification. In the future, research users should be provided with access to whole classification files for their own use as simple text files that survive system replacements and with access via the Web to interactive interfaces that give the novice user access to just the parts of the classifications needed.

Box 11.1 Key web links for Chapter 11.

http://www./caci.co.uk/	CACI's Web site referencing the ACORNa and PIN geodemographic system
http://www.eurodirect.co.uk/	Eurodirect's Web site for the CAMEOUK geodemographic system
http://www.spatialinsights.com/simosaic.htm	Describes the American version of the MOSAIC geodemographic system
http://www.mimas.ac.uk/docs/experian/gbmosiac.pdf	Documentation describing Experían's MOSAIC geodemographic system. Contact Experían Ltd, Talbot House, Talbot Street, Nottingham NG1 5HF, tel. 0115 941 0888, fax 0115 934 4905 for further information on Experían geodemographic systems.

http://census.ac.uk/cdu/Datasets/1991_Census_datasets/Area_stats/Derived_data/Area_classifications.htm	ONS ward classification: brief description and references to files available for downloading
http://census.ac.uk/cdu/Datasets/1991_Census_datasets/Area_stats/Derived_data/Area_classifications.htm	ONS district classification: brief description and references to files available for downloading
http://www.ccg.leeds.ac.uk/linda/geodem	Webpage to run GBProfiles/LGAS

Box 11.2 Additional CDS resources for Chapter 11.

ONS classifications: lookup files	
/db/census91/area_class/ward.csv	List of wards, ONS classes, and variable values in comma-separated variable format
/db/census91/area_class/ward.xls	List of wards, ONS classes, and variable values in spreadsheet format
/db/census91/area_class/district.csv	List of districts and ONS classes in comma-separated variable format
/db/census91/area_class//district.xls	List of districts and ONS classes in spreadsheet format
GB PROFILER: files associated with the original PC version	
/db/census91/profiles/Profiler.hlp	Windows help file for GB PROFILER
/db/census91/profiles/Readme.txt	Installation notes
/db/census91/profiles/data_files/P-data91.zip	The core data files without with GB PROFILER will not run
/db/census91/profiles/classification_files/Nedgb010.zip, Nedgb064.zip, Nedgb100.zip	Individual classification files
/db/census91/profiles/image_files/P-images.zip	Images that summarize different clusters within the classifications
/db/census91/profiles/latest_version/P-res22.zip	The setup files for the software

Note: The file locations refer to the *irwell.mimas.ac.uk* server. Alternatively, these files are also available on the CDS Resources CD. Users should be aware that the files may move from the irwell.mimas.ac.uk server, should this service be replaced in the future. The GB PROFILER files listed here are associated with software that no longer runs on contemporary PCs. Users should use the website given in Box 11.1 to access an on-line version of GB Profiles.

Box 11.3 Server resources for Chapter 11.

GB PROFILER: accessing the on-line UNIX version	
http://census.ac.uk/	Register to access 1991 Census SAS/LBS and DBD
http://mimas.ac.uk/	Register for a user name and password for the MIMAS irwell server
irwell.mimas.ac.uk (IP address 130.88.203.130)	Login: *user name* Password: *password*
Type the command:	*uprofiles*.
Follow the on-screen instructions.	Supply postcodes or file of postcodes.

Note: Users should be aware that the system described here may move from the irwell.mimas.ac.uk server, if this Unix service is replaced.

12

Dealing with the Census Undercount

Stephen Simpson

SYNOPSIS

This chapter reviews the nature of the people who were not counted by the 1991 Census and introduces three data sets, EWCPOP, SOCPOP, and MIGPOP, which provide estimates of their composition in terms of age and sex, social group, and migrant flows. These data sets have been produced as a result of an extensive project known as Estimating with Confidence (EwC). The estimates are provided in files in the accompanying CDS Resources.

12.1 INTRODUCTION

This chapter describes a set of 1991 population estimates for small areas—typically wards in England, Wales, and Northern Ireland, and postal sectors in Scotland—that include an allowance for those residents missed by the census. These data sets are included in the accompanying *CDS Resources*. This material should be read in conjunction with Chapter 13, which gives a review of population statistics after census years, and Chapter 21, which describes plans to deal with non-response in the 2001 Census.

Section 12.2 briefly considers the causes of census non-response, and in Section 12.3, population estimates are introduced, which refer to the number of people missed of each age and sex. Section 12.4 further disegregates these estimates by ethnic group, by employment status, and by tenure, using the consistent evidence that a census is more likely to miss some groups than others. Section 12.5 uses evidence of the number of migrants missed from the census and gives adjusted flows of migrants between districts of Britain. The data for small areas described in Sections 12.3, 12.4, and 12.5 have become known as EWCPOP, SOCPOP, and MIGPOP, respectively. The 'EWC' derives from *Estimating with Confidence*, a project spanning local authorities, academic, and government staff that undertook the research and consultation leading to the EWCPOP data set. This discussion focuses on areas smaller than local authority districts. The estimates discussed here are consistent with mid-1991 population estimates published for each local authority district area by the UK statistical offices (Heady *et al.*, 1994; Beatty and Rodgers, 1999).

Figure 12.1 'You might as well answer—we'll get you eventually with EWCPOP, SOCPOP and MIGPOP'. Source: Copyright Malcoln Campbell.

Some research has a real need for estimates of the number of *households* in an area, the *daytime population* including tourists or commuters, or other aspects of the population. These needs are addressed in Chapter 13.

12.2 A COMPLETE POPULATION ESTIMATE FOR SMALL AREAS IN 1991

Figure 12.2 shows the official estimates of the proportion of residents not counted in the 1991 Census output. The group least well enumerated by the census was male adults in their twenties, especially across London and city districts where one in five were missed. Women in their twenties and young children were also notably missed, although the evidence reviewed for official studies could not pin these down to particular types of district. Under-enumeration of the elderly was estimated mainly through careful comparison with universal pension records: it seems to have been minimal in Scotland, to have affected

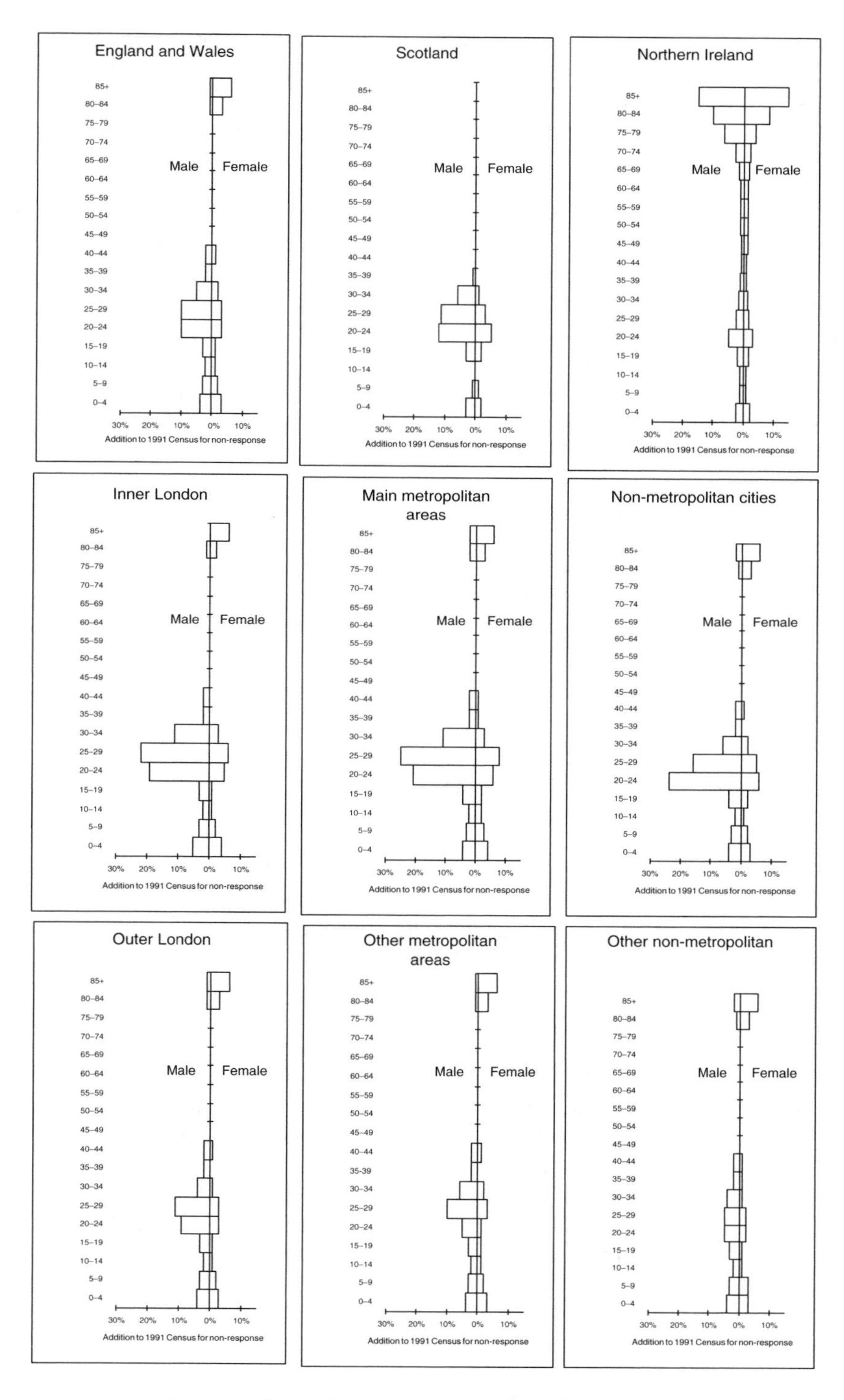

Figure 12.2 Non-response to the 1991 Census.

very elderly women in England and Wales, and to have been relatively high in Northern Ireland. Figure 12.2 should be considered in the context of a 98% coverage for the census overall as an extremely good result.

Similar demographic patterns of data quality are found in the censuses of other countries. This is partly attributed to disaffection with personal enquiries in general or government enquiries in particular. It is also attributed to people who are hard to contact and find it hard to respond because they are 'on the move'. These people may be literally in the process of moving, or in temporary accommodation, or unsettled in other ways, which make census questions less relevant to their circumstances. The great majority of those missed could also be termed *socially excluded* in some way because of poverty or lack of permanent housing, employment, or legal status (Simpson, 1999). The transfer of students from their 'home' addresses where they were counted as residents in the 1991 Census, to their term-time addresses where they are usually resident, is an additional adjustment that significantly affects population estimates in many areas. In 2001, the UK Census for the first time counted students as residents at their term-time addresses.

The following section describes mid-1991 population estimates for areas smaller than districts. They include allowances for non-response, for the transfer of students to their term-time addresses, and for the small change in time between the census date and mid-year. Because of the lack of direct evidence for the nature of census undercount in 1991, the population statistics presented in this section rely heavily on plausible estimates of non-response, validated by comparison with alternatives and extensive consultation. The methodology for the population statistics is referenced rather than explained in detail here.

12.3 EWCPOP—POPULATION ESTIMATES FOR 1991 FOR SMALL AREAS, WITH AGE AND SEX STRUCTURE

Figures 12.3 and 12.4 show example output from EWCPOP: population estimates for wards in England, Wales, Northern Ireland, and postal sectors in Scotland. A *results* file gives a summary of the adjustments made to the census, age, and sex detail of the resulting population estimates. An *adjustments* file gives the detailed adjustments made for each age–sex group. These data sets are included in the accompanying *CDS Resources*.

Figure 12.5 shows an example of how the adjustments build up the population estimate for each small area from the census results. Apart from the armed forces, the non-response estimated for districts by national statistical offices is shared among small areas in two ways: according to the number of young unemployed and according to the number of imputed households (which enumerators thought contained residents but did not return a form). The population statistics use the mean of the two estimates. Full details of the method and alternatives that were considered are given in Simpson *et al.* (1997a). The results are available in *CDS Resources*, and the files are listed in Table 12.1. The file ewctext.lst describes the field layout, which is identical for each of these files. The adjustments are available in files of the same name with the trailing letter 'p' replaced with the letter 'a', in *CDS Resources* or as described in the summary at the end of this chapter. Their field layout is described in ewcatext.lst.

Population estimates have been produced in the same way for Enumeration Districts (EDs) and Output Areas (OAs). The results are not as reliable as those for wards, as

Census county code	Area code	Area name	Local authority district 1991	County 1991	Census count of residents, all ages	Student net adjustment, all ages	Armed forces non-response, all ages	Net timing adjustment, all ages	Census modification adjustment, all ages	Residual non-response, all ages	Mid-1991 population estimate, all ages	Mid-1991 population estimate, m 0–4	Mid-1991 population estimate, m 5–9	Mid-1991 population estimate, m 10–14	Mid-1991 population estimate, m 15–19	Mid-1991 population estimate, m 20–24	Followed
8	CXFN	Ilkley	Bradford	West Y	13 534	−230	0	21	0	259	13 584	315	309	405	440	391	by the
8	CXFP	Keighley N	Bradford	West Y	15 458	−79	0	22	0	478	15 879	623	685	678	513	607	population
8	CXFQ	Keighley S	Bradford	West Y	13 287	−11	0	18	0	406	13 699	566	532	465	429	525	of other
8	CXFR	Keighley W	Bradford	West Y	16 373	−92	0	23	−1	421	16 724	684	664	631	578	644	male age
8	CXFS	Little Hor	Bradford	West Y	16 789	235	0	21	0	1 703	18 748	1 030	865	748	628	1 098	groups
8	CXFT	Odsal	Bradford	West Y	16 839	−39	0	24	0	609	17 432	619	623	600	565	718	to 85–89,
8	CXFU	Queensbury	Bradford	West Y	16 242	−40	0	21	0	532	16 754	604	569	503	541	676	and 90+,
8	CXFW	Rombalds	Bradford	West Y	15 261	−243	0	23	0	207	15 247	434	462	526	462	373	and female
8	CXFX	Shipley Ea	Bradford	West Y	14 332	−19	0	20	0	478	14 811	606	503	448	526	595	age groups
8	CXFY	Shipley We	Bradford	West Y	15 419	−131	0	22	0	360	15 669	495	535	575	542	532	
8	CXFZ	Thornton	Bradford	West Y	13 833	−55	0	18	0	496	14 292	538	467	436	471	597	
8	CXGA	Toller	Bradford	West Y	17 601	−12	0	24	−1	798	18 410	951	986	947	688	853	
8	CXGB	Tong	Bradford	West Y	14 108	−13	0	19	0	790	14 903	730	677	553	474	688	
8	CXGC	Undercliff	Bradford	West Y	15 514	−33	0	19	0	918	16 418	678	686	575	601	765	
8	CXGD	University	Bradford	West Y	18 897	1929	0	21	0	1 230	22 076	1 032	1 084	1 085	1 106	2 250	

Figure 12.3 EWCPOP results of mid-1991 population estimates for electoral wards: annotated extract from file englawsp.dat.

County code	Area code	Area name	District 1991	County 1991	Census resident counts m 0–4	m 5–9	m 10–14	m15–19	m20–24	Student net adjustment m 0–4	m 5–9	m 10–14	m15–19	m20–24
8	CXFN	Ilkley	Bradford	West Y	305	302	410	486	419	0	−1	−12	−48	−68
8	CXFP	Keighley N	Bradford	West Y	601	669	667	518	547	0	0	−2	−15	−27
8	CXFQ	Keighley S	Bradford	West Y	547	517	454	422	458	0	1	1	−1	−6
8	CXFR	Keighley W	Bradford	West Y	664	650	622	586	600	0	0	−2	−15	−31
8	CXFS	Little Hor	Bradford	West Y	954	804	708	562	688	0	1	0	20	87
8	CXFT	Odsal	Bradford	West Y	591	602	585	559	621	0	0	0	−7	−14
8	CXFU	Queensbury	Bradford	West Y	580	551	491	537	592	0	0	−1	−7	−14
8	CXFW	Rombalds	Bradford	West Y	425	457	530	503	410	0	−2	−10	−43	−70
8	CXFX	Shipley Ea	Bradford	West Y	584	486	435	519	517	0	1	1	−3	−10
8	CXFY	Shipley We	Bradford	West Y	479	523	567	560	516	0	0	−2	−24	−45
8	CXFZ	Thornton	Bradford	West Y	515	450	423	468	525	0	0	0	−8	−19
8	CXGA	Toller	Bradford	West Y	914	957	925	674	713	0	1	1	−5	−8
8	CXGB	Tong	Bradford	West Y	694	649	534	458	545	0	0	0	−3	−5
8	CXGC	Undercliff	Bradford	West Y	637	654	555	586	604	0	0	−2	−7	−10
8	CXGD	University	Bradford	West Y	976	1039	1054	896	1233	0	2	−1	180	783

Timing net adjustment m 0–4	m 5–9	m 10–14	m15–19	m20–24	Census modification m 0–4	m 5–9	m 10–14	m15–19	m20–24	Armed forces non-response m 0–4	m 5–9	m 10–14	m15–19	m20–24
1	0	2	−3	1	0	0	0	0	0	0	0	0	0	0
3	1	3	−3	1	0	0	0	0	0	0	0	0	0	0
3	1	2	−3	1	0	0	0	0	0	0	0	0	0	0
3	1	3	−4	1	0	0	0	0	0	0	0	0	0	0
4	1	4	−3	1	0	0	0	0	0	0	0	0	0	0
3	1	3	−3	1	0	0	0	0	0	0	0	0	0	0
3	1	2	−3	1	0	0	0	0	0	0	0	0	0	0
2	1	3	−3	1	0	0	0	0	0	0	0	0	0	0
3	1	2	−3	1	0	0	0	0	0	0	0	0	0	0
2	1	3	−3	1	0	0	0	0	0	0	0	0	0	0
2	1	2	−3	1	0	0	0	0	0	0	0	0	0	0
4	1	5	−4	1	0	0	0	0	0	0	0	0	0	0
3	1	3	−3	1	0	0	0	0	0	0	0	0	0	0
3	1	3	−4	1	0	0	0	0	0	0	0	0	0	0
5	1	5	−6	2	0	0	0	0	0	0	0	0	0	0

Non-response, imputation-indicated m 0–4	m 5–9	m 10–14	m15–19	m20–24	Non-response, unemployment-indicated m 0–4	m 5–9	m 10–14	m15–19	m20–24
11	9	5	7	48	7	6	3	5	31
20	17	10	14	93	18	15	9	12	80
11	9	6	8	52	21	17	10	14	93
11	10	6	8	52	21	18	11	15	97
95	79	48	66	430	48	40	24	33	216
26	22	13	18	118	23	19	12	16	104
23	19	12	16	104	20	16	10	14	89
6	5	3	4	28	8	7	4	6	37
15	13	8	11	70	23	19	12	16	104
12	10	6	9	57	14	12	7	10	66
19	16	10	14	88	20	17	10	14	93
24	20	12	17	110	41	34	21	28	185
28	23	14	19	126	37	31	19	26	170
42	35	21	29	191	33	27	17	23	150
42	35	22	29	192	60	50	31	42	274

In each section, male age groups to 80–84 and 85+ are followed by the equivalent female age groups.

Figure 12.4 EWCPOP adjustments to reach mid-1991 population estimates for electoral wards: annotated extract from file englawsa.dat.

relevant data are suppressed for a significant number of the smaller areas and institutional EDs or OAs. For further notes, the estimates and the adjustments themselves, see the summary at the end of this chapter.

12.4 SOCPOP—ETHNIC GROUP, TENURE, AND EMPLOYMENT STATUS

Differential non-response among ethnic groups, tenure categories, and employment status categories has been estimated from a review of coverage studies (Simpson and Middleton, 1997). The differentials embody the consistent evidence that minority and marginalized

	Ilkley		University	
Census count of residents		419		1 233
Adjustments for:				
Net student adjustment		−68		+783
Timing		+1		+2
Armed forces non-response		0		0
All other non-response				
(a) According to unemployment	(+31)		(+274)	
(b) According to imputed residents	(+48)		(+192)	
Mean, ½(a)+½(b)		+39		+233
Modification		0		0
EwC Population estimate		391		2 250

Figure 12.5 Examples of the construction of the population estimate from the census, males aged 20 to 24 in two wards of Bradford district.

Table 12.1 EWCPOP population estimates and filenames in CDS resources.

Population estimates	Filename
England 1991 population for wards	englawsp.dat
Wales 1991 population for wards	waleswsp.dat
Northern Ireland 1991 population for wards	niwp.dat
Scotland 1991 population for postal sectors	scotplp.dat

ethnic groups, the unemployed, and those renting private accommodation are most likely to be missed from the census. For ethnic groups, the differential non-response adjustment—the additional proportion to be added to the census count—is estimated to vary among White groups: Black groups: South Asian groups: Other groups in the ratio 1 : 6 : 4 : 3. Despite the incomplete nature of the 1991 Census Validation Survey, the same regular patterns of non-response among ethnic minorities, and particularly among the Black population, tenants (especially tenants of private landlords), unemployed residents, and students, were seen (Heady *et al.* 1994).

Population estimates for each category have been computed incorporating these ethnic, employment, and tenure differentials, as described in Simpson and Middleton (1998). A second set of estimates has been computed incorporating none of these social differentials, applying only the EWCPOP adjustment pertinent to the sex, age, and location of each ethnic group. Figure 12.6 shows an extract from the population estimates with social differentials. It shows both the estimates and the distance of these estimates from that computed with no social differentials.

Cnty code	Dist code	Ward code	Ward name	Ethnic group	Ward population estimate: 1991 Census plus non-response					
					M 0–4	M 5–9	M 10–14	M 15–19	M 20–24	Male age groups to
8	CX	CXGD	University	White	101.52	69.80	51.36	122.25	498.99	80–84 and 85+ are
8	CX	CXGD	University	Blck Car	1.13	6.55	6.33	10.91	27.67	followed by the
8	CX	CXGD	University	Blck Afr	5.63	6.55	.00	2.18	26.14	equivalent female
8	CX	CXGD	University	Blck Oth	6.76	10.92	8.44	3.27	23.06	age groups,
8	CX	CXGD	University	Indian	109.80	113.52	133.28	104.41	140.76	and a total.
8	CX	CXGD	University	Pakstani	707.37	781.11	798.67	617.06	587.95	
8	CX	CXGD	University	Bngldshi	34.84	40.62	43.06	21.93	19.93	
8	CX	CXGD	University	Chinese	2.12	4.20	1.03	3.11	69.97	
8	CX	CXGD	University	Oth Asia	25.42	13.66	13.36	19.73	49.62	
8	CX	CXGD	University	Oth Oth	32.83	35.73	23.64	23.88	22.90	

	Impact of social differentials for non-response					
	M 0–4	M 5–9	M 10–14	M 15–19	M 20–24	Male age groups to
White	−3.75	−2.03	−.91	−3.59	−60.93	80–84 and 85+ are
Blck Car	.07	.31	.18	.51	6.27	followed by the
Blck Afr	.37	.31	.00	.10	5.93	equivalent female
Blck Oth	.44	.51	.24	.15	5.23	age groups,
Indian	.32	.05	.05	.41	6.43	and a total.
Pakstani	2.06	.36	.27	2.42	26.84	
Bngldshi	.10	.02	.01	.09	.91	
Chinese	.01	.04	.00	−.01	4.59	
Oth Asia	.15	.13	.04	−.03	3.25	
Oth Oth	.20	.34	.07	−.04	1.50	

Figure 12.6 SOCPOP 1991 Census day population estimates for social groups: annotated extract from file *englawle.dat*.

Table 12.2 Topics categories and SOCPOP filenames in CDS resources.

Topic	Categories (within each age and sex group)	SOCPOP filenames
Ethnic group	White, Black Caribbean, Black African, Black Other, Indian, Pakistani, Bangladeshi, Chinese, Other Asian, Other groups.	Englawle.dat Waleswle.dat Scotde.dat
Employment status	Employed, unemployed, students, other inactive	Englawlm.dat Waleswlm.dat Scotdm.dat
Tenure	Owners, social renters, private renters, communal establishments	Englawlt.dat Waleswlt.dat Scotdt.dat

The estimates for each ethnic group are consistent with (i.e., add up to) the EWCPOP estimates. However, they do not transfer students from their home addresses to their term-time addresses (see in the preceding text). This restriction was necessary because of the lack of information about the ethnic group of students. The estimates, therefore, refer to the April 21st, 1991 Census date rather than mid-year one. Table 12.2 lists the social group estimates available in files in *CDS Resources*. The file soctext.lst describes the field layout, which is identical for each of these files.

Note that while the EWCPOP data are for all ward-level areas identified in the 1991 Census Small Area Statistics (SAS), restrictions on census data for ethnic groups necessitate that the SOCPOP estimates are only available for the (slightly fewer) wards identified in the 1991 Census Local Base Statistics (LBS) for England and Wales, only for districts in Scotland, and not at all in Northern Ireland.

Even when applying the response rate for each age, sex, and locality equally to all social groups, there are quite a few large differences in their estimated level of non-response (black bars in Figure 12.7). This is purely because White, employed, and owner residents are less likely to be in the age groups and places that Office for National Statistics (ONS) and the EWC project estimated as experiencing high non-response. The plausible

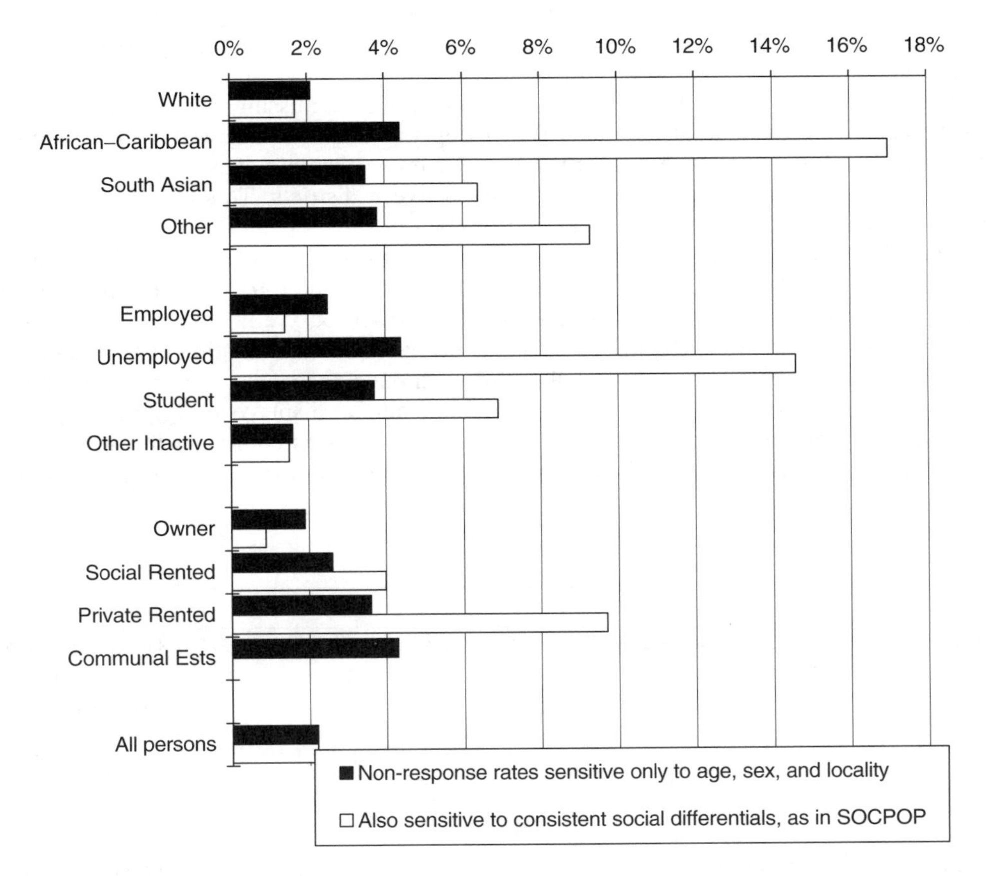

Figure 12.7 England and Wales population estimates from SOCPOP.

differentials among the response rates of the social groups used in SOCPOP make a considerable further difference (white bars in Figure 12.7). Groups other than White are estimated to number from 6 to 17% more than their census count, unemployed nearly 15% more than their census count, and private renters about 10% more than their census count. The differences make it important to use these full population estimates for purposes in which they have sufficient detail to be of use. For example, the starting point for population estimates and projections—for small areas, or with an ethnic group dimension—can now include appropriate non-response. Comparisons of unemployment rates between areas can be assessed for their robustness to census non-response. The resulting population estimates for social groups are just that—estimates rather than certain counts. The differentials are plausible but not exact. In particular, because of the lack of any relevant evidence, the same *differentials* among undercount rates for groups are applied within all areas in Britain equally and to all age groups.

12.5 MIGPOP—THE IMPACT OF NON-RESPONSE ON MIGRATION

In general, the number of migrants is boosted by an allowance for undercount by a larger percentage than is the population as a whole. This is because migrants are particularly concentrated among young adults, where non-response rates are high, and because the social groups with high non-response are more likely to be migrants, particularly the unemployed, students, and those in the rented housing sector.

The MIGPOP contains allowances for migrants among the missing million, adding 6.3% to the census count of total migrants. Estimating the origin of migrants with unknown origin adds a further 6.9%. A 10% allowance for misreporting makes a total of an extra 23% migrants to the published count of flows between districts. This total is confirmed by independent calculations in Rees *et al.* (2000).

The MIGPOP contains a single file with the complete 'population migration flows' between each of the 459 districts in Britain as constituted in 1991. The file migpop.dat included in *CDS Resources* contains estimates of the 459 × 459 flows between districts of Britain by age and sex. The file contains the origin district, destination district, and the estimated number of migrants in that flow for each male age group and for each female age group, followed by the estimated total migrant count, as displayed in Figure 12.8. The components that make up the estimates, essentially the census count and the allowances listed earlier, are available in migpopa.dat, or as described in the summary at the end of this chapter.

Estimated complete population migration								
ORIG	DEST	MIGM0004	MIGM0509	MIGM1014	MIGM1519	MIGM2024	MIGM2529	Male age groups to 80–84 and 85+ are followed by the equivalent female age groups, and a total.
08DA	08CX	63.3	39.4	26.5	47.7	236.7	278.9	

Figure 12.8 1991 migration between districts of Britain. Annotated extract from file migpop.dat: the record for migration from Leeds to Bradford.

The impact on the *pattern* of population change within Britain is of more importance than the level of movement itself. For any flow between two districts, the estimated impact of non-response recorded in these files is complex. It depends on the social and age–sex composition of the raw census flow, on the officially estimated non-response for each age—sex group of residents in the destination district, and on the non-response differentials among social groups.

However, two general patterns in the population migration flows of MIGPOP can be identified. First, the under-recording of young male adult migrants is eliminated. Second, the net outflow from the most urban city areas into more rural areas is significantly reduced. Figure 12.9 summarizes the latter impact, using official categories of district in use at the time of the 1991 Census (see also Simpson and Middleton, 1999).

Census migrant data inform regional policy and find a central role within the government's sub-national population projections. The data of MIGPOP suggest that the 'Urban

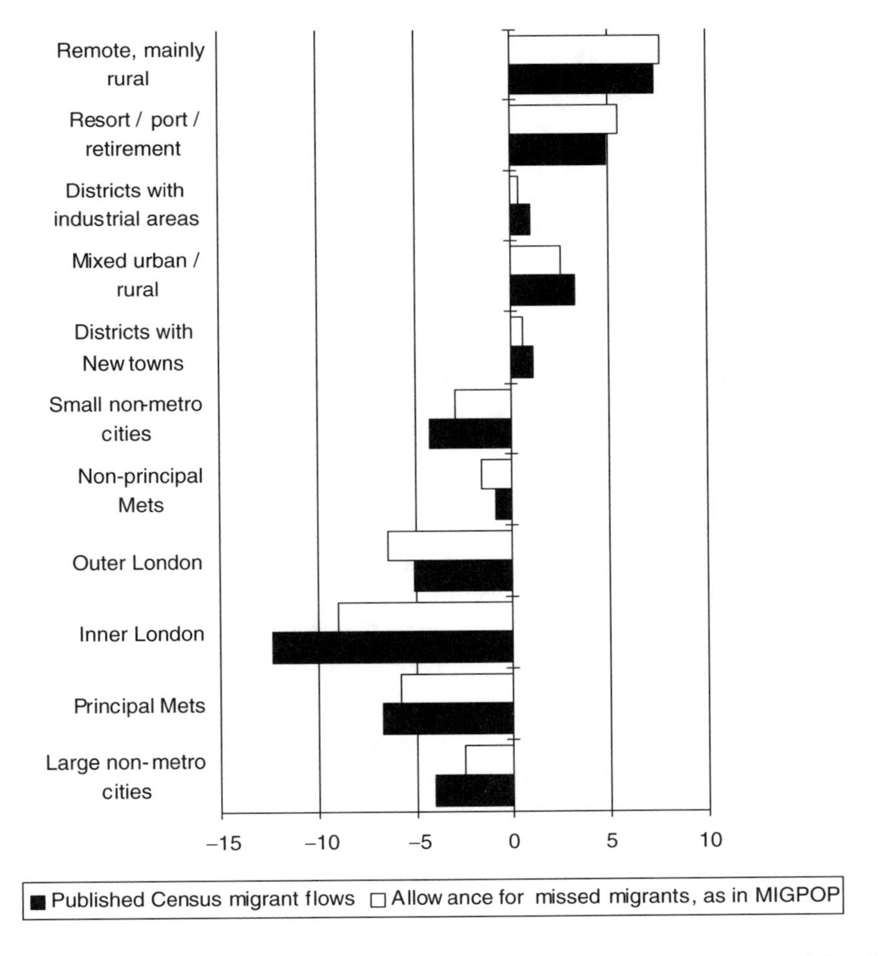

Figure 12.9 Net flow of migrants to (positive) or from (negative) each type of district in the year before the census, per 1000 population.

Exodus' of the 1980s and 1990s is not as great as measured by the census, although still dramatic and the major feature of internal migration. Over a twenty-year period, the shift from city districts to more rural districts within the 1991 counties used in government projections would be reduced by a half a million. This significant impact of better data suggests there is some urgency to use the better migrant data promised from research on patient records.

12.6 CONCLUSION

The 1991 Census enumerated 98% of all residents in the United Kingdom. Those missing were concentrated among the armed forces and in many city areas, but were also disproportionately among social groups associated with exclusion in other ways: ethnic groups other than White, unemployed and students, and households renting private landlords.

Users of the 1991 Census output should consider how the undercount may affect their application and use alternative data sets where available. For small areas, three data sets have been prepared as described in this chapter, and their key features are summarized in Box 12.1.

Acknowledgements

The data sets incorporate copyright data from the 1991 Census and from ONS mid-1991 population estimates. The data sets are by-products of two research projects funded by the Economic and Social Research Council (United Kingdom), awards H519255028 and R00023696.

Box 12.1 Summary of EWCPOP, SOCPOP, and MIGPOP features.

EWCPOP

- Mid-1991 population: five-year age group and sex.
- For SAS or LBS ward-level areas, throughout United Kingdom.
- Allowances for the missing million, student term-time address, and timing to mid-1991. Fully consistent with government, mid-1991 population estimates for Districts.

SOCPOP

- Census day 1991 population: ethnic group (10 categories), employment status (4 categories), or tenure (4 categories), as well as five-year age group and sex.
- For the 9636 LBS ward areas in England and Wales. For the 56 Districts in Scotland.
- Allowances for the missing million.

MIGPOP

- 1991 Census migration flows: five-year age group and sex.
- Within and between the 459 Districts of Britain (459 × 459 flows).
- Allowances for the missing million, recorded migrants with unknown origin, and under-reporting of migrants.

Full detail and further references are given in the Web pages of the Census Dissemination Unit at Manchester University:
http://www.census.ac.uk/cdu/Datasets/1991_Census_datasets/Area_Stats/Adjusted_data/Undercount_adjusted_census_data/

Further detail of the adjustments made to derive these data sets are freely available for academic research to those registered with the University of Manchester's MIMAS service for 1991 Census data sets, where text and SPSS formats are both available. A selection of the data sets is available in the accompanying CDS Resources CD.

Box 12.2 Additional CDS resources for Chapter 12.

EWCPOP	Guide to revised 1991 mid-year population estimates of residents and private household residents by five-year age group and sex for wards in the United Kingdom and EDs in England and Wales. See *gb.csv* for an explanation of the 1991 ward codes.
SOCPOP	Guide to revised 1991 mid-year population estimates of residents and private household residents by five-year age group, sex, and tenure, ethnic group or employment status, for wards in Great Britain. See *gb.csv* for an explanation of the 1991 ward codes.
MIGPOP	Guide to revised 1991 migration flow estimates by five-year age group and sex for districts in Great Britain.

13

Population Statistics after the Census

Paul Williamson and Stephen Simpson

SYNOPSIS

A census provides a snapshot in time that rapidly dates. This chapter outlines and evaluates methods for creating national, sub-national (district), and small-area (ward) population estimates for years between censuses. Attention is focused primarily upon estimates by age and sex, but consideration is given to a range of other measures including households, daytime population, and the size of sub-groups such as the unemployed, the disabled, and members of minority ethnic groups.

13.1 THE CENSUS AND POPULATION ESTIMATES

Population statistics underpin social research. Among many other, policy and research interests, mortality and illness rates, indices of poverty, measures of housing occupation and need, all require population statistics as a denominator. Others see the change in population in its own right as an important feature of our local social environment.

The census is the best source of population estimates each time it is undertaken nationally because small areas throughout the entire United Kingdom are enumerated with considerable accuracy. Unfortunately, the census becomes progressively less accurate as a population estimate with each year since its enumeration. By 2003, when the first outputs from the 2001 Census are anticipated, 1991 Census data will be 12 years out of date. Over this period, population counts at both the national and sub-national level can be anticipated to have changed significantly.

Figure 13.1 shows the change in population between the last two censuses for one small area, the electoral ward of Basing in Hampshire. Housing development in the 1980s added very significantly to the total population. The age structure by 1991 neither repeats the 1981 Census nor reflects a simple ageing since the 1981 Census. The type of migrants to Basing—tending in this case to be young professionals, some with young children—would have to be known before the age structure could be estimated correctly. The local population has changed again in the 1990s, no less significantly than in the 1980s, and there is every reason to expect this change to continue into the 21st century.

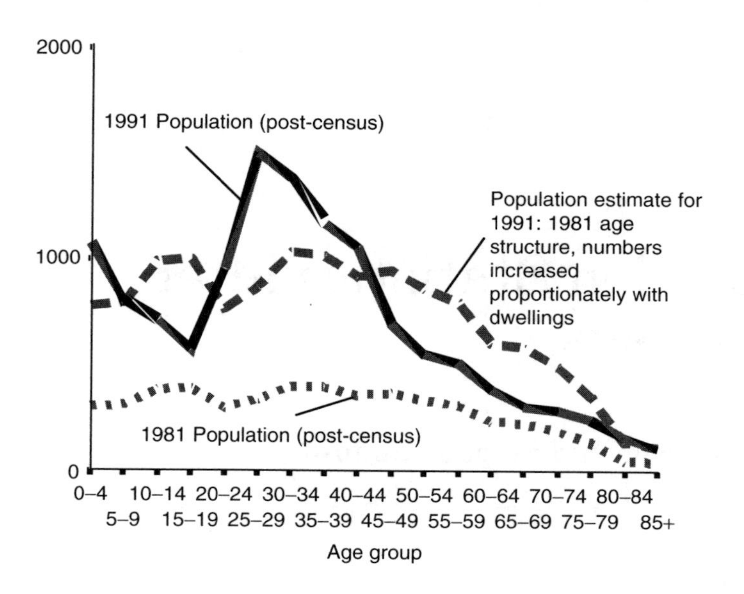

Figure 13.1 Basing ward, Hampshire, 1981 and 1991.

This chapter reviews the problem of estimating population statistics between census years. The main focus is upon estimates of residents, by age and sex, for small areas, typically wards in England and Wales and postal sectors in Scotland. Summary findings from the Estimating with Confidence Project, which evaluated the wide-range of intercensal population updating methodologies in use among local authorities, are presented in Section 13.4. Progress to date is also reviewed on a significant new initiative, the creation of quasi-official ward-level population estimates as an input to the Department of Environment, Transport and the Regions Index of Local Conditions (Section 13.5). A final section briefly considers the updating of census counts for attributes other than age and sex, including counts of households, daytime populations, and minority groups.

Importantly, all of the methodologies reviewed constrain their small-area estimates to agree with 'official' sub-national (district level) population estimates. These estimates in turn are constrained to fit 'official' national population estimates. As a preliminary, therefore, the following two sections present a brief review of the current official national and sub-national population estimation methodologies.

13.2 NATIONAL INTERCENSAL POPULATION ESTIMATES

Even at a national level, intercensal population changes can be large. Between the 1981 and 1991 Censuses, the population of the United Kingdom increased by 1 456 000 (2.6%) (ONS 1999d, Table 1.2). For constituent countries within the United Kingdom, population changes ranged from a decrease of −1.4% for Scotland through to a 4.1% increase for Northern Ireland. The latter is equivalent to an additional one person living in Northern Ireland in 1991 for every 25 persons resident in 1981.

Both the size and the diversity of these population changes emphasize the need for intercensal population estimates at a national level. For England and Wales, population estimates are prepared by the Office for National Statistics (ONS). The General Register Office for Scotland (GROS) and the Northern Ireland Statistics and Research Agency (NISRA) compile similar estimates for, respectively, Scotland and Northern Ireland. All three statistical offices adopt the same national population estimation strategy, although with minor modifications for Northern Ireland (ONS, 1999b; Beatty and Rodgers, 1999).

The basis of the ONS methodology is to roll forward the population counts from the most recent census, on a year by year basis, by ageing on the population, adding births, subtracting deaths, and adding net migration. The estimation process faces two key areas of difficulty. First, armed forces and their dependants are notoriously difficult to enumerate and estimate properly. Figures on these special populations are drawn from a number of sources including the Defence Analytical Services Agency. Second, estimating net migration flows is problematic, given the lack of a formal register of moves. In contrast, the registration of births and deaths is a statutory requirement.

Proxy migration information is drawn from a variety of sources. The main source of information for migration within the United Kingdom is the National Health Service Central Register (NHSCR). This records the previous and current addresses of all patients who register with a general practitioner outside of their previous Family Health Service Authority (FHSA) area (broadly equivalent to clusters of neighbouring local authority areas). Data from the NHSCR are used to estimate national flows between constituent countries of the United Kingdom (lagged by one month). Data from the Irish Central Statistical Office are used to estimate migration to and from the Republic of Ireland. Other international migration flows are estimated using information from the International Passenger Survey (IPS), a sample of passengers at ports of entry or departure across the United Kingdom. An allowance is made for asylum seekers, visitor switchers, and their dependants (supplied by the Home Office). Because of the lack of data, the net effect of European Union (EU) citizens switching between visitor and resident status is assumed to be zero.

The biggest problem with national intercensal population estimates is capturing the net effect of international migration flows. Critiques of the IPS abound (e.g., Champion *et al.*, 1998; Hollis, 1998), all pointing out the low sample size and the problems with accurately recording at the port of entry or departure the final intended destination of visitor switchers and asylum seekers. The NHSCR data have also come under fire (Boden *et al.*, 1992; Haynes *et al.*, 1995; Stillwell *et al.*, 1995). All moves by an individual are recorded, not just the 'transition' in address from mid-year to mid-year. There is a variable time lag between each move and re-registration with a general practitioner, and, most disturbing of all, young and fit adults, the most mobile group in society, are those least likely to register with a general practitioner after moving. Changes in the use of the NHSCR and its associated Family Health Service registers may deal with the first of these criticisms (Scott and Kilbey, 1999), but the other concerns remain.

Another problem with the national population estimates is that the age structures of the estimated migration flows are assumed to remain the same as those recorded in the 1991 Census. Apart from the fact that there is reason to suspect that the 1991 Census differentially undercounted migrants (see Chapter 12), there is also no real reason to expect that the age structure of the observed flows will remain unchanging over time. Nor are the published estimates disaggregated by population characteristics other than age and sex. This leads, for example, to an absence of vital denominators for studies of ethnic exclusion and population projections using immigrant or ethnic group-specific fertility rates. This is a point returned to briefly in Section 13.6.

On the basis of the methodology outlined earlier, the estimated 1998 mid-year population of the United Kingdom was 59 237 000, a 2.47% increase on the census-based 1991 mid-year estimate. Of this net change, 55% has been attributed to natural increase, 42% to net civilian immigration, and the remaining 3% to a combination of asylum seekers, visitor switchers, and the net movement of national and foreign armed forces (with dependants) (ONS 1999d, Table 1.7).

It is difficult to assess the accuracy of intercensal national population estimates. In theory, one opportunity would have been to compare the rolled-forward population estimate from the 1981 Census with the enumerated population total from the 1991 Census. In fact, the reverse occurred. In conjunction with a number of administrative data sources, the rolled-forward national estimate was used as a basis for identifying an estimated 2% Census undercount (OPCS, 1993a; Heady *et al.*, 1994) (see also Chapter 12). The reason for this role reversal was the close agreement of the rolled-forward estimates with other available administrative data and, hence, their perceived robustness. In any case, the estimation strategy used by the national statistical offices, under a state of constant review, has also changed slightly since that time.

13.3 SUB-NATIONAL (DISTRICT) INTERCENSAL POPULATION ESTIMATES

The UK statistical offices also publish population estimates, with detailed age–sex structure, for each local and health authority area (see respective statistical office Web sites). Virtually identical methodologies are adopted for England, Wales, and Scotland (ONS, 1999b). A broadly similar approach is used for Northern Ireland (Beatty and Rodgers, 1999).

In general terms, the statistical offices prepare population estimates for 'building bricks' that aggregate to the desired output areas (unitary, local, and health authority districts). The introduction of sub-national geography to the population estimation process increases the complexity of the estimation process, particularly in relation to migration. But first, as for national estimates, the population is aged on and births and deaths are added. At sub-national level, the ageing-on process not only excludes armed forces and their dependants, whose number and age–sex structure are estimated for national population estimates, but also boarding school pupils and prisoners. The number and age–sex distribution of these latter populations are obtained directly from, respectively, the Department for Education and Employment and the Home Office. For all three 'institutional' populations, the estimates supplied include the place of usual residence. In this way, the net effect of institutional migration flows are captured directly. This leaves the task of estimating migration flows among the remaining 'civilian' population.

International civilian migration to and from each FHSA in the United Kingdom is estimated using FHSA level indicators from the census, the IPS, and the NHSCR. Migration to and from the Republic of Ireland is distributed among FHSAs using the number of Irish-born enumerated in each FHSA in the 1991 Census. Other international migration is estimated directly from the IPS, using a weighted three-year average of flows to smooth sampling fluctuations.

The distribution of asylum seekers, visitor switchers, and immigrants uncertain of their final destination poses particular problems. Asylum seekers are first allocated 85% to Greater London, 15% elsewhere, and subsequently allocated to districts within these areas in proportion to 1991 Census counts of resident by broad country of birth. Visitor switchers are allocated directly to districts using 1991 Census country of birth data. The importance of these latter two groups has varied over time, but in 1997 to 1998 they comprised just under two-fifths of the 125 000 net in-migration to England and Wales (ONS, 1999b). Immigrants uncertain of their final destination on arrival in the United Kingdom are shared with FHSAs proportionately to those immigrants with known destinations.

The number of civilian migrants moving between FHSAs within the United Kingdom is estimated directly from NHSCR data. They are distributed to building blocks within FHSAs using recorded changes in electoral rolls, with an age–sex distribution assumed to reflect the patterns observed in the 1991 Census for each building block. International in-migrants are redistributed to building blocks within FHSAs using the 1991 Census distribution of international immigrants, while their age–sex distribution is assumed to mirror the relevant national origin–destination flow age–sex structure observed in the 1991 Census.

The use of 1991 Census data for the redistribution and disaggregation of certain elements of sub-national migration flows is clearly unsatisfactory. Migration behaviour changes over time, in terms of both volume and composition. Although there is reason to hope that NHSCR data might broadly capture the volume element of change, evidence supporting the assumption that the age–sex composition of flows remains stable over time is mixed (Bates and Bracken, 1987). There is also reason to believe that the composition of the migration flows captured in the 1991 Census are biased because of both differential undercounting (see Chapter 12) and their enumeration during an economically depressed period (Stillwell *et al.*, 1995). There would appear, therefore, to be some scope to use age–sex information recorded in the non-census data sources, such as the IPS and the NHSCR, to address this problem. In practice, however, this type of solution is restricted by the inherent problems of statistical unreliability, given the often small numbers involved.

Perhaps equally unsatisfactory is the use of information on changes in electoral roll size to model intra-FHSA population changes. Problems with electoral roll data include differential non-registration (e.g., low registration rates for young adults, recent migrants, and certain ethnic groups), double counting (students registered at term-time and home addresses and second home owners), and fluctuations due to sporadic administrative efforts to identify and remove non-existent electors (ONS, 1999a). The problems with using data from the NHSCR to estimate migration flows, already touched upon in the context of national population estimates, should also not be ignored. These problems can only grow worse as the level of spatial detail is increased.

Despite these caveats, the current sub-national estimation methodology adopted by ONS and the other UK statistical offices is widely accepted as best practice at present, given the limitations of having to produce nationally consistent estimates from nationally available data sources. The estimation strategy is also kept under constant review and revised annually. For example, improvements to the measurement of migration within the United Kingdom have been made (Scott and Kilbey, 1999) and introduced into post-1998 population estimates (Chapell *et al.*, 2000).

The accuracy of sub-national population estimates has been assessed subsequent to the 1991 Census by comparing 1981 rolled-forward estimates with 1991 Census counts revised to take account of differential under-enumeration (Armitage and Bowman, 1995). It was concluded (unsurprisingly) that the smaller the area, the greater the relative inaccuracy of the rolled-forward estimate. For standard regions, the average absolute error

Table 13.1 District-level intercensal population change.

District Top 10 gainers	Total	Natural change	Migration	District Top 10 losers	Total	Natural change	Migration
	(% change 1991–1998)				(% change 1991–1998)		
East Cambridgeshire	19.1	1.0	18.1	Isle of Anglesey	−5.8	−0.2	−5.6
Westminster	17.5	2.4	15.1	Easington	−5.7	0.6	−6.3
Kensington and Chelsea	16.2	3.1	13.1	Redcar and Cleveland UA	−5.5	1.0	−6.5
Forest Heath	16.1	3.9	12.2	City of London	−4.9	0	−4.9
Reading UA	15.6	4.1	11.5	Merthyr Tydfil	−4.8	0.5	−5.3
Milton Keynes UA	13.4	6.0	7.4	Barrow-in-Furness	−4.5	1.2	−5.7
Elmbridge	13.1	1.7	11.4	Ipswich	−4.1	1.8	−5.9
Worcester	12.8	3.7	9.1	Corby	−4.1	3.2	−7.3
North Dorset	12.5	−0.5	13.0	Liverpool	−4.0	0.4	−4.4
Bracknell Forest UA	12.0	5.2	6.8	North East Lincolnshire UA	−3.4	1.3	−4.7

Source: Authors' calculations based on Tables 6 and 10 in ONS (1999c), adjusted to remove direct effects of boundary changes, 1991 to 1998.

was only 0.4%, compared with 2.5% for local authority districts and London boroughs. This level of inaccuracy is low and probably within acceptable limits for most uses. It compares favourably, for example, with the estimated 1991 Census under-enumeration of 2.2% for England and Wales (OPCS, 1993a; 1993b).

Table 13.1 highlights the importance of producing updated district-level population counts for the years between censuses. Without updating, for example, analyses of East Cambridgeshire using 1991 Census counts would be nearly 20% out by 1998. More generally, the average absolute difference between 1991 and 1998 district population mid-year estimates is 3.8%. The error for the populations within specific age and sex groups should be expected to be considerably higher. The figures presented in Table 13.1 also serve to highlight the importance of (net) migration as the main driving force in sub-national population change.

13.4 POPULATION CHANGE BETWEEN THE NATIONAL CENSUSES FOR SMALL AREAS

For some users, updated census counts are required for areas smaller than districts, typically wards, enumeration districts, postcode sectors, or even postcode units. The *Estimating with Confidence* project established a network of local authorities and academic researchers to assess the accuracy of different methods of making population estimates for these small areas in the 1990s. This section summarizes the main conclusions of that project and suggests the way forward for researchers requiring population estimates for small areas in the years between censuses. A more detailed discussion can be found in the project's various reports (e.g., Simpson, 1998a; Simpson *et al.*, 2000).

The main estimation strategies identified, in order of increasing complexity and cost, are: do nothing; apportionment; ratio; cohort survival and local census. The 'do nothing' approach simply involves continuing to use the increasingly out-of-date information from the most recent census. Apportionment subdivides an independent estimate of district-level population change between the constituent small areas of the ward, for example, on the basis of population share at the last census, or of the share of the current electoral roll or patient registers. Ratio approaches involve increasing census counts in proportion to the size of the change in an indicator of population change (e.g., changes in numbers on the electoral roll). Cohort survival ages on a previous estimate by separately estimating changes likely to result from births, deaths, and migration, in theory, better capturing the specific population dynamics of each small area. As births and deaths are registered at ward level, the key to the success (or failure) of this approach is typically the model of migration adopted. The final option is to conduct an interim local census (e.g., as part of the electoral roll canvass).

From a survey of all local authorities, the Estimating with Confidence project managed to elicit detailed descriptions of 48 local authority estimation strategies, each representing a unique combination of one or more of the strategies outlined in the preceding text, in conjunction with the use of a wide variety of nationally and locally available data sources (Simpson *et al.*, 1997b). Figure 13.2 displays the results of an experiment using eight representative methods of population estimation for 222 wards of Hampshire county. Each method involved rolling forward 1981 small area census counts for 10 years and comparing the results with 1991 Census small-area counts, revised to take the effects of under-enumeration into account. No method estimated every five-year age group in any ward-sized area to within 10% accuracy 10 years after the last census (see also Lunn *et al.*, 1998).

All methods of providing an updated 1991 population estimate gained, on average, a better accuracy than relying on the 1981 Census with no update. There was not a great deal to choose among the six 'desktop' methods (shown with thin lines in Figure 13.2), which in various ways used statistics from administrative registers, the previous census, and government estimates of district population. The desktop methods that did worse than no update at all, at young adult ages, were those that aged on the previous census age structure. This was a particularly erroneous assumption for areas with student and armed forces establishments that regenerate a stable age structure.

However, Figure 13.2 clearly shows that a local census is more successful than other methods for estimating five-year age groups, by 5 to 10%, depending on the age group.

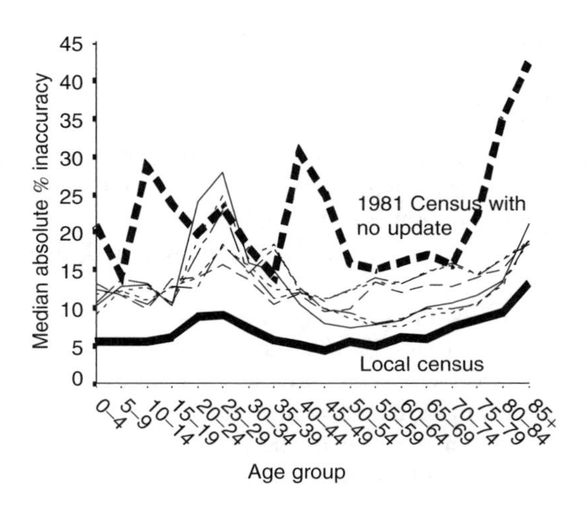

Figure 13.2 Accuracy of eight methods for estimating the 1991 population 222 wards of Hampshire.

In this case, the local census consists of a card from Hampshire Council inserted, every other year since the late 1970s, in the electoral canvass carried out by Hampshire's District Councils.

The two over-riding results from the Estimating with Confidence project on accuracy of updated population estimate were:

- An assumption that the population of small areas has not changed since the last census is rarely a reliable strategy for the following decade.
- A local census is the only clear winning strategy for accuracy and flexible detail of output, so long as it achieves at least a 70% response rate.

As in Hampshire, a local census in practice usually piggybacks electoral registration as a framework for asking demographic questions of all residents. Several local authorities in England and Scotland undertake such a census annually or each two years or in the middle year between two national censuses. The local census should be recommended and encouraged. It is capable of resolving disputes on local population.

However, unless and until a national census is taken more frequently, a next-best strategy is needed if local population estimates are to be achieved for purposes that do not command the resources needed by a local census. The evidence from the Estimating with Confidence project on a variety of desktop methods of estimating local populations suggests:

- Start simple.
- Use local knowledge.
- Suit methodology to local demographic circumstances. A mix of methods is acceptable.
- Avoid costly complexity unless the reliability of results is clearly improved.
- Reduce the largest errors. Where different methods cannot be distinguished in their accuracy, an average of methods' results will be more accurate on average and give fewer large errors than the results from a single method.

For researchers who do not have access to local data and the skills to develop population estimates themselves, there are four progressively less satisfactory options:

1. Contact the local authority for the area for which population estimates are required. A research or planning department is likely to be able to give advice and statistics freely.
2. Check the most recent government output. Development of small-area population estimates became a resourced priority for ONS in 2000. The DETR also produced estimates for all wards in England in 1999 (see Section 13.5).
3. Undertake population estimates for the local areas required, using the experience referenced in this section.
4. Contact Experian Ltd of Nottingham or CACI of London, who provide population estimates and projections for small areas that update the most recent census. They use relatively crude methods as they do not have access to relevant unpublished information for areas throughout Britain. A charge will be made.
5. Use population estimates for the year of the previous census as if the population had not changed, with awareness of the approximation involved.

One new data source appears set to help production of small area population estimates for the future decade: registers of patients held by each health authority (Health Board in Scotland and Northern Ireland). These registers contain information on age, sex, and postcode of residence for each patient, and are updated as patients move from one doctor to another, are born or die, and to some extent, as patients move address without changing doctors. Some practitioners have already experimented with the use of patient registers in population estimation (Cohen, 1998; Simpson, 1998b, includes a 'South Asian' category). A major research exercise by ONS has collated the patient registers nationally to assess their utility in estimating migration for local areas (Kilbey and Scott, 1997; Scott and Kilbey, 1999 give interim reports). Without such data, Boyle *et al.* (1998) demonstrate just how hard it is to estimate ward-level migration flows, even if relatively sophisticated modelling solutions are adopted.

13.5 'OFFICIAL' WARD-LEVEL POPULATION ESTIMATES?

Drawing in part upon the lessons of the Estimating with Confidence project, Penhale and co-workers from the Department of Applied Social Statistics and Social Research Unit at the University of Oxford consulted widely on the construction of a national set of intercensal ward-level population estimates (Penhale *et al.*, 1999a,b; Index 99 Team *et al.*, 1999). These estimates were used as denominators in the construction of the DETR Index of Local Deprivation 1999, and subsequently released as part of the ONS neighbourhood statistics. At the time of writing, they appear to constitute the nearest thing yet to a set of official ward-level population estimates, given the wide level of public consultation (and approval) involved in their construction, their general availability, and their use as the basis for an official government measure of ward-level deprivation. For a review of alternative 'unofficial' estimates, see Simpson *et al.* (2000). It is also worth noting that, subsequent to the 2001 Census, ONS aims to release its own annual series of 'official' small-area population estimates. The most significant drawback to the Index 99 ward estimates is that they are disaggregated into three broad age bands only (<16,16–59, 60+), with no account taken of sex, although the 16 to 59-year olds are subdivided into the economically active and inactive.

As with the various methods used by local authorities reviewed earlier, the Index 99 estimates are constrained to ONS district-level population estimates. Within districts, the estimated district population aged less than 16 is apportioned to wards on the basis of postcoded child benefit data, with an adjustment to account for boarding schools. The population aged 16+ is estimated by sharing the district population aged 16+ among wards on the basis of changes in the electoral roll since the 1991 Census (using Estimating with Confidence results as the basis for 1991 ward populations figures). The population aged 60+ is estimated by apportioning the district total 60+ to wards on the basis of the number of persons of pensionable age receiving state pensions and retired partners of such pensioners, or who were of pensionable age and receiving Incapacity Benefit or Severe Disablement Allowance. The working age population in each ward is calculated simply as the population aged 16+ less the population aged 60+.

The preliminary ward population estimates arising from this process were disseminated to the relevant local authorities for evaluation. Adjustments have been made to ward counts in which local knowledge has been able to demonstrate obvious under- or overestimation (e.g., owing to major housing developments or significant changes in 'special' populations such as students or armed forces). All changes made are balanced by rescaling the remaining within-district ward-level totals to the official district total.

The greatest strength of this approach is that it has been specifically designed to be nationally applicable, depending on nationally available information. In particular, the approach capitalizes on the recent release of ward-level benefits data. The main methodological problems that remain are a reliance upon electoral roll data to identify population changes within districts. For example, between elections, the number of registered voters declines. To an extent, this problem is addressed by using electoral roll data only to apportion district totals, rather than directly estimating ward-level totals. However, this solution depends on the assumption that all areas experience equal shifts in electoral roll registration rates. In fact, the decline may be differential, potentially exaggerating the gains in population for some wards at the expense of others. An alternative, or supplementary, source of information covering the whole population for future consideration is the patient register coded to ward level, as mentioned earlier.

The value of producing sub-district estimates is illustrated by the consideration of Liverpool. Between 1991 and 1998, the population of Liverpool is estimated to have declined by 3.8%. Using some preliminary results from the Index 99 project, Figure 13.3

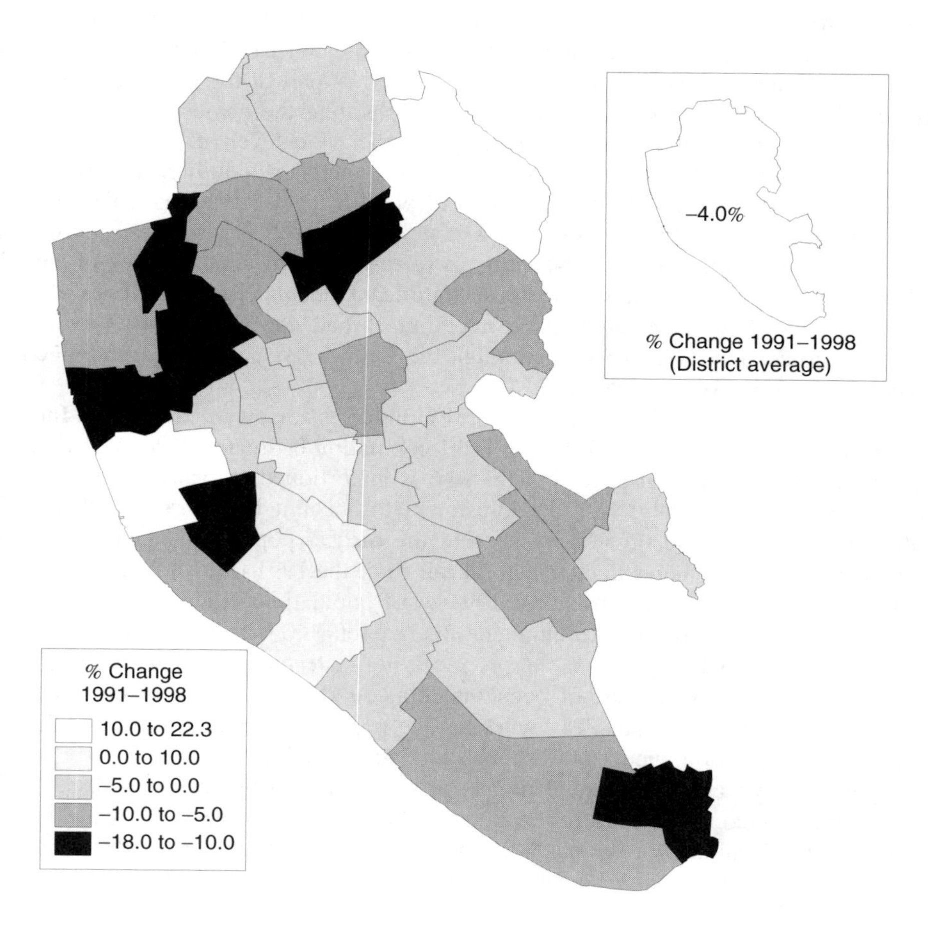

Figure 13.3 Population change, 1991 to 1998.

shows that this decline is not evenly spread across the city. In fact, five wards are even estimated to have increased in population since 1991, two by more than 20%. This underlines the fact that even within one of the fastest shrinking populations in the country, there are likely to be pockets of both growth and decline. Unfortunately, the reliability of these estimates cannot be directly tested, but average absolute error is likely to range within the ±5% range identified by the Estimating with Confidence Project for most 'deskbound' ward-level estimation strategies (Simpson *et al*., 1996).

13.6 UPDATING OTHER CENSUS COUNTS BETWEEN CENSUSES

The main need between censuses is for updated counts of residents by age and sex. However, some research has a real need for estimates of the number of *households* in an area, or the daytime population including tourists or commuters, or for estimates of small sub-populations (e.g., ethnic minorities). Brief consideration will be given to each of these types of need in turn.

13.6.1 Households

The most common way of updating household estimates between censuses is to adopt what is known as the headship rate method. At its simplest, this is effectively equivalent to calculating the last observed ratio of households to persons and assuming that it applies directly to any estimated increase or decrease in population since the last census date. Variations include calculating headship rates for population sub-groups (e.g., males aged 20–24; ethnic groups) and estimating trends in headship rate changes over time (e.g., an increasing number of households per person in response to increasing divorce rates). Every three to five years, DETR publishes district-level household 'projections'. Significantly, these projections start with a base-year intercensal estimate of the number of households in each district. DETR (1999) gives full details of the household projection methodology used to derive the 1996-based estimates and projections, at the time of writing the most recently published round of projections at a sub-national level for England. The Scottish Executive and Welsh Assembly provide similar household estimates and projections for districts in Scotland and Wales. In contrast, official household estimates for Northern Ireland are based on housing stock estimates as household projections are not currently produced, a situation that is under review.

13.6.2 Daytime Populations

The 1991 Census recorded the location of the population over the night of April 21 to 22, 1991. Within hours, this count was outdated by the mass movement of population from home to work. In contrast to the official census count of the 'population present' overnight, the population usually present in an area during working hours is often referred to as the daytime population. The size of the daytime population can be important for a number of purposes, including planning for the provision of sufficient emergency service cover. The use of daytime population figures can also sometimes be a more appropriate denominator for use in calculating rates, such as the incidence of crimes against the person and exposure to environmental pollution.

A crude initial estimate of the daytime population for an area can be made simply by calculating the net flow of commuters to or from an area, and adding this amount to the usual resident population. This is possible for areas as small as wards using the Special Workplace Statistics gathered in the 1991 Census. By this method, the April 1991 'daytime' population of Liverpool can be estimated to have been nearly 12% greater than its resident night-time population, equivalent to a net daytime influx of more than 50 000 persons. At ward level, the scale of change is even greater (see Figure 13.4). Within Liverpool itself, more than half of all wards experienced a net daytime population loss of 10% or more, whereas three wards experienced a net gain of more than 50%. For many districts and wards, the difference between the usually resident and daytime populations will be significantly greater than the change in population that occurs between censuses.

The Special Workplace Statistics also record the composition of work flows by age, sex, occupation, and hours worked (see Chapter 19). This allows the estimation of daytime

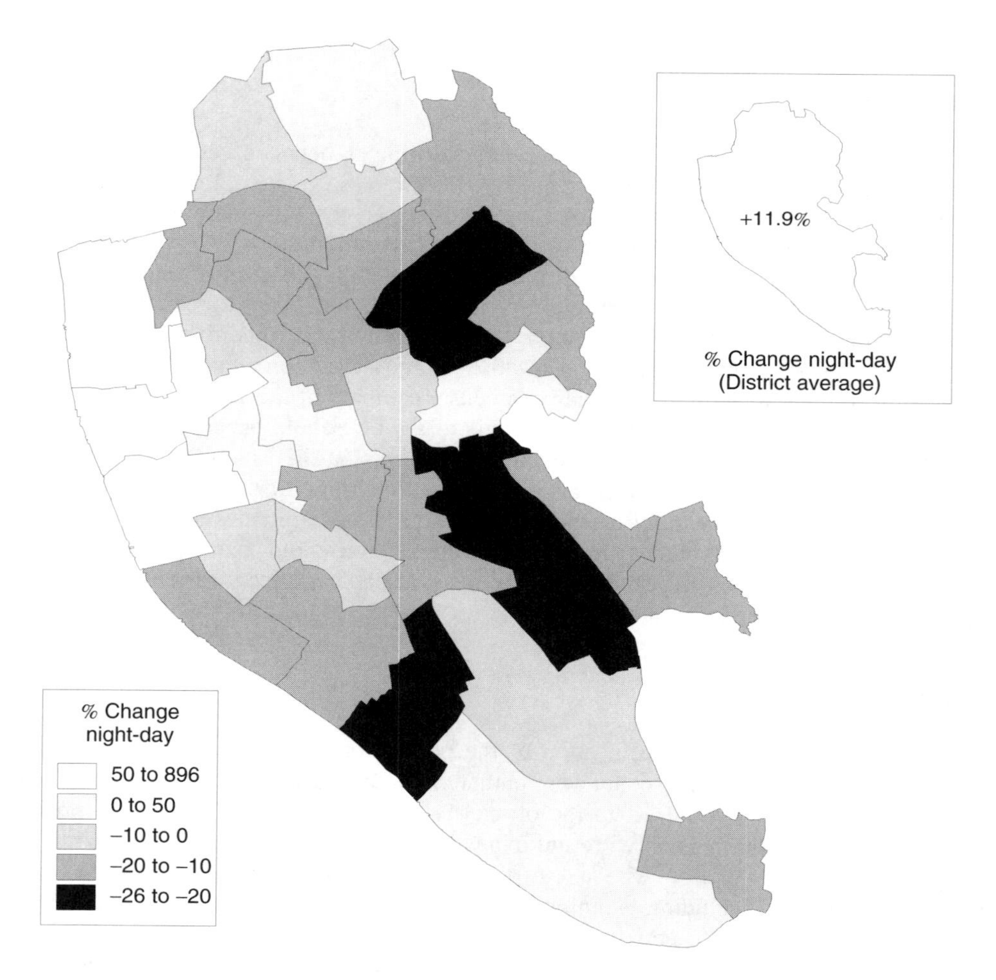

Figure 13.4 Population change, night-day (1991).

populations disaggregated by similar characteristics. To update the estimated daytime population to a subsequent intercensal year, some form of modelling is required. For example, estimated intercensal changes in the economically active population for each ward in a labour market area could be used to reapportion observed 1991 workplace flows. A more sophisticated approach is outlined by Akkerman (1995).

What the census fails to capture are three additional and important population flows. First, no record is made of flows of pupils from place of residence to place of study. Second, no attempt is made to capture the far more volatile flow of shoppers, which can be expected to contribute particularly to a significant increase in weekend daytime populations in some areas. Third, no account is taken of the seasonal variations in population that arise from tourism. Faced with a lack of 'official' information on any of these three types of population flow, the only way to include any of them in a daytime population estimate would appear to be some form of specially targeted local census or survey.

13.6.3 Estimates of Population Sub-Groups

Some groups of residents give special difficulty in many estimation procedures, in particular students, prisoners, armed forces, and generally residents in institutions. These are often either missing or unusually complete in registers, as their inclusion depends on the management of the institutions involved as much as on the individuals concerned. The quality checks on the 1991 and previous censuses did not include institutions because the difficulty of accurate re-enumeration was considered too great to justify the information that might be gained. These special groups are not only hard to measure, but they are not always clearly required within a population estimate. Their measurement is not specifically discussed here but is treated in some detail by the Local Authorities Research and Intelligence Association (LARIA) (Simpson, 1998a).

Alternatively, interest may be focused upon specific sub-groups of the population such as members of the economically active, ethnic minorities, the disabled, and the disadvantaged. The methods of small area estimation are many, summarized for a statistical audience by Ghosh and Rao (1994). In practice, in Britain, they include:

Direct measures—

- national surveys such as the Labour Force Survey (LFS) provide estimates for relatively large areas; see Labour Market Trends for summaries of the LFS, and Schuman (1999) for its use to update estimates of ethnic minority populations.
- administrative data provide counts of those in receipt of benefit (Penhale *et al.*, 1999a), and distinctly named populations (Simpson, 1998b).

Cohort survival methods—

- the distinctive age structures of each ethnic group make cohort survival a suitable method of estimation for years since the most recent census (Storkey, 1998; Simpson, 1996).

Prevalence ratio methods—

- the application of the prevalence of a condition is often applied to a demographic estimate to disaggregate it to sub-groups. For example, disability prevalence at each age, or participation rates in voluntary organisations, each estimated from a national survey.

- the method of household projection described earlier uses headship rates estimated from previous censuses and from the LFS subsequent to the last census as a prevalence ratio.

Often the sub-groups concerned comprise only a small proportion of the national population, and measures of their change over time exist only for areas larger than those desired. For example, the LFS captures changes in economic activity over time, but reliably disaggregated only to county level. Apportionment or ratios can be used to apply these higher level (changes in) rates to the populations of smaller areas to produce (admittedly crude) sub-national estimates. An example of this practice is the estimation of the number of economically active within wards by the Index 99 team, applying ward-level 1991 Census-based economic activity rates to the current estimated population of working age, and subsequently scaling these estimates to county level estimates of economic activity derived from the LFS.

Although undeniably crude, such a methodology will at least capture broad economic effects. If possible, disaggregating and applying rates for a greater number of population sub-groups (e.g., age by sex) could add some further precision and credibility to these estimates. But combining data from a variety of sources and spatial scales moves us into the realm of synthetic population estimation, a topic discussed at greater length in Chapter 17.

Box 13.1 Key web links for Chapter 13.

Population estimates	
http://www.statistics.gov.uk/statbase/mainmenu.asp	ONS's StatBase contains free downloadable national population estimates and information about the availability of sub-national estimates for England and Wales. The most recent methodological revisions may be requested from *popinfo@ons.gov.uk*
http://www.nisra.gov.uk/	The Web site of the Northern Ireland Statistics and Research Agency includes free downloadable national and sub-national population estimates for Northern Ireland
http://www.open.gov.uk/gros/	The Web site of the General Register Office for Scotland includes free downloadable national and sub-national population estimates for Scotland
http://www.statistics.gov.uk/neighbourhood/ and *http://index99.apsoc.ox.ac.uk/*	Total population for wards in England, as estimated by the Department of Applied Sociology, Oxford University for the DETR (now DTLR) and distributed by the ONS Neighbourhood Statistics Service
Household estimates	
http://www.housing.detr.gov.uk/research/project/	DETR 1996-based household projections: free downloadable national and sub-national household estimates and projections for England, and documentation of methodology

http://www.scotland.gov.uk/	Scottish Office 1996-based household projections. On-line documentation and national/sub-national estimates and projections. Similar 1998-based estimates and projections due for release mid-2000, produced by the Scottish Executive (post-devolution equivalent to Scottish Office). Search for 'household projections'

Part IV

Microdata

14

Microdata from the Census: Samples of Anonymised Records

Angela Dale and Andy Teague

SYNOPSIS

This chapter presents the innovative and widely used microdata resource generated for the first time in the United Kingdom from the 1991 Census. A thumbnail sketch of the household and individual Samples of Anonymised Records (SARs) is provided together with pointers to harmonized UK SARs built from common variables in the Great Britain and Northern Ireland SARs. The fourth section of the chapter shows how new variables have been added to the SAR, the most important of which are two measures of income. A final section of the chapter describes the detailed work carried out over the past few years to assess the risks of and to justify the procurement of enhanced microdata samples from the 2001 Census.

14.1 INTRODUCTION

The SARs were a breakthrough in terms of products from the 1991 Census. For the first time in the history of the UK Census, samples of microdata were released to the research community. The decision to release these anonymized samples was made by the Office of Population Censuses and Surveys (now the Office for National Statistics) following extensive research by an ESRC Working Group, headed by the late Professor Cathie Marsh. The group demonstrated not only the research potential of these files but also the negligible risk to confidentiality that they posed (Marsh *et al.* 1991).

Two SARs have been extracted from the 1991 Censuses for England and Wales, Scotland, and Northern Ireland:

- A 2% sample of individuals in households and communal establishments; this comprises 1.1 million records containing information on all the topics asked in the census and limited information about other members of the household. The geographical areas identified in this file have a minimum population size of 120 000 with sub-threshold areas grouped with neighbouring areas.
- A 1% hierarchical sample of households and individuals in those households. This comprises 215 000 households and the 542 000 individuals enumerated in them. The

full range of census variables is available, with standard region (England) as the lowest level of geography.

The data for England, Wales, and Scotland are combined into Great Britain files, whereas data for Northern Ireland are held in separate files, reflecting the slightly different schedule that was used and the different processing procedures.

The nature of microdata means that there is very considerable scope for the addition of derived variables, either through combinations of existing variables or through linking additional external information. These derived variables have provided a greatly enriched data set that increases the research potential of the data in a number of ways, described in the following text.

14.2 THE OPPORTUNITY OFFERED BY THE TWO SARs

The 1% Household SAR has a hierarchical structure with details of all members of the household linked, so that one can identify relationships both within the household and also within the family (Figure 14.1). Using this information, a large number of variables that describe the structure of the household and family have been derived—discussed in more detail later. The restricted geography of the Household SAR means that it contains more detailed information than the Individual file, particularly with respect to occupation, industry, and qualifications. For example, there are 358 occupational categories at the level of unit SOC (SOC has 371 categories but a few have been collapsed for confidentiality reasons).

The 2% Individual SAR is a flat or rectangular file containing one record per person. It is not, therefore, possible to link information between household members and

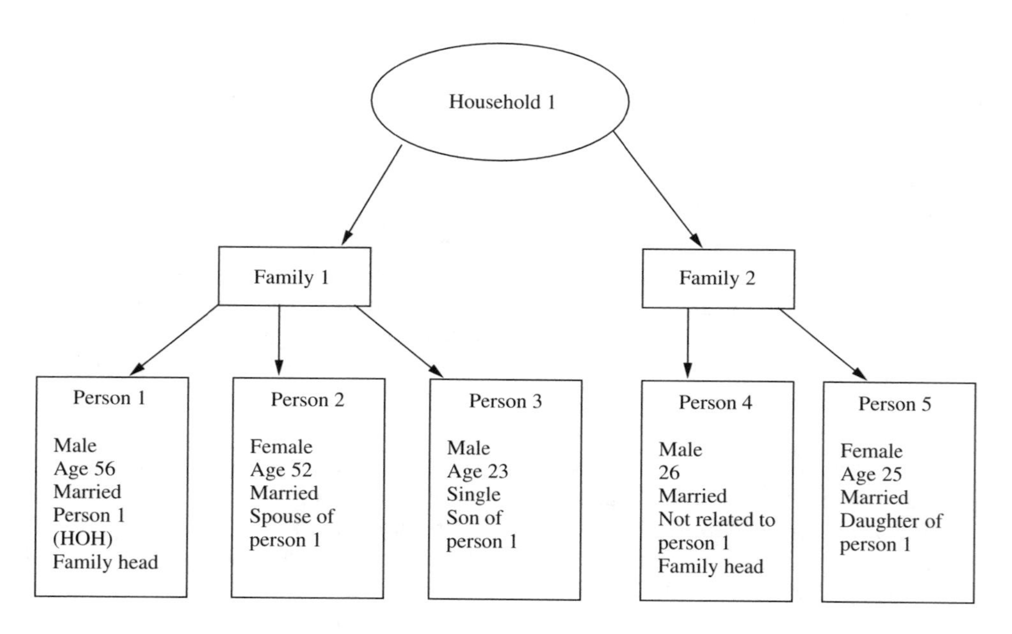

Figure 14.1 The hierarchical structure of the Household SAR.

information about other members of the household is very limited. Therefore, there are fewer opportunities to add derived variables. However, the great strength of this file is the degree of geographical detail that it contains. The geography of the Individual SAR is based on 1991 local government districts, with areas grouped together where their population falls below the threshold of 120 000. This results in a total of 278 SAR areas in Great Britain.

In the following sections, we outline some of the additional value that has been added to the 1991 SARs in the years since their release.

14.3 A UK SAR

The availability of data for Northern Ireland provided the possibility of making comparisons across the countries of the United Kingdom and therefore a UK SAR was derived, which maximized this comparability. This file loses some of the detail available on the Great Britain SARs, for example, in terms of family composition. It also loses the greater detail available in the Northern Ireland files, for example, on number of children ever born. However, as the UK file is *additional* to the separate GB and NI files, this harmonization does not lose any data for the countries individually. Further details are given on Web site *http://les.man.ac.uk/ccsr/cmu/*.

14.4 ENHANCING THE RESEARCH VALUE OF THE SARs: ADDING MORE VARIABLES

14.4.1 Measures of Socio-Economic Status

The detailed information in both the SAR files on occupation and employment status has provided a basis for deriving a number of variables related to socio-economic status. Two key variables, both providing a continuous measure of socio-economic status, are the Cambridge Occupational Score (CAMSCORE) and mean hourly earnings derived from the New Earnings Survey (NESSCORE). Both are discussed in the text that follows.

The CAMSCORE is a continuous measure based on occupation and employment status that provides an alternative to social class (Prandy, 1990; 1992). It is designed to measure social advantage and is based on the assumption that there is social interaction between those with similar lifestyles. The scale was derived using multi-dimensional scaling methods based on data on friendships and marriages between members of different occupational groups. It therefore provides a finely graded hierarchy, rather than a structure of discrete and homogenous classes. A manual explaining the derivation of the scale in further detail and providing the score value for each SOC unit is available from the University of Cambridge (Prandy, 1992).

Income estimates have been derived from the New Earnings Survey (NESSCORE) to overcome one of the shortcomings of the 1991 Census, that is, the lack of any income information. The situation for 1991 has been partially remedied by using mean earnings information from the New Earnings Survey (NES) to attach to individuals in the SARs. Despite strong lobbying for inclusion in the 2001 Census, an income question was not included and similar methods will need to be used to add imputed income to the 2001 SARs. Mean hourly earnings (excluding overtime) have been derived from the New Earnings Survey for 1991. The NES is a large survey (sample size about 160 000) conducted annually by government and collects information from employers about the earnings of their employees. The data are collected under statute and only aggregate (tabular) data are released. Any cell that contains fewer than three individuals is suppressed. In order to attach aggregate earning information to individuals in the SARs, hourly earnings (excluding overtime) in 1991 were obtained by the following characteristics:

London and the south-east; rest of GB
Male; female
Full-time; part-time
16 to 19; 20 to 29; 30 to 49; 50+
Occupation: level of minor SOC—73 categories

The mean hourly earnings can be seen as representing the earnings potential of someone in the relevant occupation and in the particular age, sex, employment status, and regional group. There are many other factors, including amount of overtime worked, that affect actual individual earnings. The NES information has also been added to people not currently in work, using information on occupation and employment status related to their last occupation (employee/self-employed status from the SEG of the last occupation), but with age and geographical location related to the 1991 situation.

A comparable variable (LFSSCORE), based on earnings from the Labour Force Survey, has been derived by Drinkwater and Leslie (1996). The variable is computed by extracting the coefficients estimated from two wage equations, one for males and the other for females, using Labour Force Survey (LFS) data from the fourth quarter of 1992 up to the second quarter of 1995, to the individual data contained in the SARs. The NES scores provide an indication of average earnings (within age–sex and full/part-time categories) for a given occupation (at minor SOC level). All within-occupation variation is absent, as is any individual variation by family status, level of qualification, or ethnic group. The LFS scores depend less heavily on occupational information but incorporate much more detail concerning family, ethnic group, and qualifications. The LFS earnings scores are available for both the 1 and 2% SARs.

14.4.2 Additional Variables Describing Socio-Economic Status

The Household SAR has greater individual detail than the Individual SAR. This means that variables such as occupation are coded to the full 371 unit SOCs (with a few exceptions). Because of this, a fuller range of special classifications have been added, including several designed to provide international comparability—for example, the International Standard Classification of Occupations.

International Standard Classification of Occupations (ISCO-88)

Developed by the International Labour Office, this has been supplied by ONS and matched onto the SARs.

Standard International Occupational Scale (SIOPS)

This scale was originally constructed by Treiman in 1977 by averaging results of prestige evaluations carried out in approximately 60 countries. Occupational titles from national and local prestige studies were matched to the three-digit version of ISCO-68, now updated to ISCO-88 (Ganzeboom and Treiman, 1996).

International Socio-Economic Index of Occupational Status (ISEI)

ISEI measures the attributes of occupation that convert a person's education into income. In constructing the scale, occupation was scaled to capture the indirect influence of education on income as much as possible (Ganzeboom *et al.* 1992).

The Goldthorpe Class Scheme (GCLASS)

The Goldthorpe classification was originally developed as a seven-category system, which was specifically British. More recently, it has been developed by Erikson and Goldthorpe (1993) into an international scheme and the number of categories extended. Goldthorpe class categories distinguish positions in the labour market and in addition, reflect skill levels and sectoral differences. They have been found to be a powerful tool for analysis of inter-generational occupational mobility.

1991 Occupational Segregation Variables

Derived by Catherine Hakim of the London School of Economics, these are three variables that measure occupational segregation and have been created from 10% census statistics on occupational structure of the 1991 Census (Hakim, 1996). Each variable distinguishes male occupations, mixed occupations, and female occupations. The three variables differ only in the sex ratios used to define the mid-point in the classification and in the width of the middle Integrated occupations band. More details are given in SARs Newsletter No. 8, accessible from the CCSR Web site (*http://www.ccsr.ac.uk/cmu/*).

14.4.3 Adding an Area-Level Classification

The census is unique in providing coverage of the entire population and summary statistics for very small areas—in 1991, Enumeration Districts and electoral wards were the smallest areas with an average size of about 180 and 2000 households, respectively. However, confidentiality prevents this amount of detail being released on microdata files. As a way of partially overcoming this problem, an area-level classification has been attached to individual records in each of the SARs. A different classification at a different level has been added to each SAR. The ONS ward-level classification has been added to the 1% Household SAR and an ED-level classification, GB Profiles has been added to each record in the 2% Individual SAR. For both files, the classifications had to be added by ONS as the matching required access to the detailed geographical location of each individual and households.

The ONS Classification (ONSCLASS)

The ONS classification provides continuity with district level classifications produced for earlier censuses and represents the 'official' classification, which is widely used by local authorities and health authorities. It is restricted to a ward scale of geography rather than

ED or postcode, which is used by most commercial and academic research applications. A value has been attached to every individual in the Household SAR. This provides a descriptor of the ward in which the household is located. Individuals in the same household share the same ONSCLASS value. The ONS ward classification assigns wards in England and Wales and postcode sectors in Scotland to one of 14 Groups and 43 Clusters according to their characteristics based on 1991 Census data. However, because of ONS confidentiality requirements, some amendments have been made to the standard classification to reduce possible disclosure risks. The derivation of the classification is described in Wallace, Charlton and Denham (1995) and some information is provided in Wallace and Denham (1996). Full details are given in SARs Newsletter No. 8, accessible from the Web site: *http://www.ccsr.ac.uk/cmu/*.

The GB Profiles ED-level Classification

The classification added to the 2% Individual SAR was GB Profiles, developed by Openshaw, Blake and Wymer (1994). It is freely available for teaching and academic research purposes in the United Kingdom. For addition to the SARs, ONS confidentiality criteria required that each SAR area (278 areas in total) should contain either zero or more than 10 EDs in any one category on the classification. In order to meet this confidentiality requirement, Professor Openshaw produced a specially optimized version of GB profiles with a total of 49 clusters, on the basis of 80 census variables. Full details of the derivation of the classification and descriptions of the make-up of each of the 49 clusters is given at: *http://les.man.ac.uk/ccsr/cmu/data/gbprof.htm*. The GB Profiles classifications are further discussed in Chapter 11.

Research Value of the Area-level Classifications

The addition of an area-level classification to the SARs is of considerable academic value. In particular, it greatly enhances the scope for modelling multilevel area effects. In analyzing unemployment, the addition of an area-level classification provides a much better indicator of the locality than can be obtained from knowledge of either the SAR area (large LA) in the 2% sample or the region in the 1% sample. Similarly, in the analysis of long-term limiting illness, it is important to be able to include in a multilevel model, characteristics of an area at a fine level of geography. An area-level classification provides the means for doing this while protecting information about the actual ED or ward of residence. It also enhances the value of the SARs to the commercial sector and to health authorities, both of whom make considerable use of such classifications.

14.4.4 Population Weights in the Individual SAR

The SARs were drawn from the 10% sample of the 1991 Census. This omitted imputed wholly absent households as well as those individuals and households missed altogether by the Census. The SARs, therefore, are drawn from an enumeration of approximately 96% of the population. Often, however, the user will wish to make an inference from the SARs to the full population on census night in 1991. In order to compensate for this under-enumeration, which varies by age, sex, and geography, weighting factors have been added to the SARs, which allow adjustment to the mid-1991 population estimates.

These are specific to age, sex, and SAR area and are in the form of an additional variable (POPWGHT). It is important to note that population estimates are based on residents, and therefore visitors should be excluded from analysis when applying population weights.

14.4.5 Maximizing the Value of the Hierarchical Structure of the 1% Household SAR

One of the strengths of the Household SAR is its hierarchical structure (Figure 14.1). This allows a very large number of variables to be derived describing the characteristics of the household (e.g., number of adults in the household, number of pensioners, age of youngest dependent child) and the characteristics of the household head. Over 78 variables have been derived on the basis of characteristics of the household or the household reference person. Most of these variables were derived in accordance with the variables in the SAS and LBS, thereby providing comparability for researchers moving from one data source to the other. The SARs code book and glossary for derived variables describe these fully (*http://www.ccsr.ac.uk/cmu/data/hdvar.html*).

The hierarchical structure of the household SAR has also allowed two new levels of analysis to be derived—the family unit and the Minimal Household Unit (MHU). This information has been used in the SARs to derive a set of variables that parallel those for the household. Although only about 15% of households contain more than one family unit, the ability to conduct analysis at the family level opens up important research opportunities not previously available with the census. This is particularly important for those sub-groups of the population—for example, lone parents and some minority ethnic groups—that are most likely to live in shared households.

The MHUs are a reduced form of family unit and were first derived by Overton and Ermisch for the then Department of Environment (Overton and Ermisch, 1984). The most obvious difference between MHUs and family units is that unpartnered adult members of a family are allocated to a separate MHU. Variables have been derived that describe, for each individual, the type of MHU to which they belong, their position in the MHU, and the combination of MHUs in the household. These are very valuable in any detailed analysis of household composition.

The MHUs may also be used to identify the position of individuals within household units and are therefore more versatile than other household and family classifications available for the SARs. The algorithms used to create the variables are based on the SAR family type variable (FAMTYPE), which was derived by ONS from the question on relationship to household head.

There are three MHU variables: MHUTYP, Type of minimal household unit, MHUPOS, Position within minimal household unit, and MHUCOM, Combination of minimal household units.

The classification of MHUs adopted for the SARs distinguishes between cohabiting and married couples:

MHU type 1: Unmarried adult (including non-dependent children and lone parents with non-dependent children)
MHU type 2: One-parent family with dependent children
MHU type 3: Married couple with no dependent children
MHU type 4: Cohabiting couple with no dependent children

MHU type 5: Married couple with dependent children
MHU type 6: Cohabiting couple with dependent children

Minimal Household Units are important because they enable the identification of individuals within families. This has allowed comparison of the position of women within minimal household units by age in four ethnic groups: White; Black (including Black Caribbean, Black African, and Black Other); Indian; and Pakistani and Bangladeshi combined (Holdsworth, 1995). For all four ethnic groups, the majority of women in their late teens live with their parents as non-dependent children and are therefore classified as in MHU type 1—'unmarried adult'. However, women in their twenties and thirties have very different living arrangements across the four ethnic groups. In particular, the Black group has the highest proportion of women heading one-parent families, whereas the lowest proportion occurs for Indian women. The larger proportion of Black women heading one-parent families is offset by a much smaller proportion of Black women in their twenties and thirties living with partners and children. In contrast, the majority of Asian women in this age group live with dependent children and partners. At older ages, there is a higher proportion of White women living with a partner and no dependent children than in the other three groups. This may be explained by a greater proportion of Asian women with dependent children at older ages, a higher rate of widowhood among Asian women, and a reduced partnership rate among Black women.

The MHU variables may therefore be used both to examine patterns of living arrangements and characteristics of specific family members. They may also be used to exclude certain family members from analyses, for example, children, or to restrict analyses to a particular group, such as lone mothers.

The 1991 SARs have been extensively used across a wide range of subjects and disciplines. Areas of particular interest are: ethnicity (listed by more than half of users), demography, household or family, and labour markets. A list of publications based on the 1991 SARs is regularly updated and can be accessed from *http://les.man.ac.uk/ccsr/cmu/jointpub.html* or *http://les.man.ac.uk/ccsr/publications/sarpub.htm*.

14.5 SARs FOR 2001

There is a clear demand for SARs from the 2001 Census. However, a wide consultation of users and non-users conducted by the CMU (Brown and Dale, 1997) identified some areas where changes and amendments to 2001 SARs could be beneficial to users and some reasons for the lack of use of the 1991 SARs.

Some users found that the coarseness of the SARs' geography impeded their research. For a number of respondents, the SAR areas were too big or did not provide the required boundaries. Health Authorities, in particular, needed a geography that reflected their boundaries. Local Authorities expressed the need to be able to work at the level of their authority. Thus, for small authorities that were grouped with neighbouring authorities, analysis was problematic. Some respondents found the sample size a problem when working with sub-groups of the population; others alluded to the crudeness of some coding schemes (e.g., the bands on distance to work). Local Authorities (LAs) generally wanted a larger sample size to improve the precision of their within-authority estimates, and there was also a requirement to conduct analysis at a sub-LA level. These and other comments led to the following list of user requirements:

- a reduction of the threshold of the Individual SAR;
- an increase in the sample size of the Individual SAR;
- an increase in detail on some variables in the Individual SAR;
- an additional SAR with more geographical detail but less individual information; and
- larger sample size and finer geography for the Household SAR

Having established user preferences, the next step was to establish the risk to confidentiality if these changes were implemented.

14.5.1 Confidentiality Assessment of Increasing the Size and Content of 2001 SARs

We began by working through the assessment made by Marsh *et al.* (1991) for the 1991 SARs and concluded that the overall risk of identification of any one individual in the SARs is unlikely to be greater than the figure suggested by Marsh *et al.* (1991) and may well be less. In the rest of the work, we therefore used a 2% sample, at a minimum geographical level of 120 K, as a base line against which to measure change in risk of identification. This work is reported in Dale and Elliot (1998, 2001).

A Reduction in Geography and an Increase in Sample Size for the Individual SAR

We estimated that only 39% of local government areas in 2001 would meet the existing population threshold of 120 K, 67% would meet a 90 K threshold, and 90% would meet a 60 K threshold. We therefore assessed the disclosure risk of a reduction in the population threshold. We found that, on the measures used to assess risk, moving from a 2% sample with a 120 K threshold to a 3% sample and either a 90 K or 60 K threshold would lead to only a very modest increase in risk.

Adding Extra Variables to the 2% SAR for Head of Household or Other Household Characteristics

This extra information was requested by a number of users. However, there was a significant increase in disclosure risk if one assumed that the extra variables formed part of an identification key and therefore the benefit did not outweigh the risk.

Enlarging the 1% Household File

Requests were made for a larger sample size and a finer geography for the Household SAR. The structuring of the household file increases the disclosure risk at smaller geographical areas. However, the risk of a larger 3% sample without geographical definition was similar to that for the existing 1% sample with regional geography. Consultation with users has suggested that this trade-off would not be widely welcomed.

14.5.2 Small Area Microdata: A Third SAR

There were a number of requests for a third SAR at a much smaller level of geography. The benefits of a SAR with a 7 to 10 K population threshold are:

i. It can serve as a building block for aggregation to old 1991 SAR boundaries and also new 2001 local authority boundaries, thus providing a link between censuses.
ii. These building blocks can be grouped to other areas—for example, health authorities—or can be combined to provide local areas of homogeneity within larger areas.
iii. They provide the potential for modelling local area effects.

Users have made powerful cases for all these aspects. An assessment of confidentiality suggests that the risk of releasing data with a much lower population threshold can be reduced to the baseline measure on the 2% SAR by grouping categories on key variables (Tranmer *et al.* 2000). The Office for National Statistics will decide whether or not to accept these proposals during 2001–2002.

Acknowledgements

We are grateful to Harry Ganzeboom of Utrecht University for supplying us with the algorithms to create SIOPS and ISEI from ISCO-88 codes.

Box 14.1 Key weblinks for Chapter 14.

http://les.man.ac.uk/ccsr/cmu/ or *http://www.ccsr.ac.uk/cmu/*	Web site for the Census Microdata Unit (CMU) giving access to comprehensive information on the SARs.
http://les.man.ac.uk/ccsr/index.html	Web site for the Cathie Marsh Centre for Census and Survey Research (CCSR). The CCSR hosts the CMU (see next link).
http:// /les.man.ac.uk/ccsr/cmu/sarsdata.html	The SARs code book and glossary for original and derived variables can be accessed via this link.
http://les.man.ac.uk/ccsr/cmu/data/gbprof.htm	Full details of the derivation of the classification and descriptions of each of the 49 GB Profile clusters attached to records in the Individual SAR
http://les.man.ac.uk/ccsr/publications/sarpub.htm	Information on publications that use the SARs.

15

On-Line Tabulation for the Samples of Anonymised Records

Ian Turton

SYNOPSIS

This chapter describes a software tool developed to make it easier to begin analysis of census microdata. The Unix tabulation tool for the Samples of Anonymised Records (USAR) gives the user the freedom to design multiway tables from the individual or household SAR, either on-line or via a version that can be downloaded for PC use. USAR enables the user to recode variables during the tabulation process and to generate tables very swiftly. The chapter concludes with a discussion of how such a tool might be adapted for the SARs from the 2001 Census.

15.1 INTRODUCTION

The release of the Samples of Anonymised Records (SARs) from the 1991 Census of Population in 1993 was a major milestone in census analysis, for the first time freeing users from the tyranny of fixed tabulations of census data. The SARs opened the way to many new and exciting types of analyses that could not have been contemplated before. However, for these promised riches to be achievable, users needed an easy way to access the data set. This chapter describes one of the methods developed to allow the easy access needed. The USAR system is described and some of the lessons learnt during its development are discussed. Finally, some suggestions are made for a similar package to be developed following the 2001 Census.

15.2 BACKGROUND

A major UK Census milestone was reached in 1993 with the release of Britain's first ever sample of anonymized census microdata. This amounted to a 3% sample of the 1991 Census being released in an anonymized form. The data was anonymized by recoding the geography to large areas and grouping variables with large numbers of classes (such as age and occupation). The two SAR files were: a 2% sample of individuals and a 1% hierarchical sample of households and individuals within those households. The SARs were, and still are, an important data set for social scientists because they offer census users an opportunity to design their own tabulations free of the many restrictions that apply to other census data sets.

As Marsh (1993b, p296) puts it,

> 'Users can explore relationships on the sample data, interacting until they reach the table that they feel gives the best information. They may still decide to commission a special tabulation, perhaps to get a level of geography specificity not available in the SAR or because small numbers demand a 100% run, but they will be me much more confident than in the past that they will be getting what they want'.

However, it soon became clear that the SARs were much more than a source of enhanced tabulations. They allowed users to explore and analyse census data in ways that had previously been restricted to other kinds of survey sample data such as, for example, analysis of variance and regression at the individual level free of any danger of the ecological fallacy.

The SAR sample is sufficiently large to investigate subpopulations in society and permits the hierarchical study of aspects of household structure. Equally important, the flexibility of the sample allows users to define their own variables; for example, measures of deprivation, head of household, and household composition. Now, no variable definition needs to be fixed, but can reflect what users really need.

New uses that were made possible by the introduction of the SARs include multilevel modelling of health variations (Gould and Jones, 1996), investigations of relationships to unemployment (Fieldhouse, 1996), and household and family structure (Dale, 1994). Users were also able to explore relationships that were previously difficult or impossible to explore as a result of the limited number of the small-area tables in which a variable appeared. Researchers in the field of ethnicity benefited particularly from this: see, for example, Rees (1995b) and Owen and Johnson (1996). Other fields also made good use of this new data source—for example, more information could be found on migrants, allowing new areas of research to be followed (Boyle 1998).

The problem was to develop delivery methods that allowed the typical social scientist to access this rich data source. This was more complex than simply providing a custom tabulation service as the SARs were too large and complex for the majority of users to be sure of the table that they actually required. Turton and Openshaw (1995) drew attention to the problems of this: 'Left to their own devices, there is a danger that instead of practising their cross-tabulations on a small survey of 100 or so households, the availability of the SAR will merely result in many social scientists practising their cross-tabulations on the SAR.' Fortunately, the release of the SAR coincided with an explosion in desktop computing power, which encouraged quantitative social scientists to make much more use of the SAR than was feared.

15.3 USAR—CUSTOM TABULATION FOR BEGINNERS

The SAR release was considered at the time to be a large and complex data set, the original files being approximately 126 Mb. Today this does not seem to be very large, especially with the growth in disk and memory sizes in standard desktop machines. Before the SARs were released, much thought and preparation were carried out as to the best means of dissemination and delivery of this large and complex data set. Turton and Openshaw (1995) noted 'The original intention in 1991 was to load the SAR datafiles onto a large mainframe at Manchester Computing Centre (MCC) to provide a table output service for users, by using the same relational database package (Model 204) as used by the UK census agencies in processing the 1991 census.'

In 1992, the Census Microdata Unit was established at Manchester University to provide three routes of dissemination for the SARs:

- an on-line service, available over the network using database software such as Quanvert and SIR and statistical packages such as SPSS and SAS
- distribution of the raw data files
- a customized tabulation service (based on Model 204).

Turton and Openshaw (1995) describe a fourth distribution route. A program called USAR was developed under an Economic and Social Research Council (ESRC) grant (1992–1994). The aim was to make the program freely available over the Internet for any user with a Unix workstation to download. This would allow users an alternative method to access the SARs without the need to register on a remote machine with an operating system that they may not have met before. Registration to use the SARs is still, at the time of writing, required, but after 2001 may no longer be needed (see Chapter 22).

As noted earlier, the SAR is a complex data set; there are potentially more possible tables available than there are stars in the known universe. Clearly, there was a need to provide novice users with as much help as possible, and when the SARs were released all users were novices. Therefore, USAR was developed as a menu-driven program (a relative novelty at the time) rather than using a command file as was more common. However, it was also built in such a way that, as users became more experienced, they were helped by the interface rather than defeated by it, as is too common in novice-friendly interfaces.

The basic function of USAR was to produce tables because this was what the majority of census users wanted (or thought they wanted). The USAR leads the user through this process using a series of menus to select the variable(s) required for the rows and columns of the table. The user could then preview the table before reading the whole data file. Subsequently, they could either make changes to the table, so that it showed what they required or proceed to make the table. For many users, this was as much as they initially needed; in some cases, this was the limit of exploration the SAR.

The USAR also provided a set of standard recodes such as, for example, the number of dependent children in a household, the number of adults in a household, and standard industrial or occupational codes. These recodes are provided in a transparent way to the users so that they appear to be standard variables and can be used in an identical way to build tables. The USAR also allowed the user to recode variables in ways that only they or a small group of users required, for example, a household-type indicator relevant to migrant households. These recodes could be developed and subsequently saved to allow their reuse in later sessions. To speed this process, USAR allowed the user to develop the recodes using a subset of the data set for greater speed. As many of the target user group were novices in designing cross-tabulations, USAR provided a visual warning of unreliable values in the tables created. This was essential as it was very easy to produce tables that were very sparse or entirely populated with small numbers. The USAR highlights any cell that falls within two standard errors of zero, once sampling uncertainty is considered. The user can then regroup the classes of some or all of the variables in the table with the USAR instantly redisplaying the table until a majority of cells in the table are significantly different from zero.

15.4 ADVANCED EXPLORATION OF THE SAR

The previous section discussed how users could build simple, but safe, tables with little prior experience of table design. However, many users wanted to go beyond this simple analysis to investigate how variables interacted with each other for specific groups found in the SAR. The USAR provided the tools to allow these users to do this. The first was a simple exploration method that allowed a user to select a single variable and a specific value of that variable or a group of variables and values. The USAR then constructs a linked list of AND filters for each variable (selections of the same variable used an OR join) and selects each record from the SAR that fulfils the filter conditions. For each record that fits the selection, each variable is examined and a counter for that variable case is incremented. These results are subsequently displayed for the user to examine, sorted in order of size. The user can then select more pairs of variables and values and repeat the search or construct tables on the basis of the filters used. This allows the user to discover the variables that are associated with a subpopulation of the SAR.

The USAR could also handle a 'fuzzy' query for users who were unsure of the exact variables that were most likely to be associated with a subpopulation. A fuzzy query in the context of the SAR can be considered as the possibility of a record matching a given 'select if' criteria even when it would fail if the criteria were applied in a deterministic manner. This allows the user to be a little uncertain about precisely what is required or to count cases that nearly meet the requirements. This is not uncommon, as census analysis is still more art than science. The USAR implements this function by allowing the user to specify a template or set of ideal values for a set of variables and then requests the identification of records that meet a certain fraction of the variables (i.e., four out of six), but not necessarily all of them. This fuzzy query can also be regarded as providing another form of categorical data exploration. The selected records might be regarded as providing clues about the multivariate structure present in the SAR. Any interesting result might then be used to specify tables. Openshaw and Turton (1996) demonstrate the use of these features to present a new analysis of lone parents in the United Kingdom.

15.5 DEVELOPMENTS OF USAR

The USAR was extended for selected users (who had the necessary data set registration) to handle other survey data sets. Much of the time in starting to use a new data set is spent in learning the software that is used with the data set and in exploring the data set to discover any interesting relationships hidden within the data. Thus, by making use of a common interface that has built-in data exploration tools, it is possible to begin substantive research much more quickly than is normal.

As development of USAR progressed, PCs became more powerful and it became apparent that more users wanted to access the SARs via their desktop machine. Despite the obvious advantages that a Unix workstation had as a research machine, many users were using an Intel-based PC. As USAR had been developed to be portable, it was easy to recompile it for a PC and release an identical version of the program for PC users. This also had the advantage that if a user wanted to carry out an analysis that was too large or complex to be completed in a reasonable time on their desktop PC, they could run the job using exactly the same commands on a central Unix mainframe.

15.6 LESSONS LEARNT IN 1991

The USAR was a success because it was free and available on almost any platform the user required. It was also fast and easy to use; a novice user could produce a meaningful substantive table within less than an hour of starting. The inclusion of data exploration tools within the program allowed users to discover interesting features of the data set and to develop this exploration into tables suitable for publication.

However, if users wanted to carry out any analysis of the data beyond tabulation, they were forced to move to another program. For this reason, USAR provided an easily understood data export format that allowed users to import results of a USAR run into spreadsheets and graphical programs. This allowed USAR to remain small and simple as it concentrated on being good at a specific task. The addition of spreadsheet or graphical functionality would have resulted in a larger and more complex program. This would have been worse at drawing graphs than specialized programs were, and so advanced users would still have exported the results and novices would have presented poor graphics.

15.7 SOFTWARE FOR SARs IN 2001

When the 2001 Census is processed, there will be a release of SARs that will be at least comparable to the 1991 SARs (see Chapter 14). There is also a possibility that more samples will be produced to meet some of the concerns of the user community. This clearly will lead to a continued requirement for access software for the SARs. Some have argued that modern databases on personal desktop computers will be able to handle these data sets. However, this may be possible only in a purely technical sense, that is, the SARs will fit in to a relational database, but the fact remains that the majority of the social scientists lack the necessary skills to develop and make full use of a relational database. There are also types of query that users require that are hard to deal with in a simple database. In addition, the proponents of this solution often feel that providing a solution for Microsoft Windows' users meets the requirements of the whole user base. However, many users still make use of Unix-based systems both for research and teaching, and even with the increase in computing power that has occurred between censuses, there are still types of analysis that require the power of a larger (possibly parallel) computer.

With these restrictions in mind, it is clear that a custom software package designed from within the user community will still be required in 2001. Ease of use must be the most important design criterion; more users are joining the quantitative social sciences and making use of census data all the time. It is very important not to place any unnecessary barriers in the way of these new users. If a police force needs to provide base statistics for its crime analysis, for example, then collecting these data should be easier than obtaining their crime statistics, as that is their job, and not census analysis. This leads to a second design requirement that it must be easy to add different data sets to the program allowing users to compare or combine elements of their own data sets with the SAR.

Portability is still one of the most important requirements as users will want to be able to move seamlessly from one system to another as they move from home to office to a remote site depending on where and what they are working on. Obviously, a user will not want to learn a new system simply to complete a more complex analysis more quickly. Portability is a free benefit if the package being developed is written in Java. The development can be carried out on a range of machines and operating systems and the final product can be used on any system for which Sun Microsystems (or other developer) has produced a virtual machine.

15.8 CONCLUSION

Thus, it can be seen that computing is a fast moving arena, and the only certainty is that whatever you plan for will have changed in the two to three years between planning and release of a software product. Ease of use and the special features required for census analysis require the program to be developed in conjunction with the user community that has gained 10 years of expertise using the SARs. Any attempt to shoehorn the SAR and its diverse users into a commercial package will lead to a product that certainly satisfies only a few of the users and risks alienating them all. In addition, the system developed should concentrate on being good at accessing the SARs; additional functionality can be provided by the existing programs that the individual user is already familiar with. By combining this specialized functionality with a portable system to meet the needs of all the potential users of the SARs, there seems no reason for the release of the 2001 SARs not to have as great an impact on census analysis as the release of the 1991 SARs did.

Box 15.1 How to access and use USAR.

To access USAR, you need to register with the MIMAS service of Manchester Computing for a user name and password for the *irwell.mimas.ac.uk* system and as a SAR user with the Census Microdata Unit, CCSR, Faculty of Economic and Social Studies University of Manchester. See Web pages *http://census.ac.uk/cdu/Registration/How_to_register.htm*
You then need to access the server using a telnet command or other suitable interface
Type telnet *irwell.mimas.ac.uk*, after which you will be prompted for a user name and password
After entering these, you will be logged in and presented with a command line
Type usar
The program then runs in an interactive mode, prompting you to select an SARs data set, variables for rows and columns in your table. The USAR allows you to design the classification scheme for the variables. When instructed, the USAR program will compute the sample counts that occupy your table design. You can view and correct this table and ask for it to be written to a file
Here is a simple example of a table created using the Northern Ireland Individual SAR, which tabulates Religion against Age Group for children and adolescents. The output file, containing counts separated by tabs, has been converted into a table
Table created with USAR Version 2.0b
Data file niperson.in
No filters

	AGE	0–6	7–13	14–19
RELIGION				
RC		1681	1762	1412
Presb		596	615	622
CofI		547	540	486
Methodist		102	99	88
Other		237	241	250
None		148	116	105
NS		361	272	210
Indefinite		11	2	5
Crown Copyright				
Further instructions on how to use USAR are given in *http://les.man.ac.uk/ccsr/cmu/analysis/usar.html* or in Turton and Openshaw (1994)				
Both Unix and PC (DOS) versions of USAR are available from the author to registered users				
E-mail *i.turton@geog.leeds.ac.uk*				

16

The ONS Longitudinal Study: Linked Census and Event Data to 2001

Rosemary Creeser, Brian Dodgeon, Heather Joshi, and
Jillian Smith

SYNOPSIS

This chapter provides an account of the Longitudinal Study (LS) of England and Wales, which links a 1% sample of the lives between the censuses of 1971, 1981, and 1991 and adds vital events (e.g., births or deaths of sample members). The chapter discusses the preparations being made to add a link to the 2001 Census to the LS. When this is in place, users will be able to track a variety of demographic, social, and spatial behaviours of the English and Welsh population across three intercensus periods. Arrangements for access to the LS are described. Because of their confidential nature, the data themselves cannot be released; instead, researchers arrange for access through a support unit, the staff of which is approved by Office for National Statistics (ONS) for hands-on extraction of aggregated tables and analyses on behalf of approved user projects. Details of the Web site and how to access the data dictionary are provided.

16.1 INTRODUCTION

The ONS LS is a record linkage study of census and vital event information. A 1% sample of census records was initially selected from the 1971 Census by taking all individuals born on four specific dates across the calendar year. This covered about 500 000 people, or 1% of the population of England and Wales, including those in both private and communal households. At each census, a sample has been drawn on the same basis and linked at individual level to the LS. The database now includes three census samples: 1971, 1981, and 1991, and plans have been made to link the 2001 Census sample. The entire census record for the LS member and all members of that person's household are entered into the LS. This contrasts with other census data sets, which are adjusted, aggregated, or otherwise limited to safeguard confidentiality. The LS is protected by security measures that provide a safe setting for its use.

Vital event information, drawn from the ONS registration systems, has been linked over the period of the study to the individuals in the sample. Two types of entry events replenish the sample: births on the four LS dates and immigration of people with LS dates of birth. Exit events linked to the study are deaths and embarkations. Records of people who have left the study continue to be held for analysis. Other major events that are linked include births and infant deaths registered to women in the study, cancer registrations, and widow(er)hoods. Figure 16.1 shows the structure of the data and the number of events available for analysis.

The National Health Service Central Register (NHSCR) is used to facilitate the linkage of events to the LS, which itself does not carry identification information such as name and address. The linkage of events is achieved either through routine notifications direct to the NHSCR or by regular statistical processing from the ONS registration systems. The LS master database is updated several times a year.

The LS data are maintained in a 32-file database in Model 204. This is surrounded by extensive security measures to provide a safe setting for use. Data are extracted from the database into subsets of information tailored for particular analyses that are completed on-site at the ONS. Users are able to receive their work in several forms, ranging from summary data sets for further analysis to final analyses from a range of software packages.

An important feature of the LS is the provision of facilities to support research within the safe setting. The LS is distinguished by a high level of user support, providing skilled analysis expertise to assist users wishing to use these complex data. The result is a developed and supported research environment in terms of computer facilities, documentation, quality information, and practical data preparation and research skills. The support work is shared through a partnership between the ONS, the Economic and Social Research Council (ESRC), and the Joint Information Systems Committee (JISC) of the Higher Education Funding Councils, who funded staff in the 1991–2001 period at the Centre for Longitudinal Studies (CLS) at the Institute of Education, London University and at the ONS. The staff supported the analyses of a wide range of LS users, including academics, government departments, and individuals working in the voluntary sector. New support arrangements are being introduced in 2002, which are briefly described in section 16.5.

The LS is among the largest longitudinal data sets in the country. This makes it particularly suitable for studying small groups, change over successive time points, geographic breakdowns, migration, and mobility. It complements other national studies that are mostly smaller and may be more specialized, such as the British Birth Cohorts (see Ferri, 1993; Bynner *et al.*, 1997), or use different design criteria covering shorter periods in more detail, such as the British Household Panel Study (Buck *et al.*, 1994). When these data sets are used together, they help alleviate individual deficiencies, such as the 10-year gap between census data in the LS. On the other hand, the long time span covered by the LS allows it to be used to check how far other studies suffer from biased attrition.

The LS is collected with no direct respondent burden. It only draws upon data sets initially collected for other purposes, many of which are required by law. While the LS is unique in the United Kingdom in providing linked census and event data, some other countries have, or can collate, similar information. This enables the United Kingdom to take part in important collaborative international analyses and contribute, for example, to the policy agenda on public health (Fox, 1989).

16.2 THE 1991 CENSUS DATA AND BEYOND

Since 1994, the research potential of the LS has been considerably enhanced by the inclusion of data from the 1991 Census and events such as births, cancer, and deaths

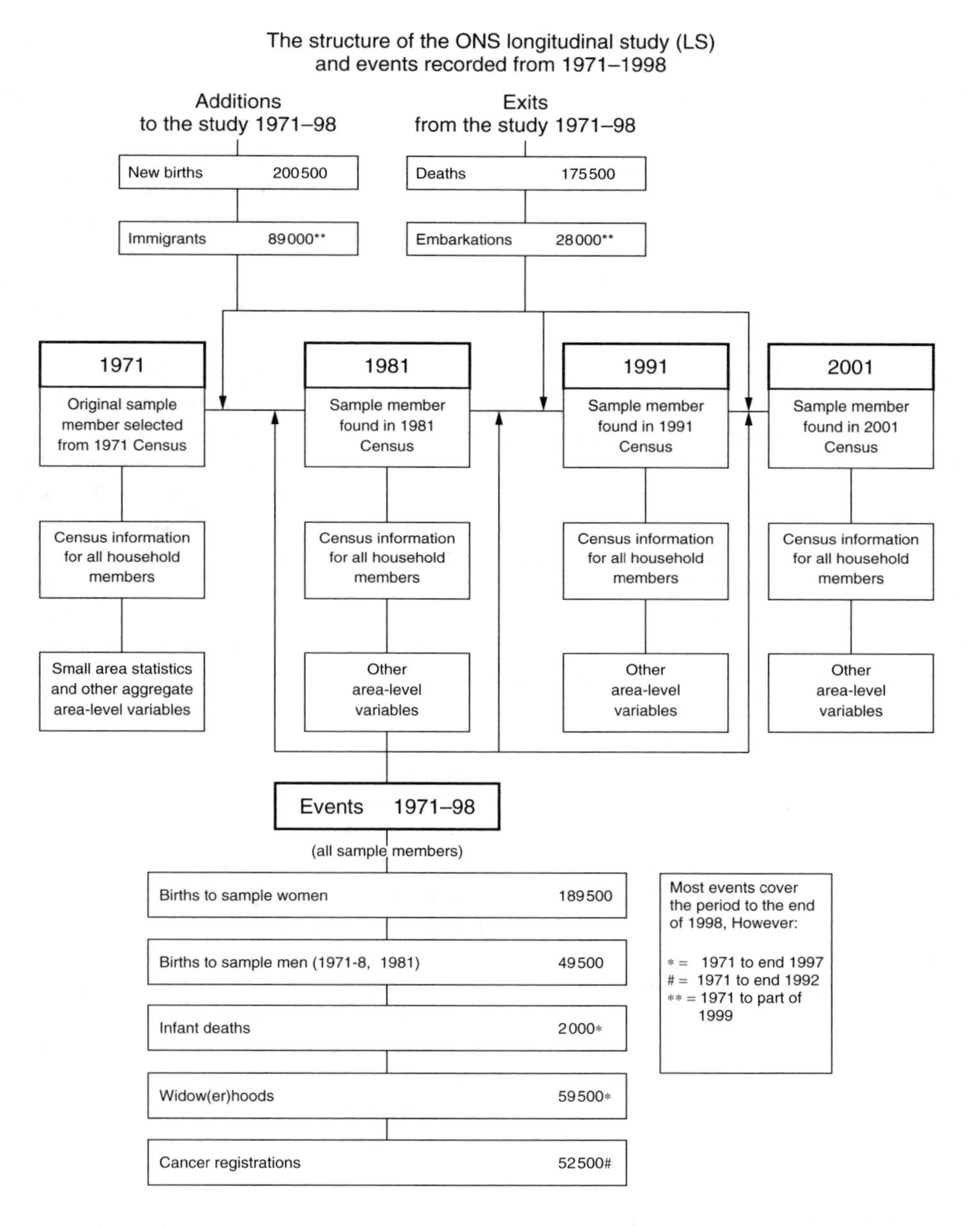

Figure 16.1 The structure of the longitudinal study and the events available for analysis.

have been registered after 1991 (see Figure 16.1). The dynamic nature of the database means that new events are added at regular intervals. An up-to-date version of this figure is available via the LS Web site *http://www.cls.ioe.ac.uk/Ls/intro.htm.*

Longitudinal census data covering the period 1971 to 1981 and 1981 to 1991 are available for in excess of 400 000 members of the study. For LS members enumerated in ‘private households’, a separate census record also is available for every individual household member. These are often referred to as *non-LS members*. It is important to highlight that the LS does not routinely follow up these ‘non-LS’ individuals from census to census: longitudinal census data for non-LS members will only be available where study members have been living with the same person(s) at consecutive censuses. Although these data are indispensable for household and family research, the scope for studying many ‘new’ or ‘non-standard’ living arrangements is unfortunately restricted by the questions asked at previous censuses and by the way in which they have been coded. For example, the way in which the 1991 Census recorded household relationships makes it difficult to distinguish stepchildren or ‘concealed families’, where there is no direct relationship to the first person on the census form.

The 1991 Census LS data are not restricted to individual-level records. The LS database also includes a substantial subset of aggregate-level data summarizing the socio-economic characteristics (demographic composition, housing stock, local labour market) of the neighbourhood in which LS members were living in 1991. The basic ‘building block’ used to create these Small Area Statistics (SAS) is the enumeration district (ED), which aggregates up to ward, local authority district, and district health authority level.

For LS members enumerated in 1991, other aggregate-level data are also available. These include: the 1991 Travel-to-work Areas (TTWAs), the 1991 Carstairs Deprivation score (Carstairs and Morris, 1989), the 1991 ONS Area Classification (Wallace *et al.*, 1995), and a subset of variables summarizing population and road density at ward and ED level. Some corresponding aggregate-level data are also available for the 1971 and 1981 Censuses.

The Department of SAS, London School of Hygiene and Tropical Medicine, provides the data on road density. They were derived using census data and information on motorways, A and B roads, supplied by the Automobile Association (AA). In certain cases, ‘road density’ gives a better measure of the urbanization of an area than population density.

Since the 1991 Census LS link was achieved, data on the ‘vital events’ of the post-1991 Census, routinely recorded by the ONS, have been added every year to the LS database. The most significant of these events—both in terms of the analysis potential of the study and of the sheer number of events—are birth and death registrations. There is a time lag of approximately two to three years for the inclusion of the majority of vital events recorded in the LS. The one exception is cancer registration data—at the time of writing, complete cancer registration data were available only in the study upto 1990. Some of the limitations of the National Cancer Registration System, coupled with the fact that cancer registration is not compulsory, account for this delay. The LS Web site shows the number of vital events currently available for analysis (*http://www.cls.ioe.ac.uk/Ls/intro.htm*).

16.3 THE 2001 CENSUS LINK

Every ten years since 1981, a major project is mounted to add the new LS census sample to the LS database. The LS sample criteria, that is, four dates of birth, are used to draw the LS sample from the census. This sample is matched to existing LS records using the NHSCR as the intermediary linkage mechanism. Whether or not a match is successfully achieved, the census data for the LS member and that person’s household are added into the LS database. A high level of linkage is vital to the continued quality of the LS and the future capture of event information; thus, extensive efforts are made to reduce non-matches to a minimum. The detective work made possible by the comprehensive registration systems held by the NHSCR allowed linkage rates of 94% for the 1981 Census and 90% for the 1991 Census.

At the time of writing, plans are in progress for the next LS census sample to be matched into the database. When this work is complete, users will be able to draw upon four census data points and all the intermediary events.

16.4 ANALYSIS POTENTIAL OF THE LS 2001 LINK

Even without the incorporation of another census, the potential list of possible studies is far from being exhausted, and grows as evidence accumulates from the ongoing addition of event data from the mid-1990s, not to mention additions of other contemporary data such as contextual geographical variables and the development of computing facilities for more complex analysis.

The topics that can be studied are anything upon which census questions are asked and those for which vital registration contributes additional data, principally death, bereavement, cancer, birth, and international migration. Because the LS is based on individual-level data, several topics are normally combined in any one analysis. Evidence is normally taken from more than one point in time. This illustrates to what extent people retain the same state from census to census and how far there is mobility in social space. The study can also be used to detect stability and mobility over geographical space and relate the values of one variable at a point in time with circumstances that precede or follow it.

Some change over time is inevitable—individuals increase in age over time—and other changes are more or less expected as part of the life course—graduation from schoolchild to employee or from a family with dependent children to an 'empty nest'. The LS can describe how the nature and timing of these life-course transitions may vary between social or ethnic groups and between the different intercensal periods. Other changes recorded by individuals between censuses may be less routine or expected, and hence be of particular interest in their own right. A few examples of this might be changes in housing tenure, geographical moves, occupational mobility, becoming or ceasing to be a lone parent, moving out of employment into permanent sickness or limiting longstanding illness, and (for older people) moving into residential accommodation. Census to census comparison might also involve looking at area-level variables. Successive SAS could be used to identify, for example, wards that have undergone gentrification, while the files on LS members may highlight whether there has been upward mobility of continuing residents, in-migration of persons above average social class, or downward selective out-migration. However, it should be stressed that it is not possible to identify the particular small areas where the LS members reside, but it is possible to identify the generic type to which the small area belongs.

Information from the event files can be compared with data from previous or following censuses. The classic example of the former demonstrates the study's original purpose, which is to study the social differentials in mortality (Fox and Goldblatt, 1982). The reverse comparison can be made if the details of a child's birth are matched to its circumstances at a subsequent census (Brown, 1986). Looking both ways, the census could be used to describe the housing or employment status (say) of people who have a registered cancer before the event, and for survivors, at the next census. Event data can also be compared with other events, such as the length of a previous birth interval or survival after cancer registration; but even here, it would often be useful to bring in some census data about the person's location or social class, for example.

Most of these forms of analysis would be possible with only two censuses linked. The third has permitted comparisons over 20 as well as 10 years. This has opened up the possibility of comparing intercensal changes across two decades, particularly for two cohorts experiencing the same age transition at two periods. At the same time, it is possible to see whether transitions in a second decade are conditioned on the state at the beginning of the first, which is important if observed transitions are to be used in forecasting household size, for example.

Bringing a fourth census into the data set enhances the possibilities enormously. Perhaps the most obvious innovation is that the 2001 Census LS link will enable data spanning 30 years to be used. Intervals of this length and longer are useful for the study of occupational mobility, return migration, and change of household composition. The interval begins to be long enough to analyse the adult outcomes of young LS members—qualifications, employment, housing, by the circumstances of their families of origin. The birth cohorts for which this would be feasible are those young enough to be still living with their parents in 1971, that is, those born between 1956 and 1971. These cohorts happen to encompass the two British Birth Cohort Studies of 1958 and 1970. As indicated earlier, the LS could be used to supplement these on topics such as attrition bias, and the cohort studies could be used to amplify certain topics (e.g., school achievements) not present in the census.

Four censuses do not have to be compared in the same analysis. They offer the possibility of three sets of intercensal comparisons, or two sets of three. These increase the scope for seeing how life-cycle transitions may differ across three periods. The LS also offers the scope for investigating whether the pattern of social differentials in mortality or cancer has widened in the 1990s compared with the 1970s and 1980s (see Harding *et al.*, 1997). Indeed, as event data accumulates into the first decade of the twenty-first century, it will become possible to compare differentials in a fourth decade.

The extended follow-up of event data can also be used to prolong the period in which survival (or cancer recurrence) can be observed, and increases the chances of observing return immigration. The birth histories are of major significance in the accumulation of event data. From 2001, there begin to be cohorts of LS women whose entire childbearing age span is covered by registration data in the LS. These complete (and near complete) birth histories will be useful for studying and forecasting fertility. They permit a quantification of birth rates by parity and the interval since last birth, and also for analysis by socio-economic factors. Another advantage of accumulating event history records is that they are adding to the small number of records on infant deaths and stillbirths, which should make meaningful analysis more possible.

There is another dimension in which the 2001 Census will augment the LS: the addition of new variables. The ONS included a number of additional topics. These include religion, general health, voluntary care, qualifications gained at school, but not banded income, which was strongly supported by users but felt by the ONS to create difficulties for census-taking. These new variables will be available for analysis with data from earlier censuses and events. The religion question, for example, may help differentiate some of the minority groups in 1991, and indeed earlier years, and data on qualifications can also be projected backwards. Better information on relationships within the household resulting from the 'matrix' question defining the relation of each person in a household to every other person (up to six members) will be very helpful in the construction of variables about changing household composition and the identity of the people living with the LS member.

Finally, let us note what the LS can bring to the rest of the census. It should throw some light on the 1991 undercount, insofar as LS members who were not enumerated in 1991 may reappear in the 2001 link. Also, there may be some missing information on persons enumerated in 2001 if, for example, automated coding of occupation leaves some cases unresolved, particularly of the elderly. The LS will be able to offer alternative information on which to assess a social grading and inform other census users of any likely bias in the uncoded material. It should also be possible to assess the degree to which individuals change their reported ethnic identity between 1991 and 2001.

The LS as an extract and extension of the census has already offered a wide range of research opportunities. There is evidence of these in the LS publications available via the Web (*http://www.cls.ioe.ac.uk/Ls/pub.htm*) or from the LS User Support Programme at the Institute of Education. As data and access diversify, and as international comparisons are made, the potential for further work continues to increase.

16.5 ACCESS TO THE ONS LONGITUDINAL STUDY

Access arrangements for academic use of the LS are in transition in 2001–2002. From December 2001, the contact to support use of the LS moved from the Institute of Education to the London School of Hygiene and Tropical Medicine, to an LS Support Team directed by Professor Emily Grundy. In this section, generic access arrangements are described, while section 16.6 provides information about the Institute for Education LS web pages, which will remain accessible during a transition period to the new service.

Researchers are asked to submit applications to an LS research board, giving details that allow an assessment of the scale of work and the confidentiality implications of their projects. Each project is allocated a support person who will help with applications, liaise for the duration of the project and, in many cases, complete the hands-on analysis. To ensure that confidentiality is maintained, the ONS or LS Support Team staff clear all the LS outputs before publication or release.

Access to the LS is governed by several constraints, which distinguish it from other census products. The three C's: complexity, confidentiality, and computing environment are central.

The data set's complexity means that most researchers need to consult ONS or the LS Support Teams for advice before embarking on a project. The LS Support Team runs workshops, in which researchers can learn the intricacies of the data, familiarize themselves with the data dictionary (see Section 16.4), and experiment using a test data set.

The issues of confidentiality and the computing environment are closely linked. Because the LS is a completely disaggregated census data set, with no data-blurring or other adjustments, researchers may not access the data directly. Until recently, all analysis had to be done on the ONS mainframe, with the LS Support Teams acting as intermediaries to run analyses specified by researchers in SAS or SPSS. The outputs would be summary statistical analyses such as linear regression or survival analysis, or tabulations in printed or machine-readable form (column or spreadsheet format, with a limited number of variables/cells to guard against disclosure of individual information).

This is still the main approach, but recently it has become possible for large subsets of LS data to be moved to NT servers in the 'safe environment' of the ONS London office. This enables the underused potential of the data set's complexity to be exploited by allowing analysis with many other software tools such as STATA, MLn, MAPINFO, ARCINFO, EGRET, and others.

NT analysis is still normally performed by the LS Support Teams, but in approved circumstances, it is possible for researchers themselves to visit the ONS London office and work on the NT servers with the supervision and advice of support staff.

With the fourth time-point provided by the 2001 Census link, the LS will become even more complex. It is likely to be maintained in a transformed computing environment. A server will almost certainly supersede the mainframe, with the Model 204 database also being replaced. Increased possibilities for complex analyses inside the 'safe environment' are clearly feasible.

16.6 THE LS WEB SITE AND ACCOMPANYING CD-ROM

The enclosed CD-ROM includes a copy of Version 5 (December 1998) of the LS IDEALIST Data Dictionary. The LS Data Dictionary provides information on approximately 3000 variables. This includes details of the range of each variable and references in the 'LS Technical Volume' (Hattersley and Creeser, 1995) and LS User Guides to the source and quality of the data. Version 5 includes information on the post-1991 event variables (births, cancers, deaths) that were unavailable in previous versions. A concerted attempt has been made to reference the post-1991 event variables with the nearest meaningful equivalents for 1971 to 1981 and 1981 to 1991 (and vice versa). The LS data dictionary incorporates hypertext links to more than 100 census and event data appendices.

The CD-ROM includes a single self-extracting file (LSDICT98B.EXE) for creating the data dictionary, instructions for installing it (install.doc), and a copy of LS User Guide 15 (lsug15.doc). The User Guide describes the contents of the data dictionary and how to search for relevant variables, using the main search functions of FIND, WIDEN, NARROW, and EXCLUDE. A read-only version of the IDEALIST database management system is provided free with the LS Data Dictionary. This means there is no need to obtain any special database software. Six megabytes of space on the hard disk is required to install and run the data dictionary. As there are no licence restrictions for the read-only version of IDEALIST, the data dictionary can be copied to more than one platform.

The information held in the data dictionary is revised at regular intervals. A copy of the most recent version may be downloaded directly from the LS Web site (*http://www.cls.ioe.ac.uk/Ls/ddict.htm*).

The LS Web site includes summarized information on the LS and the LS User Support Programme based at the CLS, Institute of Education. The LS homepage (*http://www.cls.ioe.ac.uk/Ls/lshomepage.html*) includes links to:

- an MS PowerPoint introduction to the LS;
- an annotated copy of the LS Undertaking Form—the form that all prospective LS researchers must complete before any analysis may be carried out;
- summaries of LS projects that have recently started or due to start soon;
- back issues of some LS newsletters and LS User Guides as. pdf (portable document format) files; and
- an up-to-date list of all publications using the LS by broad subject area.

Box 16.1 Key web links for Chapter 16.

http://www.cls.ioe.ac.uk/Ls/lshomepage.html	Home page of LS on CLS Web site
http://www.cls.ioe.ac.uk/Ls/intro.htm.	Up-to-date version of Figure 16.1
http://www.cls.ioe.ac.uk/Ls/pub.htm	List of publications using the LS
http://www.cls.ioe.ac.uk/Ls/ddict.htm	Web page from which latest version of the LS Data Dictionary can be downloaded
http://www.celsius.lshtm.ac.uk/	Web page describing the new LS Support Service: Centre for Longitudinal Study Information and User Support (CeLSIUS). Takes over fully from the Centre for Longitudinal Studies in April 2002

Box 16.2 Additional CDS resources for Chapter 16.

LS Data Dictionary, Version 5	This lists all the variables available on the LS and the categories/classes used in categorial variables in searchable IDEALIST format

17

Synthetic Microdata

Paul Williamson

SYNOPSIS

Population microdata comprise lists of individuals, each with an associated set of personal characteristics (e.g., age, sex, occupation). In order to protect respondent confidentiality, publicly available survey microdata typically suffer from top-coding, the collapsing of response categories, and suppression of spatial detail. This chapter outlines a range of strategies that have been adopted to overcome these data shortcomings, leading to the creation of 'synthetic' data sets. Some consideration is given to reweighting and imputation, subjects already well covered in the literature. However, the main focus of the chapter is on the retrieval of lost spatial detail. Four competing approaches are outlined and critically reviewed. Ongoing work in this area is also highlighted.

17.1 INTRODUCTION

Population microdata comprise lists of individuals, each with an associated set of personal characteristics (e.g., age, sex, occupation). For consideration of the widest possible range of social policy issues, it is usually necessary for these individuals to be nested additionally into families and/or households. There are two key advantages to representing a population in this way. First, for data sets comprising many variables per individual, storage of population as a list is far more computationally efficient than representation of the same data in a tabular form. Second, population microdata offer flexibility of data aggregation, avoiding the constraints placed on analysis by the prior aggregation of microdata into fixed tabulations before release.

In Britain, the two largest, readily accessible (and non-commercial) survey microdata sets are the 2% individual and 1% household Samples of Anonymised Records (SARs) from the 1991 Census (see Chapter 14). Since their release, they have both proved very popular with researchers for precisely the reasons outlined earlier. Unfortunately, the types of analysis achievable using the SARs are limited in a number of ways. These limitations include the relatively restricted range of questions asked in the census: the restriction of sample size, which limits the reliability of in-depth analyses that may be undertaken and the collapsing of response categories to protect confidentiality, in particular a reduction in spatial detail. In an attempt to overcome these shortcomings, researchers often draw upon external data sources to modify the raw microdata in some way, either through reweighting or imputation.

In the broadest sense of the term, 'synthetic' microdata comprise all microdata that have been modified in this way. This follows the dictionary definition of 'synthetic' as the combination of parts (different data sources) into a whole. However, in practice, use of the term *synthetic microdata* appears so far to have been restricted to describe microdata in which the spatial coding has been greatly enhanced (e.g., Birkin and Clarke, 1988; Williamson *et al.*, 1998). Duke-Williams and Rees (1998) and Beckman *et al.* (1996) use the equivalent term *synthetic population*, whereas Mitchell *et al.* (1998) refer to *synthetic EDs*. In a number of these examples, microdata are not so much modified as estimated from scratch. The term *synthetic microdata*, therefore, may be viewed as covering a continuum in which data sets range from spatially coarse microdata that have been barely modified through to spatially detailed microdata that are entirely 'synthetic'.

In the light of this definition, the aims of this chapter are threefold. First, to consider briefly the range of possible techniques that may be used to enhance existing publicly available microdata, under the broad headings of reweighting and imputation. Second, to review the current state of the art with respect to the estimation of geographically detailed 'small-area' synthetic microdata. Third, to review a range of concerns that pertains to all modified microdata, at whichever end of the continuum they lie.

17.2 WEIGHTING

Weighting survey microdata has now become such a time-honoured practice that arguably no survey statistician would regard data so modified as synthetic. Yet, weighting almost invariably involves drawing upon one or more external sources of data, even if only to calculate survey design factors. For example, the individual SAR, released via the Census Microdata Unit (CMU), includes a set of weights designed to correct the errors introduced both by the sampling method used to obtain the SARs and by census under-enumeration (CMU, 1994; see also Chapter 14). These weighting factors, disaggregated by age, sex, and SAR area, are based on the Registrar General's 1991 mid-year population estimate. To call such weighted data *synthetic* might seem like mere pedantry. However, there is reason to believe that, even having taken account of age, sex, and SAR area, the extent of census under-enumeration varied considerably between population sub-groups and, possibly, between geographic areas (see Chapter 12 for a fuller discussion). In consequence, weighting the SAR only by age, sex, and SAR area may significantly distort the relationships between SAR variables from 'true'. Labelling weighted data as 'synthetic' does at least flag up a warning to analysts: the weighted data are only a *best estimate* of the population distribution, not the distribution itself.

A variety of schemes exist for weighting microdata (Elliot, 1991). The simplest hinge around the idea of partitioning the population into smaller sub-groups before calculating and applying appropriate weighting factors. Alternative approaches include raking and model-based solutions. Model-based weighting methods use data collected during the survey itself to estimate the response rate of different types of individuals captured in the survey. These estimates are converted into target totals for population sub-groups, thereby avoiding the need to weight the externally supplied totals. Raking (also known as iterative proportional fitting) similarly weights microdata to given (external) constraints, but minimizes the overall distortion introduced by the weighting process to unconstrained distributions. Thus, raking is a way of 'avoiding the extra variability caused by spreading the sample thinly across the weighting classes' (Elliot, 1991: 27). Raking can also be used when only marginal rather than full multi-way target population distributions are known, although it is worth noting that raking-based weighting procedures that use only marginal distributions can still have unintended side effects on the joint distributions of variables in the data set.

Merz (1986) reviews a range of alternative techniques for reweighting survey samples to agree with marginal totals, while Merz (1994) and Gomulka (1992) propose a raking method for weighting survey microdata that satisfies simultaneous household and individual level constraints, on the basis of the Minimum Information Loss principle. Chambers (1996) outlines an alternative, non-iterative method, on the basis of ridge regression. All of these solutions discard the integer weighting of the initial microdata. This can become problematic if large multi-way tabulations are desired because in a sparse matrix, many of the weighted cell counts will be close to zero. Rounding in this instance will result in an apparent loss of population, although rounding thresholds can be iteratively determined to minimize data loss. Alternative integer weighting strategies are discussed by Brown *et al.* (undated).

More generally, no weighting strategy can fully compensate for the fact that the characteristics of survey respondents not controlled in the weighting process may differ substantially from those of survey non-respondents, or from survey respondents drawn from another geographic area. The effect of this will be to reduce the variability among the households or individuals within population sub-groups; it may also lead to actual bias away from the target 'true' distribution.

17.3 IMPUTATION

A second way in which census microdata may be modified to enhance data quality is imputation. Imputation involves 'filling in missing data with plausible values' (Schafer, 1997: 1). The alternative, case deletion (of cases with missing variables), fails to take account of the response bias that typically arises from non-respondents sharing characteristics in common (Kalton, 1983). Imputation also allows the addition of variables not included in the original microdata, the disaggregation of variables aggregated for confidentiality reasons, and even alternative adjustments for differential under-enumeration.

In nearly all imputation methods, 'strength is borrowed' from auxiliary data sources and combined with local information to produce the best estimate of a missing value. The main methods of imputation include replacement of missing values with a mean value, typically calculated after partitioning the data into population sub-groups; regression; Monte Carlo sampling from conditional probability distributions; maximum likelihood–based approaches including logistic regression, empirical Bayes, and multilevel modelling; exact matching (data fusion), as exemplified in the Longitudinal Survey; statistical matching (data merging); and neural networks. Statistical matching techniques include multiple imputation, a form of sensitivity analysis for synthetic estimates, and 'hot decking'. As the hot deck procedure was used to impute missing values in the 1991 Census (Mills and Teague, 1991; Vickers and Yar, 1998), it could even be argued that the post-edit version of the 1991 Census is itself a set of synthetic microdata.

A more detailed review of the various imputation methods available may be found in an appendix to this chapter supplied on the accompanying *CDS Resources*, including an extended bibliography on the problem of small-area estimation. General reviews of imputation are also provided by Kalton (1983), Government Statistical Service (1996), and Ghosh and Rao (1994). Crucially, every imputation method depends on the Conditional Independence Assumption (CIA). Briefly stated, if an individual's income is imputed, given occupation alone, the (implicit) assumption is made that the relationships between income and other variables in the same data set, such as tenure, are entirely explained by income given occupation; that is, given tenure, income, and occupation are independent. An alternative is to include tenure in the definition of yet smaller population sub-groups. However, some degree of violation of the CIA is really an inevitable aspect of all forms of both weighting and imputation. In fact, the assumption that a sub-area/domain/group shares the same characteristics as larger group is cited by Gonzales (1973) as the definition of a synthetic estimate.

17.4 ADDING SPATIAL DETAIL

Over time, there has been increasing interest in estimating and using small-area data for a range of public and private policy analysis purposes (c.f. Brackstone, 1987; Ghosh and Rao, 1994; Fotheringham, 1997). For example, in the United Kingdom, lack of information on the local distributions of income is perceived as a major gap in official statistics (Bramley and Lancaster, 1998; Dorling, 1999). However, most attempts to remedy the lack of small-area data hinge upon the estimation of univariate measures (e.g., Isaki, 1990; Fotheringham *et al.*, 1998; Mugglin *et al.*, 1999). At first sight, the lack of small-area population microdata appears to preclude investigation of these sorts of problems from a microdata perspective. A second glance, however, reveals that there is no need to be restricted to the use of only that spatial coding inherent in available microdata. A number of alternative strategies for estimating spatially detailed population microdata already exists. These approaches are outlined later, classified under four main headings.

17.4.1 Stratified Sampling or Geodemographic Profiling

A number of geodemographic profiling products exist, including ACORN, SuperProfiles, PiN, and MOSAIC (Brown, 1991). The common feature of all of these products is that they classify areas on the basis of their demographic, socio-economic, and housing characteristics, mainly drawing upon variables captured in the census. The ONS ward classification and GB Profiles ED classification are further examples of this genre, based entirely on census data (see Chapter 11 for details). These two classifications have been added to records in the household and the individual SAR, respectively, allowing analysts to assign each individual or household to a geodemographic area type, without knowing their precise spatial location. It is possible, therefore, to construct synthetic populations from the SAR consisting of individuals and households that come from the same area type. Unlike a sample of records taken from the SAR at random, this approach is thought to have the advantage of capturing some of the internal homogeneity of wards and enumeration districts.

More sophisticated sampling strategies could be envisaged, in which this stratified sampling from the SAR is extended to include other known characteristics of the local area, such as the age–sex profile or the tenure and dwelling type composition. In this case, the synthetic populations generated might be viewed as a plausible representation of the likely population in an area, rather than simply as representative of the type of individuals or households likely to be found within a given area type. If other publicly available microdata were made available with similar area classifications attached, the potential for data matching between surveys would be increased, while still circumventing data confidentiality concerns.

The application of geodemographic profiles to small-area estimation problems is already widespread. For example, Madden *et al.* (1996) develop a small-area electricity demand forecasting system on the basis of small-area classifications, while Batey *et al.* (1999) identify the profile of university student intakes on the basis of the types of areas from which the students were drawn. The adoption of a microdata approach might help overcome one of the major criticisms of geodemographic profiling, which is that it masks known within-area heterogeneity. For example, in one commonly used geodemographic profile, only 10% of the between–area proportion of women working full-time was 'explained' by the geodemographic classification (Voas and Williamson, 2001). Synthetic population microdata created as outlined in this section would at least allow some of the potential heterogeneity within small areas to be recaptured, although a note of caution has to be sounded. It is quite conceivable that the populations within specific small areas are, in fact, more homogeneous than random sampling from geodemographic strata within the SAR might imply, suggesting a need for a greater number of constraints in the stratification process. Methods for selecting a combination of individuals or households from the SAR that conform to a larger number of constraints than can be dealt with by simple stratified sampling are outlined in Section 17.4.3 that follows.

17.4.2 Data Fusion and Data Merging

Data fusion or data merging could, in theory, be used to produce highly spatially detailed population microdata by combining, for example, data from census returns with data collected by credit agencies, vital registration statistics, and so on. Data fusion works by linking together records from different data sets that share a unique identifier, such as National Insurance number or address, while data merging links records from different data sets if they share a core set of common characteristics.

In practice, data confidentiality, commercial sensitivity, and civil liberty issues mean that neither such approach is likely to be pursued widely within the United Kingdom. Disregarding a number of commercial data sets that link together mainly financial information from a range of sources, the major UK data set created by data fusion is the Longitudinal Survey (LS). Although providing a rich seam of data for analysts interested in changes over time, the LS is hampered in terms of spatial resolution by being only a 1% sample of the UK population. In contrast, data merging appears to have been used primarily to construct databases for use in central government tax-benefit models, which operate at only a crude regional or district level of spatial resolution (e.g., POLIMOD—Redmond *et al.*, 1998). In other countries, particularly those operating a population register, the situation can be strikingly different (Holm *et al.*, 1996).

17.4.3 Iterative Proportional Fitting or Reweighting

Simpson and Middleton (1998) describe the use of iterative proportional fitting (IPF) to estimate unknown small-area tabulations by scaling known national level tabulations so that they are consistent with known but less detailed small-area tabulations that act as marginal totals. Although this is strictly speaking a technique for estimating unknown small-area tabulations, it is a trivial matter to convert the estimated small-area tabulation into an appropriately weighted set of small-area microdata. In principle, the approach described by Simpson and Middleton could be extended to include a wide range of variables through the use of additional marginal totals. Beckman *et al.* (1996) describe an example of such an approach being used to estimate synthetic microdata for census tracts in the United States. However, Gomulka (1992) points out that iterative proportional fitting is really just a special case of the more general problem of reweighting or grossing up sample survey microdata. The only difference to be borne in mind here is that, whereas the usual situation involves the sample population being reweighted upwards, the likely excess of the sample size over the actual small-area population being estimated typically leads to a situation in which the sample is 'factored down'.

A further variant of reweighting is to select that combination of households from the available microdata that, when tabulated, best fit known constraints for the selected small area. A highly simplified example of this combinatorial optimization approach is presented in Figure 17.1. In effect, this is equivalent to an integer weighting scheme in which the majority of households in the survey sample are weighted to zero. The difference lies in the heuristic adopted to find the best 'weighting' scheme. The number of small-area tabulations that can be used as constraints is limited only by the number of tables available and the amount of computing time taken. Williamson *et al.* (1998) describe this approach in more detail and offer some preliminary evaluations of its performance in replicating known small-area distributions.

Mitchell *et al.* (1998) combined elements of a stratified sampling approach with combinatorial optimization. They created synthetic ED populations by randomly drawing households from the SAR subject to stratification by household type (15 types based on a k-means cluster analysis of age, sex, socio-economic group, family type, number of rooms, tenure, and household space type). The approximate number of households of each cluster type present in an ED was determined using a specially trained Artificial Neural Network, on the basis of published small-area tabulations. For each ED, 100 different combination of households were randomly drawn, and the combination that best fitted selected local area constraints was retained. Their initial results appear promising, although again a more thorough evaluation remains to be undertaken.

17.4.4 Synthetic Reconstruction or Imputation

A final approach to small-area population estimation is to reconstruct the population by repeated Monte Carlo sampling from a chain of linked conditional probabilities. For example, one published census tabulation reports the number of household heads by age, sex, and marital status in each small area. After converting this table into a list (microdata), the age of each household head's spouse (if any) can be determined by Monte Carlo sampling from the conditional probability of spouse's age, given age, sex, and marital status of partner. As a next step, the economic activity of each adult in the household could be determined by sampling from a conditional probability of economic activity, given age, sex, and marital status (see Figure 17.2). In similar fashion, a wide range of additional census and non-census based characteristics, including number and age of children and other adult household members (if any) can be reconstructed, leading eventually to a set

Step 1: Obtain sample survey microdata and small area constraints

Survey microdata

Household	Characteristics: size	adults	children
(a)	2	2	0
(b)	2	1	1
(c)	4	2	2
(d)	1	1	0
(e)	3	2	1

Known small area constraints [Published small area census tabulations]

1. Household size (persons per household)

Household size	Frequency
1	1
2	0
3	0
4	1
5+	0
Total	**2**

2. Age of occupants

Type of person	Frequency
adult	3
child	2

Step 2: Randomly select *two* households from survey sample [(a) & (e)] to act as an initial small-area microdata estimate

Step 3: Tabulate selected households and calculate (absolute) difference from known small-area constraints

Household size	Estimated frequency (i)	Observed frequency (ii)	Absolute difference | (i)-(ii) |
1	0	1	1
2	1	0	1
3	1	0	1
4	0	1	1
5+	0	0	0
		Sub-total:	4

Age	Estimated frequency (i)	Observed frequency (ii)	Absolute difference | (i)-(ii) |
adult	4	3	1
child	1	2	1
		Sub-total:	2

Total absolute difference = 4 + 2 = **6**

Step 4: Randomly select one of selected households (a or e). Replace with another household selected at random from the survey sample, provided this leads to a reduced total absolute difference

Households selected: (d) & (e) [Household (a) replaced]

Tabulate selection and calculate (absolute) difference from known constraints

Household size	Estimated frequency (i)	Observed frequency (ii)	Absolute difference | (i)-(ii) |
1	1	1	0
2	0	0	0
3	1	0	1
4	0	1	1
5+	0	0	0
		Sub-total:	2

Age	Estimated frequency (i)	Observed frequency (ii)	Absolute difference | (i)-(ii) |
adult	3	3	0
child	1	2	1
		Sub-total:	1

Total absolute difference = 2 + 1 = **3**

Step 5: Repeat step 4 until no further reduction in total absolute difference is possible:

Result: Final selected households: (c) & (d)

Household size	Estimated frequency (i)	Observed frequency (ii)	Absolute difference | (i)-(ii) |
1	1	1	0
2	0	0	0
3	0	0	0
4	1	1	0
5+	0	0	0
		Sub-total:	0

Age	Estimated frequency (i)	Observed frequency (ii)	Absolute difference | (i)-(ii) |
adult	3	3	0
child	2	2	0
		Sub-total:	0

Total absolute difference = 0 + 0 = **0**

Figure 17.1 A simplified combinatorial optimization process.

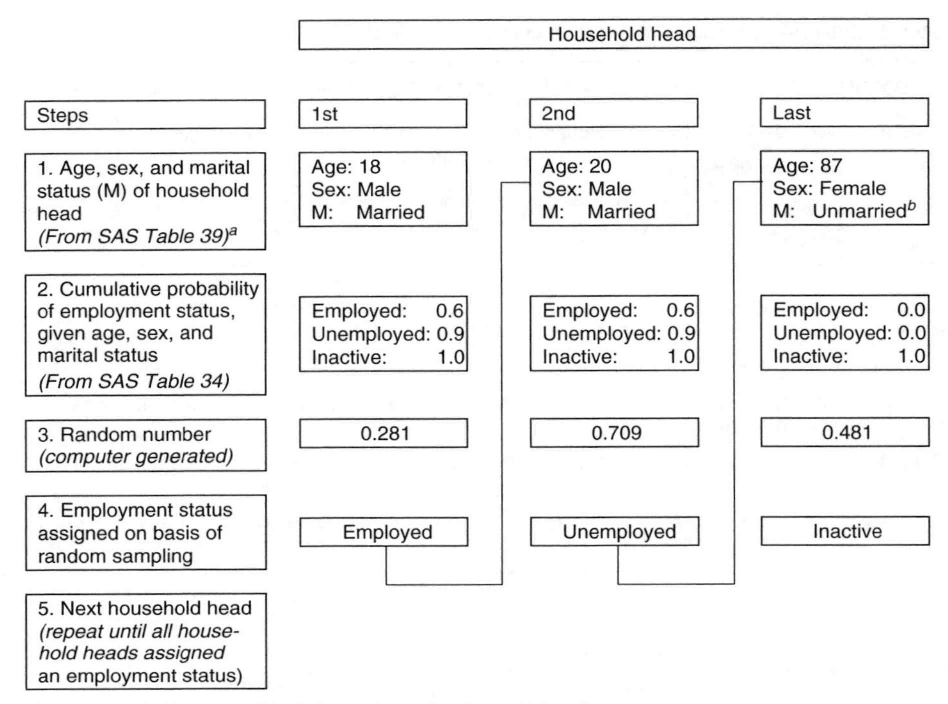

Figure 17.2 A simplified synthetic reconstruction procedure (after Clarke (1996, Figure 1)).

of fully specified small-area microdata. The requisite small-area specific conditional probabilities can be calculated either directly, from published small-area census tabulations, or indirectly, using published small-area census tabulations constrained to more detailed higher level tabulations using techniques such as iterative proportional fitting.

From this description, it may be apparent that synthetic reconstruction is really just an extended application of the sorts of data imputation techniques commonly used when preparing survey data for analysis. And, indeed, there is no reason for the sampling from conditional chain-probabilities not to be replaced or supplemented by one or more of the data imputation techniques discussed in Section 17.3. The difference lies in the extent of the imputation and, currently, in the use of small-area specific probabilities. Examples of this approach include SYNTHESIS, a model developed by Birkin and Clarke (1988) to estimate small-area income distributions, OLDCARE, a model developed by Williamson (1992, 1996) to model community care services for the elderly, and UPDATE, a fully dynamic microsimulation model developed by Duley (1989) to update small-area populations between censuses. A variation on this theme has recently been developed by Wanders (1998). Profiles of likely migrant characteristics, created by k-means cluster analysis of regional microdata, are used to impute small-area migrant characteristics, given distance migrated.

17.5 AN EVALUATION OF THE ALTERNATIVES

17.5.1 Computational Advantages

Stratified sampling offers the single quickest route to creating synthetic small-area microdata. However, unless the number of constraints used to stratify the sampling procedure is large, the resultant microdata are likely to represent relatively crude estimates of the actual underlying population distributions. Conversely, as the number of constraints increases, the speed and simplicity of stratification decreases to the point at which some form of reweighting or combinatorial optimization–based approach becomes more appropriate.

After stratified sampling, reweighting is computationally both the simplest to program and the quickest to compute, but results in fractions of households and individuals. At higher spatial scales, this problem can be addressed by rounding of results with little loss. However, as the spatial unit of interest decreases in size, tabulations become sparser, so that the rounding up or down of results can lead to serious population underestimation. In addition, weights have to be stored for every household in the original survey microdata for each small-area population estimated.

Synthetic reconstruction overcomes the problem of rounding, as every individual and household in the population is represented as a separate record. However, this gain comes at some cost. Synthetic reconstruction programs have to be specifically tailored to fit available national and small-area data sets. As a result, program development typically takes a considerable period of time (person months), although final run-times should be more or less equivalent to those needed for sample reweighting. The list representation of every individual and household in an area also carries a relatively high storage overhead, although well within the capacity of modern personal computers.

Combinatorial optimization falls somewhere between the two previous approaches. The data storage requirement is minimal, being limited to a list of the unique IDs of the sample survey households identified as best representing each small-area population. In addition, there is no reason for reprogramming to incorporate a different set of constraints to take any longer than that needed to reprogramme the constraints for use in a reweighting algorithm. However, these gains come at the expense of greatly extended computing times.

In all cases, it should be pointed out that those interrelationships between individuals and households not directly constrained in the population estimation process cannot by any means be assumed to be 'well-behaved'. However, this is no different to the situation that pertains when reweighting or imputing sample survey data in an aspatial manner, and is only to be expected of any process that attempts to estimate or approximate an unknown distribution. In addition, it should be noted that, for the reweighting and combinatorial optimization–based approaches, an acceptable solution might not be achievable if an appropriate mix of households is not available in the sample being 'weighted'. The appropriate method to adopt, therefore, depends on the task in hand.

17.5.2 Performance Advantages

More important, of course, is the degree to which each approach produces accurate and reliable spatially detailed microdata. Unfortunately, for all of the approaches outlined in the preceding text, only very tentative and preliminary evaluations have been published to date. In fact, the problem of evaluating synthetic data can appear intractable. By their very nature, such data are an attempt to fill gaps in knowledge. Therefore, there is no publicly available 'ground truth' against which to test them. Often the best that can be done is simply to indicate the general robustness of the estimates made, and their sensitivity to breaches of underlying assumptions. In response to these problems, an ongoing ESRC-funded project is currently attempting a thorough evaluation of two of the approaches (combinatorial optimization and synthetic reconstruction) as the precursor to the 'creation of a validated set of national small-area population microdata'. The results from this project will be widely disseminated, including the deposition of a set of validated synthetic small-area microdata at the Data Archive. Progress to date may be checked by accessing the project Web site (see Box 17.1). For combinatorial optimization–based approaches, unresolved issues at present include the number of constraints to be used, whether or not it is better to limit household selection to households that come from the same region as the small-area population being synthesized. For synthetic reconstruction, the number and order of the conditional probability chains sample from is unclear. In all cases, appropriate measures of goodness-of-fit are debatable (see Voas and Williamson, 1998) and remain a focus for further work.

17.6 CONCLUSION

The increasing interest in spatially detailed synthetic microdata looks set to continue into the future. As the sophistication of survey analysis also continues to increase, so does the acceptability of the various data imputation and reweighting techniques that contribute to the estimation of synthetic microdata. For example, plans for a 'one number census' in 2001 (see Chapter 21) suggest that the 2001 Census will rely upon such imputation techniques even more heavily than the last one. However, a note of caution is still appropriate. In 1993, the US Census bureau had their official 'synthetic' estimates of area-specific head counts overturned in the law courts as legally unsupportable modification of the raw data (Breiman, 1994). This decision had a number of roots. First, there was perhaps a lack of understanding and, therefore, trust by the courts of the statistically sophisticated techniques the bureau used—the problem of the 'black-box'. Second, a number of academics argued against these synthetic estimates on statistical grounds (as reviewed in Belin and Rolph, 1994; Freedman and Wachter, 1994). They argued either that alternative data adjustment techniques should have been used—the problem of sensitivity to initial choices—or that the quality of the resultant data could not be readily evaluated.

In fact, this debate rather neatly crystallizes the wider debate about the virtue of, and need for, synthetic data of any description. Publicly available microdata suffer from lack of spatial detail, aggregated response categories, restricted sample size, and question range. Synthetic microdata, which attempt to redress shortcomings by statistical means, are never going to entirely solve these problems. In particular, the resulting microdata are likely to understate inter-area heterogeneity (and intra-area homogeneity). However, the contention remains that these synthetic estimates are likely to be better than the alternative solution, which is to do nothing. This is effectively the same philosophy that underlies the revised small-area population estimates of demographic, ethnic, and other social counts arising out of the Estimating with Confidence project (see Chapter 12) and the use of imputation methods in Census Offices worldwide. Exactly how much better, and in what ways, is the subject of the ongoing ESRC project mentioned earlier.

Box 17.1 Key web links for Chapter 17.

http://pcwww.liv.ac.uk/~william/microdata	Population Microdata Unit, Department of Geography, University of Liverpool: Progress to date on ESRC-funded program to create a national set of validated small-area population microdata
http://www.cs.york.ac.uk/euredit	European Union–funded program for the development and evaluation of new methods for editing and imputation

Box 17.2 Additional CDS resources for Chapter 17.

Item imputation	Appendix to chapter dealing with item imputation

Part V

Interaction Data

18

Migration Data from the Census

Phil Rees, Frank Thomas, and Oliver Duke-Williams

SYNOPSIS

The chapter makes available to the census user a comprehensive inventory of the migration statistics output from the 1991 Census together with an account of the shape of the enhanced set of migration data likely to be produced from the 2001 Census. Migration statistics are provided at national, regional, local authority and small area scales in flow summary tables. Special Migration Statistics (SMS) hold data on origin to destination flows, while information on the relocation of students between parental domicile and term-time residence is available in a special table. The proposals for the 2001 Census envisage production of enhanced migration, commuting and journey to school statistics. The academic community provides the best means of accessing these complex data on-line via bespoke or adapted general software. Details of how to access these extraction engines are given in the chapter.

18.1 INTRODUCTION

The importance of studying intra-national migration has long been recognized. Migration is now a more important element in sub-national population redistribution across regions and localities than differences in natural change in the United Kingdom. Migration is one of the key processes through which cities and regions respond to the changing geography of production and the new spatial division of labour. Consequently, there is a strong demand for detailed and accurate migration estimates for small areas, which play a vital role in sub-national population forecasting, that to date has been provided only from the decennial population census.

The Office for National Statistics (ONS) and the General Register Office Scotland (GROS) provide information on migration between censuses, using patient re-registration data from the National Health Service Central Register. However, the spatial units employed are large: successively Family Practitioner Committee areas, Family Health Service Authority (FHSA) areas, and Health Authority areas, which correspond closely with shire counties, metropolitan districts, and groups of London boroughs. More recently, ONS has developed a new method of generating migration information from the National Health Service patient register (Chappell *et al.*, 2000), while GROS are harnessing the same data to provide inputs to small area estimation. Downloads of the individual postcoded records from FHSAs at the end of July in one year are compared with those at the end of July in the next year to generate counts of migrants between local government districts. In principle, this system could generate migrant flow statistics for any area that can be built from postcodes. The new system is being used in the production of mid-year population estimates for local government and health authorities from mid-1999 onwards. The publication of migration statistics from the system will, however, only occur after comparison in 2003 of the results of the FHSA download method with the results of the 2001 Census, which will remain the gold standard. Only the census will provide information about migrant characteristics beyond those of age and gender for the whole population.

The aims of this chapter are as follows:

- to outline the characteristics of census migration data and the issues to be addressed when using those data (Section 18.1);
- to review the migration data sets from the 1991 Census including user-enhanced data, and the improved data sets to be produced from the 2001 Census (the first parts of Sections 18.3–18.7); and
- to describe the methods for accessing the 1991 Census migration data and the improved methods being prepared for accessing 2001 Census data (the second parts of Sections 18.3–18.7).

18.2 THE CENSUS MIGRATION QUESTION AND THE INFORMATION IT PROVIDES

We begin by looking at the nature of the migration question used in the census. The 1991 Census asked a simple question about migration (Figure 18.1), and that used in the 2001 Census is very similar (Figure 18.2). However, the simplicity of the question hides three groups of issues: measurement concepts, population coverage and missing information.

Usual address one year ago

If the person's usual address one year ago (on the 21st April 1990) was the same as his or her current usual address (given in answer to question 7), please tick 'Same'. If not, tick 'Different' and write in the usual address one year ago.

If everyone on the form has moved from the same address, please write the address in full for the first person and indicate with an arrow that this applies to the other people on the form.

For a child born since the 21st April 1990, tick the 'Child under one' box.

Same as question 7 ☐ 1
Different ☐
Child under one ☐ 3

If different, please write the person's address and postcode on the 21st April 1990 below in BLOCK CAPITALS

Post-code

Source: OPCS and GROS (1992a), p.88

Figure 18.1 The migration question used in the 1991 Census.

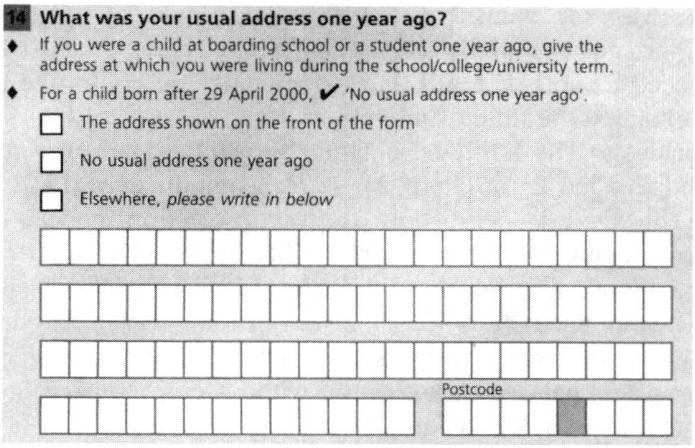

14 What was your usual address one year ago?

- If you were a child at boarding school or a student one year ago, give the address at which you were living during the school/college/university term.
- For a child born after 29 April 2000, ✔ 'No usual address one year ago'.

☐ The address shown on the front of the form

☐ No usual address one year ago

☐ Elsewhere, *please write in below*

Postcode

Source: ONS/GROS/NISRA (2000c: Annex A)

Figure 18.2 The migration question used in the 2001 Census.

18.2.1 Measurement Concepts and Issues

Usual residence definition. Underpinning migration statistics produced in the census is the notion of usual residence: the address where a person habitually lives or lives most of the time. With the expansion of second home ownership and of vacationing, less time is being spent at the habitual residence. Particular groups, such as students or the military, may spend large fractions of a year at more than one address. Rules are suggested to the census respondent about the address that should be regarded as the usual residence.

Time interval and demographic accounting concept used. Both 1991 and 2001 questions ask respondents for their usual residence one year previously, if they have moved. The census migration question thus applies to a fixed time interval, and measures transfers of residential location between the start and end of that interval. In demographic accounting terms, the census questions measure exist-survive transitions in location between two fixed points in time. The question is asked of all residents except children under one, born in the interval, whose migration status is undefined. The census question also fails to count the migrants dying during the year and the migrants who leave the country. The question counts 'migrants' (persons) rather than 'migrations' (events). Annual birth and death rates are about 10 per 1000 for the United Kingdom or about 1% each. Estimates by Boden *et al.* (1992, Table 2.4) for the period 1980 to 1981 suggest that migrations exceed migrants by about 8%. So we should expect a 10% difference between a count of exist-survive migrants and a count of all migrations in the same interval, although this difference will differ by scale of the migration. In using census migration data in either population estimation or projection, these attributes must be borne in mind. For the purposes of population estimation and projection, the difference between migrant counts and the number of migrations does not matter, as long as the appropriate model of population change is used (Rees, 1985). However, the census question covers only 1 year in 10 and must be used in conjunction with annual statistics on migration.

Age classification. Migrant counts from the 1991 Census and 2001 Census are reported by age at the time of the census, both by single years of age and by 5- and 10-year age groups. However, users need to note that this age classification refers to age at the end of the interval and not to age at the time of migration. Figure 18.3 shows two typical situations encountered in analysis. The left-hand diagram shows the age–time space occupied by migrants classified as aged 25 on April 21, 1991. Migrants aged 25 at the census could have migrated when aged 24 or 25. For use in a projection model, the migrant count should be used to compute the transition probability of the migration for persons aged 24 at the start of the interval. The right-hand diagram shows that persons aged 40 to 44 at the time of the census were aged 39 to 43 at the start of the interval. Any intensity measures computed from either of these situations cannot be directly compared with intensities computed from registers or used in projection without proper interpolation based on the use of the Lexis diagram (Bell and Rees, 2000).

Geographical coding. Census migrants are required to supply address details, including postcode, for their usual residence one year ago. There is no difficulty in geographical coding if the respondent provides a full postcode of the previous address. Using a directory of postcodes linked to the different census output zones, the responses can be coded to the administrative areas used in census geography (enumeration districts, output areas, wards, districts, counties, regions) or to the larger zones used in postal geography (postal sectors, postal districts, postal areas). Many respondents, however, fail to provide a full postcode. The full postcode can often be deduced from the partial postcode and address information.

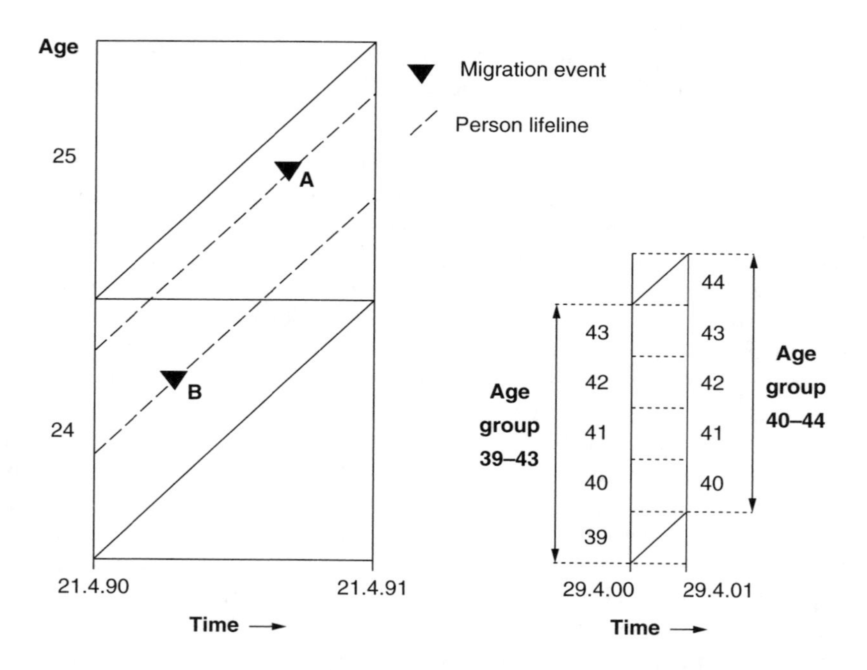

Figure 18.3 Time—age graphs (Lexis diagrams) showing the meaning of age-classified census migrant counts.

But in many cases there is insufficient information to derive an origin location and the previous address is coded as 'origin not stated'. This category made up 6% of all those counted as migrants in Great Britain (Champion *et al.*, 1998: 10). Researchers make a decision whether to ignore these migrants or to assign them, using an assumption, to origins. Usually, they are allocated in proportion to the number of fully recorded migrant origins. In the consultations prior to the 2001 Census, research has been undertaken by Andrew Mortimer and Keith Whitfield (GROS/NISRA/ONS 2000, Annex B) into the feasibility of imputing usual address one year ago using the census microdata records containing partial or wholly missing information. The idea is to apply a common technique employed in item imputation. Another record, near to the record that has missing migrant origin information, is sought that matches that record on a set of key attributes, and then the known migrant origin is borrowed to fill the gap in the record with the origin missing. The research used a sample of records, in which the origins were known, and compared imputed and real locations. The match was not overly precise but the structure of migration (e.g., distances moved) was reasonably reproduced. The Census Offices have committed themselves to develop the process of imputing missing postcodes for workplaces and migrant origins in the outputs from the 2001 Census.

Moving group classification. The analysis of migration by household structure is relatively undeveloped. The 1991 Census reported the migration of 'wholly moving households' and of households where the head has migrated. However, there are other types of migration: individual migrants moving into households with non-migrant members and migrants moving between, into, or out of communal establishments. Several migration researchers proposed classifications of migrants into 'moving units', which place all persons into a group with the same migration characteristics, but none are very satisfactory. The strategy used in other household tables to overcome this difficulty of potentially different behaviours for members of the household is to focus on the behaviour of household reference persons. Researchers interested in understanding the links between migration and household dynamics will generally need to use microdata (the SARs from the 1991 and 2001 Censuses, the Longitudinal Study (LS) linking 1971, 1981, 1991, and 2001 Censuses or the British Household Panel Study) rather than aggregate statistics. In the 2001 Census, moving groups within households will be identified.

18.2.2 Population Coverage Issues

Student migrants. One migrant group, students at higher education institutions, is not tracked well in the 1991 Census. There is a strong tradition in England of student migration to a non-local university or college at age 18, and students spend a larger part of the year at their term address than at their parental address (about 36 weeks versus 16 weeks). However, in the 1991 Census, the usual residence of undergraduate students was assumed to be their parental domicile. Neither movement between parental home and university term-time residence nor movement between term-time residence and first residence after graduation was recognized as a migration (but see Section 18.8 for information on Table 100 about student relocation). This leads to difficulties in making population estimates or projections for local government or small areas, because the concept used for population estimates includes students at their term-time residence. Many estimation and projection models for the 1990s (e.g., Rees, 1994) started with a population base that included students at their term-time locations, but the migration inflows and outflows to student towns or wards were insufficient to prevent a steady ageing in place of the base student age population. The English Sub-national Model (MVA Systematica with London Research Centre, 1996) adopted the expedient of separating out the student population and maintaining a constant age structure through the projections, ignoring the resulting inconsistency in specifying the demographic rates in the main model using populations at risk, which included students. The Geodemographic firms Experían and CACI produce small area population projections that apply either 'ageing on' or a fixed age structure to small area populations dependent on a prior analysis of their population composition (Experían, 2000). The 2001 Census question (Figure 18.2) will be an improvement on the 1991 Census question in that it will ask students for their term-time address as usual residence and so be able to capture migration between parental domicile and higher education residence. However, the migration after graduation may not be captured unless former students apply the same rule of using their term-time address as usual residence one year ago.

Infant migrants. In the 1991 Census, no information about the migration experience of children under one was provided. The user was left to estimate the number of infant migrants. Several techniques have been employed to fill this information gap. The number of infant migrants can be estimated as half of that of persons aged one at time of census or half the probability of migration can be applied to the number of births during the one-year interval. Or else, the number of babies born in the interval and who migrate after birth can be estimated by multiplying the number of women in the childbearing ages by the age-specific fertility rate and half of the migration probability. User lobbying in the consultation process (Rees, 1997b, 1997c) failed to persuade the Census Offices to extend the migration question for children aged under one to ask about the place of usual residence of the mother at time of maternity. However, the 2001 Census will offer an improvement on the 1991 Census information. The category of child under one year of age will still be used, but when processing is carried out such children will be assigned the usual address one year ago of their mother or father or guardian or nearest relation, where this is possible. There will be a small number of infants for whom it will not be possible to impute usual residence at the time of birth. These individuals will be included in the category 'no usual address one year ago'. The migration tables will contain the classification '0 years' (e.g., Table S05 in the Standard Tables as outlined in ONS/GROS/NISRA, 2000a, 2000b). How this information should be used will depend on the application. For use in population estimation and projection, the assumption might be made that half of those infants under one year of age were born prior to the mother's migration and are therefore themselves migrants, whereas the other half were born after the mother's migration and so are non-migrants.

18.2.3 Missing Information Issues

Undercounting. It is officially estimated that the 1991 Census count of the Great Britain's population of 54.890 millions was 1.209 millions or 2.2% lower than the most likely figure. This undercount poses a particularly serious problem for migration analysis because the majority of the undercount fell into the most migratory ages (20–29), because males were more subject to undercount than females and because the undercount was geographically concentrated in major metropolitan centres. Users employing migration and population data classified by age can correct this undercount using the general inflation factors published by OPCS/GROS (1993a, 1994a). Or else, they can use factors based on the revised estimates for small-area populations prepared by the Estimating-with-Confidence project (Simpson *et al.*, 1996), applying the ratios of census populations to revised estimates allowing under-enumeration to the census migration data. In Section 18.7.3, we describe the enhanced interdistrict migration estimates that allow under-enumeration prepared by Simpson and Middleton (1999). The One Number Census (ONC) process applied to the 2001 Census (see Chapter 21 for details) will impute records to account for the number and characteristics of missing people, including migrant status and origin.

Misreporting. The Census Validation Survey (Heady *et al.*, 1996: 14) suggests that 9.8% of respondents reported that they had not migrated, who in fact had moved. This means that, although the census recorded 5.35 million migrants resident in GB, the actual figure should have been around 5.87 million (Champion *et al.*, 1998: 9). The Census Coverage Survey associated with the 2001 Census will count migrant status and so provide information on misreporting. The 2001 Census form included the category 'no usual address one year ago', which may attract more responses than the strict definition adopted for current usual residence would warrant. Users should bear this in mind when using the migration information of the 2001 Census.

Characteristics of migrants at the start of the interval. There is little information captured about the status of the migrant at the start of the year at the origin location, apart from those characteristics that can be easily inferred (e.g., age one year ago) or which do not change (e.g., country of birth). For example, the important relationship between employment status and migration cannot properly be measured because only status after the migration is known.

18.2.4 Conclusion

Even given the issues outlined above, migration data from the census provide an excellent opportunity to study some of the most important migration research questions; the information is comprehensive and allows migrant characteristics to be identified. In fact, much of the migration data available to academics from the 1991 Census has been only lightly used. There are a number of reasons for this. The interaction data are probably the most complicated census data output and, in 1991 and 1981, the extraction of these flow data was extremely complicated. No group was funded to provide specialist academic advice and support. Knowledge of the different data sets was not widespread. In the remainder of this chapter, we aim to fill this gap.

18.3 MIGRATION DATA SETS: OVERVIEW

Six different data sets from the UK 1991 Census provide information about migration. The equivalent 2001 Census data sets are indicated in parentheses, where known.

- National Migration Statistics
- Regional Migration Statistics
- Local Base Statistics (LBS) (2001 Census Standard Tables)

- Small Area Statistics (SAS) (2001 Census Area Statistics)
- Special Migration Statistics (2001 Census Origin-Destination Statistics: Migration Tables)
- Samples of Anonymised Records (2001 Census Samples of Anonymised Records)
- Longitudinal Study linking the 1971, 1981, 1991 Censuses (Longitudinal Study extended to 2001).

National Migration Statistics have been produced from the 1961, 1971, 1981, and 1991 Censuses and *Regional Migration Statistics* were published from the 1971, 1981, and 1991 Censuses, although in 1991 they were produced only as machine-readable tables. The form that these national and regional statistics will take in the 2001 Census was, at time of writing, yet to be determined. Some migration data are available in the Small Area Statistics (SAS) from the 1971, 1981, and 1991 Censuses and in the Local Base Statistics (LBS) in the 1991 Census. The current proposals for the 2001 Census envisage much better migration being provided in the *Census Area Statistics* and *Standard Tables*. The 2001 Census processing strategy makes it possible to generate out-migration and in-migration totals for areas. Detailed flow data were provided in the SMS from the 1981 and 1991 Censuses and improved flow data are proposed for the 2001 Census *Origin-Destination Statistics*.

Much recent analysis on migration has taken advantage of the data provided in the *Sample of Anonymised Records* (SARs). These individual level data were especially useful for allowing migrant characteristics to be determined. However, the data were frustrating in some ways because migrant origins were coded to only nine standard regions in England together with Wales, Scotland, and Northern Ireland, whereas migrant destinations are available for 278 destinations. In the 2001 Census SARs, some improvement of origin coding is planned either with finer distance or area-type coding. The *Longitudinal Study* (LS) (for England and Wales) provides information about migration over 25 years by recording migrant locations at seven points in time: 1966, 1970, 1971, 1980, 1981, 1990, and 1991, to which will be added 2000 and 2001 when the new census is linked. Examples of migration analysis are given in Creeser and Gleave (2000). In discussions between the Census Offices and census users, there has been a strong demand for *a flexible tabulation system*, which users can employ to produce their own designed tables. This desire applies with particular force to migration statistics. Fast software (the SuperStar suite) required to generate the tables has been licensed by the Census Offices for use with the 2001 Census. Users will be able to order bespoke tables from the 2001 Census much more easily than in the past.

18.4 THE NATIONAL MIGRATION STATISTICS (NMS)

18.4.1 Nature of the NMS from the 1991 and 2001 Censuses

The NMS form part of the programme of published Topic Reports laid before Parliament after each census. They provide the most detailed cross-classifications of migrants for countries, regions, and a set of zones covering the whole of Great Britain. The 1991 Census NMS are organized in two parts occupying three bulky published volumes of nearly 1200 pages of statistics (OPCS and GROS, 1994a, 1994b, 1994c) for Great Britain with a separate volume on Northern Ireland (CONI, 1994). Integration of these into a single volume of text covering the United Kingdom, with accompanying CD containing detailed tables, is being considered for the 2001 Census.

Table 18.1 Contents of the 1991 Census NMS, Part 1 (100%).

Table	Title	Contents: cross-tabulation variables	Spatial units	Migrant count	Files [a]
1	Origins and destinations	*Sex by area of origin by area of residence at census*	Set One[b]	Residents	nm1a.csv to nm1 g.csv
2	Age and sex	*Sex by area of origin by age by area of residence at census*	Set One	Residents	nm2a.csv to nm2 g.csv
3	Marital status by age	*Age by sex by marital status by type of move*	Set One	Residents	nm3a.csv to nm3e.csv
4	Single years of age and marital status	*Age (single years) by sex by marital status by type of move*	Countries[c]	Residents	nm4a.csv to nm4 c.csv
5	Economic position	*Sex by economic position by age by type of move*	Countries	Residents 16+	nm5.csv
6	Migrant households	*Type of move (for migrant and wholly moving households) by economic position of household head, amenities, car availability, and persons per room by tenure, not in self contained accommodation and households with at least one child under one*	Countries	Households with residents	nm6a.csv to nm6b.cvs
7	Migrants in households	*Type of move by economic position of migrants aged 16 and over, car availability, and persons per room for all migrants, by tenure, not in self contained accommodation*	Countries	Residents in households	nm7a.csv to nm7b.csv
8	Composition of wholly moving households	*Type of move by household composition*	Countries	Wholly moving households	nm8.csv
9	Migrants in wholly moving households	*Type of move by age and limiting long-term illness*	Countries	Residents in wholly moving households	nm9.csv
10	Distance of move	*Sex by distance of move by age*	Set One	Residents	nm10.csv
11	Ethnic group	*Sex by age by ethnic group by type of move*	Countries	Residents	nm11.csv
12	Communal establishments	*Sex by type of establishment by limiting long-term illness by age by type of move*	Countries	Residents (non-staff) in communal establishments	nm12.csv

Source: OPCS/GROS (1994a, 1994b).

[a] The files are located on server *irwell.mimas.ac.uk* under directory */db/census91/topic/* with file names nm1a.csv to nm12.csv. The number refers to the table in the NMS. The letters a to g refer to table parts. The parts need to be integrated by the user into one text or spreadsheet file.

[b] Set One: Great Britain, England and Wales, England, regions, metropolitan counties, Inner London, Outer London, other main urban centres, regional remainders, Wales, Cardiff district, remainder of Wales, Scotland, main urban centres, remainder of Scotland (32 zones plus aggregations).

[c] Countries: Great Britain, England and Wales, England, Wales, Scotland.

The table contents of the NMS (Great Britain) are set out in Tables 18.1 and 18.2. It is useful to make a distinction between *Migration Area Tables* and *Migration Flow Tables*. *Migration Area Tables* classify residents in an area who report migration in the year prior to the census by a variety of attributes. Migration Area Tables use the *Type of move (TYMO)* classification to organize migrants by general flow categories within which

Table 18.2 Contents of the 1991 Census NMS, Part 2 (10%).

Table	Title	Contents: cross-tabulation variables	Spatial units	Migrant units	Files[a]
1	Origin and SEG	*Sex by area of origin by area of residence by economic position and SEG of employees, self-employed and unemployed*	Countries[b]	Residents 16+	nmt1.csv
2	Age and SEG	*Socio-economic group by sex by age by type of move*	Countries	Residents 16+ employees, self-employed, and unemployed	nmt2a.csv to nmt2b.csv
3	Economic position and employment status	*Sex by economic position by employment status by type of move*	Countries	Residents 16+	nmt3.csv
4	Occupation	*Sex by standard occupational classification (sub-major groups) by type of move*	Countries	Residents 16+ employees, and self-employed and unemployed	nmt4.csv
5	Industry	*Sex by industry divisions by type of move*	Countries	Residents 16+ employees and self-employed	nmt5.csv
6	Distance of move and SEG	*Sex by distance of move by economic position and socio-economic group of employees, self-employed, and unemployed*	Set One[c]	Residents 16+, economically active	nmt6a.csv to nmt6b.csv
7	Distance of move and occupation	*Sex by distance of move by standard occupational classification (sub-major groups)*	Set One	Residents 16+ employees, and self-employed and unemployed	nmt7.csv
8	Distance of move and industry	*Sex by distance of move by industry division*	Set One	Residents 16+ employees, and self-employed and unemployed	nmt8.csv

Source: OPCS/GROS (1994c).

[a]The files are located on server *irwell.mimas.ac.uk* under directory */db/census91/topic/* with file names nmt1a.csv to nmt12.csv. The letter 't' after 'nm' refers to the 10% sample data used to construct these data. The number refers to the table in the NMS. The letters a to g refer to table parts. The parts need to be integrated by the user into one text or spreadsheet file.

[b]Countries: Great Britain, England and Wales, England, Wales, Scotland.

[c]Set One: Great Britain, England and Wales, England, regions, metropolitan counties, Inner London, Outer London, other main urban centres, regional remainders, Wales, Cardiff district, remainder of Wales, Scotland, main urban centres, remainder of Scotland (32 zones plus aggregations).

they are further classified. The type of move classification varies with the geographical scale of the focus (called area x in the following text) but is basically of the form

a. Migrants resident in area x
b. Migrants moving within area x
c. Migrants moving within sub-areas of area x
d. Migrants moving into area x from the rest of the country
e. Migrants moving into area x from outside the country
f. Migrants moving into area x from origin not stated
g. Migrants moving from area x to the rest of the country.

Migration Flow Tables provide interaction information about both origins and destinations. Flow tables can as a result be very large. Only three of the twenty NMS tables contain flows. Flow tables from the 2001 Census will be supplied in computer readable form with summaries provided in any published report.

Two geographical classifications are used in the NMS. In 13 tables, a simple classification into component countries within Great Britain is used. In 7 tables, a 32-zone classification system (Rees, 1989) is used that identifies large urban centres and region remainders using selected districts and counties. This classification enabled researchers to examine the migrant flows into and out of large urban centres, although many researchers preferred to use their own classifications built up from ward and district migration matrices (Champion and Dorling, 1994; Rees *et al.*, 1996). From the 32 by 32 matrices in the NMS, researchers can extract inter-region matrices and inter-country matrices, but not matrices for inter-county flows or flows between other comparable classifications The SMS have to be used to generate those flow tables (see Section 18.7).

18.4.2 Access to the NMS

The 1991 Census NMS are available in computer format for manipulation and analysis. The academic community copies are stored on the Unix server *irwell.mimas.ac.uk* administered by Manchester Computing (University of Manchester) as a set of files under directory */db/census91/topic* with names in the generic form *nm*.csv*, where csv stands for comma-separated variables. The final columns in Tables 18.1 and 18.2 list the file names. The *csv* files load easily into spreadsheet packages. Spreadsheet files are also available from ONS Census Customer Services (*census.customerservices@ons.gov.uk*).

18.5 THE REGIONAL MIGRATION STATISTICS (RMS)

18.5.1 Nature of the RMS from the 1991 and 2001 Censuses

Tables 18.3 and 18.4 provide information on the content and geography of the Regional Migration Statistics from the 1991 Census. Table 1 in the RMS, Part 1 (100%), provides migration flows between and within districts within each region. Table 2 holds flows from origins outside the region to districts within the region, using the 32-zone system. Table 3 gives flows to destinations outside the region, using the 32-zone system, from districts within the region. The other 100% RMS parallel the NMS for those of the 32 zones within the region. The 10% Regional Migration Statistics also parallel the National Migration Statistics using the 32-zone system.

18.5.2 Access to the RMS

The Regional Migration Statistics from the 1991 Census were originally to be published as a set of printed volumes, one per region (as in the 1971 and 1981 Censuses), but because of cost they were published only as computer readable tables. The copies are stored on the Manchester Information and Associated Services (MIMAS) system of Manchester Computing as a set of files under directory */db/census91/regmig* with names in the generic form *xy*.csv*, where the letters *xy* are replaced by a two-letter abbreviation for the region and *csv* stands for comma-separated variables. Each table, however, spreads over several files and the user has to merge and edit these to obtain a convenient spreadsheet. Alternatively, spreadsheet files are also available from Census Customer Services of the ONS. Figure 18.4 describes the steps needed to access national or regional migration statistics from the 1991 Census in as generic a form as possible.

Table 18.3 Contents of the 1991 Census RMS, Part 1 (100%).

Table	Title	Contents: cross-tabulation variables	Spatial units	Migrant units	Files[a]
1	Migrants within [region]	*Sex by area of origin [in the region] by area of residence[in the region] at census*	Regions, counties, districts	Residents	xy1a.csv to xy1c.csv
2	Origins	*Sex by area of origin [outside region] by area of residence at census [in the region]*	Regions, counties, districts	Residents	xy2a.csv to xy2c.csv
3	Destinations	*Sex by area of origin [in the region] by area of residence at census [outside the region]*	Set One[b]	Residents	xy3a.csv to xy3g.csv
4	Age and employment	*Sex by age and employment status by type of move*	Counties, districts	Residents	xy4.csv
5	Economic position	*Sex by economic position by age by type of move*	Set One	Residents 16+	xy5.csv
6	Migrant households	*Type of move (for migrant and wholly moving households) by economic position of household head, amenities, car availability, and persons per room by tenure, not in self contained accommodation and households with at least one child under one*	Set One	Households with residents	xy6a.csv to xy6b.csv
7	Migrants in households	*Type of move by economic position of migrants aged 16 and over, car availability and persons per room for all migrants, by tenure, not in self contained accommodation*	Set One	Residents in households	xy7a.csv to xy7f.csv
8	Composition of wholly moving households	*Type of move by household composition*	Set One	Wholly moving households	xy8.csv
9	Migrants in wholly moving households	*Type of move by age and limiting long-term illness*	Set One	Residents in wholly moving households	xy9.csv
10	Ethnic group	*Sex by age by ethnic group by type of move*	Set One	Residents	xy10.csv
11	Communal establishments	*Sex by type of establishment by limiting long-term illness by age by type of move*	Set One	Residents (non-staff) in communal establishments	xy11.csv

Source: OPCS/GROS (1991, 1995a).

[a] The files are located on server *irwell.mimas.ac.uk* under directory */db/census91/regmig* with names in the generic form *cc415.rmwxyzh*, where *wxyz* is replaced by a four-letter label for each standard region and *h* stands for 100% data. There are also comma-separated versions of the 100% tables with generic name *xy*.csv*, where the letters *xy* are replaced by a two-letter abbreviation for the region and *csv* stands for comma-separated variables. The * is replaced by the table number.

[b] Set One: Great Britain, England and Wales, England, regions, metropolitan counties, Inner London, Outer London, other main urban centres, regional remainders, Wales, Cardiff district, remainder of Wales, Scotland, main urban centres, remainder of Scotland (32 zones plus aggregations).

Table 18.4 Contents of the 1991 Census Regional Migration Statistics, Part 2 (10%).

Table	Title	Contents: cross-tabulation variables	Spatial units	Migrant units	Files[a]
1	Origin and SEG	*Sex by area of origin by area of residence [in the region] by economic position and SEG of employees, self-employed and unemployed*	Set One[b]	Residents 16+ employees, self-employed and unemployed	xy101.csv
2	Destination and SEG	*Sex by area of residence [outside the region] by area of origin [in the region] by economic position and SEG of employees, self-employed and unemployed*	Regions, Wales, Scotland	Residents 16+ employees, self-employed and unemployed	xy102.csv
3	Economic position and employment status	*Sex by economic position by employment status by type of move*	Set One	Residents 16+	xy103.csv
4	Occupation	*Sex by standard occupational classification (sub-major groups) by type of move*	Set One	Residents 16+ employees, self-employed and unemployed	xy104.csv
5	Industry	*Sex by industry divisions by type of move*	Set One	Residents 16+ employees and self-employed	xy105.csv

Source: OPCS/GROS (1991, 1995b).

[a] The files are located on server *irwell.mimas.ac.uk* under directory */db/census91/regmig* with names in the generic form *cc415.rmwxyzt*, where *wxyz* is replaced by a four-letter label for each standard region and t stands for 10% data. There are also comma-separated versions of the 10% tables with generic name *xy10*.csv*, where the letters *xy* are replaced by a two-letter abbreviation for the region and *csv* stands for comma-separated variables, except that 'wal' is used for Wales and 'n' for North. The * is replaced by the table number with the South East (se) table being divided into parts a and b.

[b] Set One: Great Britain, England and Wales, England, regions, metropolitan counties, Inner London, Outer London, other main urban centres, regional remainders, Wales, Cardiff district, remainder of Wales, Scotland, main urban centres, remainder of Scotland (32 zones plus aggregations).

18.6 AREA MIGRATION STATISTICS (AMS)

18.6.1 Nature of the AMS from the 1991 and 2001 Censuses

There is some migration information in the 1991 Census LBS and SAS. Tables L15, L16, and L17 from the LBS and Tables S15, S16, and S17 from the SAS report on migrants resident in Local (districts, wards) or Small areas (enumeration districts, output areas). Tables L15 and S15 provide an age by sex by type of move classification. Tables L16 and S16 report on wholly moving households by household composition and type of move. Tables L17 and S17 simply report numbers of migrants among residents in areas by ethnic group.

The type of move (TYMO) classification is developed more elaborately at smaller scales. In the LBS (Table L15), the following TYMO classification is used

a. Moved within wards
b. Between wards but within district

Step	Description
(1)	Register for a username and password on the *irwell.mimas.ac.uk* server. See http://www.mimas.ac.uk/registration/ for details. You will also need to register as a user of the 1991 Census statistics. See http://census.ac.uk/cdu/ for details.
(2)	Activate the telnet interface on your PC desktop.
(3)	Login to *irwell.mimas.ac.uk.*
(4)	Change directory to */db/census91/topic.*
(5)	Copy the relevant file to your home directory on *irwell.mimas.ac.uk.* e.g., cp nm5.csv ~username/censuswork/nm5.csv.
(6)	Activate the file transfer protocol interface on your PC desktop.
(7)	Connect to the *irwell.mimas.ac.uk* server.
(8)	Set up the relevant directory on your PC desktop
(9)	Click on the file on *irwell.mimas.ac.uk* and drag and drop it into your PC directory.
(10)	Activate your spreadsheet software.
(11)	Open the file and load it into your spreadsheet.
(12)	Proceed with analysis of the statistics.

Note: These arrangements and those for the equivalent 2001 Census data may change in the future, in terms of either server or data unit or file formats or interface or registration. However, what will not change is the storage of the data at an on-line facility, to which academic users will have access.

Figure 18.4 How to access and transfer 1991 Census NMS or RMS files.

c. Between districts but within county
d. Between counties but within region
e. Between regions or from Scotland
f. From outside GB
g. Between neighbouring districts
h. Between neighbouring counties or Scottish regions.

The classification does not cover 'Migrants from the area of residence to the rest of GB' (out-migrants from the area) and there are no statistics on migrants with origin not stated, as these have been merged into the 'Moved within wards' category. It is not possible to measure the balance of in and out internal migrants properly and so the tables cannot be used easily in population change analysis or population estimation.

Considerable improvements are planned in the provision of migration information from the 2001 Census in the Standard Tables (wards and districts) and in the Census Area Statistics (output areas). Output statistics from the 2001 Census will be produced only when the national data set is declared complete after all returns has been processed and the first census imputation process (described in Chapter 21) has been successfully completed. This means that out-migrants can be counted for all areas, including those at the smallest scale. The description here of 2001 Census outputs is provisional and may be revised before the release of the statistics in 2003 to 2004.

The planned 2001 Census Standard Tables provide better migration tables for all wards and districts in the United Kingdom by including both in-migration and out-migration flows, together with immigrant flows from outside the United Kingdom. Residents in households are classified into the following migrant categories (ONS/GROS/NISRA, 2001d). This classification is to be used for Standard Table S08, for example, for geographical levels from ward to United Kingdom.

	Table category	*Interpretation*
a	Lived at same address	Non-migrants
b	Lived elsewhere one year ago, within same area	Within-area migrants
c	No usual address one year ago	Will need redistribution
	Inflow	
d	Lived elsewhere one year ago outside present area but within district (UA)	Internal in-migrants (local)
e	Lived elsewhere one year ago outside district (UA), but within United Kingdom	Internal in-migrants (not local)
f	Lived elsewhere one year ago outside the United Kingdom	Immigrants
	Outflow	
g	Moved out of area, but within district (UA)	Internal out-migrants (local)
h	Moved out of district (UA) but within United Kingdom	Internal out-migrants (not local)
	Balance	
i	Net Migration	Net internal migrants $= (d + e) - (g + h)$

This classification provides information on inflows and outflows within the whole United Kingdom for any area. The missing flow, which cannot by captured in the 2001 Census, is that of emigrants (moved out of the area to the rest of the world). There will be a demand for an estimate of the number of emigrants from areas in the year before the census and a method for redistributing the category 'No usual address one year ago'.

In the Census Area Statistics, the migration classification, as currently proposed, is a simplified version of that for Standard Tables (ONS/GROS/NISRA, 2000c, 2000d). However, users have requested that this classification be revised to that used in the Standard Tables in order to distinguish immigrants, internal in-migrants, and within-area migrants.

The Standard Table S05 records Gender and Age by Migration. The population base for Table S05 consists of people in households. Table S70 records migration statuses for people in communal establishments but not by age (though disaggregation has been requested). Table S06 tabulates Age (of Household Reference Person) and Dependent Children by Migration (of Households) and S07 gives Household Composition by Migration of Households. Theme Table T08 classifies people by migration status and, separately, by sex, age (12 groups), family status (9 categories), ethnic group (5 groups), illness (2 categories), and economic activity (11 statuses). Theme Table T09 provides information on the migration status of households and gender and new socio-economic classification of the household reference person.

18.6.2 Access to the AMS from the 1991 and 2001 Censuses

Access to these statistics is via the data extraction packages, SASPAC (London Research Centre, 1992) and CASWEB (Chapter 8).

18.7 THE SPECIAL MIGRATION STATISTICS (SMS) AND THE ORIGIN-DESTINATION STATISTICS (ODS)

18.7.1 The Nature of the 1991 Census SMS

The SMS form one of the specialist data sets produced from the 1991 Census of Population. They consist of matrices of migrant flows classified by origin area and destination area. The SMS used in the academic community were purchased by ESRC/JISC using the system developed by OPCS/ONS to produce flow statistics at user request. A similar system was in place for extracting SMS from the 1981 Census. These data are not currently on-line and must be obtained from Data Archive. An Economic and Social Research Council 2001 Census Development project directed by Boyle (University of St. Andrew's) has restored some of these data (simple ward to ward flows) and converted them to 1991 ward definitions.

Full details of the organization and content of the SMS are given in OPCS/GROS (1993b, 1993c, 1993d), in Flowerdew and Green (1993), and in Rees and Duke-Williams (1995a). Table 18.5 summarizes the information available for interward/postcode sector and interdistrict flows in the 1991 Census SMS. The 1981 and 1991 SMS proved to be large and complex data sets subject to considerable suppression. Most users found the statistics very difficult to use, although the 1991 situation was improved compared with the 1981.

The SMS from the 1991 Census as purchased by ESRC/JISC for academic research can be regarded as a pair of three-dimensional arrays (Figure 18.5). The array dimensions are, respectively, origins, destinations, and attributes, while the cell contents are counts of migrants falling in the origin-destination-attribute combinations.

1. SMS1 provides migration flow statistics between and within wards (England and Wales) and postcode sectors (Scotland).
2. SMS2 provides migration flow statistics between and within local government districts in Great Britain.

The *Set 1* array has a very large origin-destination face or matrix. The origins are composed of four types: (1) 9930 wards in England and Wales, (2) 1003 pseudo-postcode sectors in Scotland, (3) 99 areas outside Great Britain (96 individual countries or groupings of countries or other areas such as Northern Ireland plus categories for the 'Rest of the World' and 'Elsewhere'), and (4) an origin-not-stated category. The destinations are made up of the first two types of area. The third dimension is made of broad age–sex groups (Table SMS M01) and the categories of wholly moving households and residents in such households (Table SMS M02).

The *Set 2* array has a much smaller origin-destination face or matrix. The origins consist of three types of area: (1) 459 local government districts in Great Britain, (2) 98 areas outside Great Britain (as in Set 1), and (3) an origin-not-stated category. The attribute dimension can be divided into two parts. The first part consists of three tables: Table M01 classifying migrants by broad age groups and sex, Table M02 counting wholly moving households, and Table M03, which provides a more detailed age–sex classification (19 age groups by sex). These data are published without modification and are extremely valuable for demographic change analysis, population estimation, and forecasting.

Table 18.5 Tables available in the 1991 Census SMS for interarea flows.

Table	Areas in GB	Population base	Rows (no. of categories)	Columns (no. of categories)
M01	Wards/postcode sectors & districts	All migrants	Ages (5)	Sex (2)
M02	Wards/postcode sectors & districts	Wholly moving households	(1)	Counts of households, residents (2)
M03	Districts	All migrants	Ages (19)	Sex (2)
M04	Districts	All migrants	Sex (2)	Marital status (3)
M05	Districts	All migrants	(1)	Ethnic group (4)
M06	Districts	All migrants	Household status (2)	Limiting long-term illness (2)
M07	Districts	All migrants 16+	(1)	Economic position (7)
M08	Districts	Wholly moving households	(1)	Tenure (3) or (4 in Scotland)
M09	Districts	Wholly moving households	(1)	Sex (2); economic position (5)
M10	Districts	Residents in wholly moving households	(1)	Sex (2); economic position (5)
M11	Districts	All migrants	Gaelic speakers (1) /Welsh speakers (1)	Count (1)

Source: OPCS/GROS (1993c).

The second set of attributes by which the Set 2 district migration flows are classified consist of a further 38 counts organized in eight tables for Great Britain (Table 18.5). These socio-economic classifications were regarded by the Census Offices as a threat to the confidentiality of individual census microdata. The data were suppressed when the number of migrants in a flow fell below 10. Of the 210 222 interdistrict flows (459^2-459), some 135 916 were suppressed in the migrant tables and 110 268 in the wholly moving household tables.

18.7.2 The 1991 Census SMSGAPS Enhanced Data Set for Inter-District Migration

The difficulties posed by this wholesale suppression in the second part of the SMS Set 2 challenged Rees and Duke-Williams (1995b, 1997) to 'reverse engineer' the suppression to reconstruct the flow array for Tables M04 to M10 (Table 18.5). Three techniques were used: *logical data patching* (searching for rows, columns, or layers in the data in which only one element was suppressed and could be recovered by subtraction of all unsuppressed flows from the unsuppressed total), *iterative proportional fitting* (applied to arrays and marginals reduced by subtracting known terms), and *controlled rounding* of the remainders left after using the integer parts of the IPF solution. This new version of the SMS Set 2 has been made available for general use on the MIMAS service of Manchester

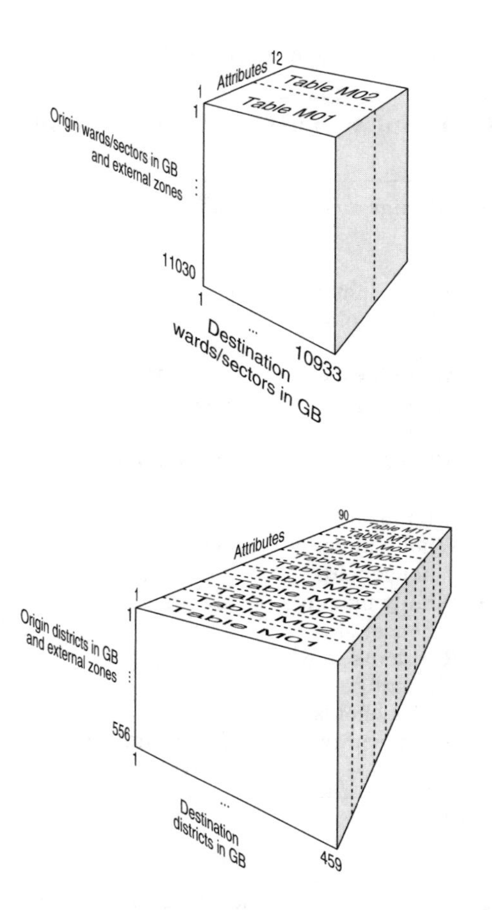

Figure 18.5 SMS represented as a pair of arrays.

Computing with the SMSTAB interface software. The new flow data have been used, for example, to analyse the pattern of interarea migration by a broad ethnic group (Rees and Duke-Williams 1995b). The SMSGAPS data array is the default interdistrict data set when using the SMSTAB extraction software.

18.7.3 The 1991 Census MIGPOP Data for Inter-District Migration Adjusted for Undercount

The SMSGAPS data set fills the holes in the interdistrict array left by official suppression but makes no adjustment for unit non-response (missing households and individuals). Simpson and Middleton (1999) make estimates of the undercount of migration in the 1991 Census. Building on the SMSGAPS data set produced by Rees and Duke-Williams (1997), Simpson and Middleton estimated new numbers of migrants to include non-respondents for each of the 459 × 459 flows within and between all districts of Britain. The estimates were made in two stages: (1) they calculated non-response factors by sex and age for each local authority district and by social group and (2) they used the non-response estimates to adjust the 459 × 459 interdistrict flows within Great Britain. Both stages involve iterative proportional adjustment of arrays of migrants classified by age and social group from the 2% Individual Sample of Anonymised Records to the full census information, adjusted for undercount, as marginals.

The Simpson and Middleton (1999) analysis of their adjusted migration estimates emphasizes the greater impact of the undercount on migrants (13% missed) than on residents in general (2% missed). The considerable increase in the size of migrant flows has differential impacts: it 'repopulates' the cities, reduces the net exodus from city to countryside, and eliminates the apparent excess of young female migrants in the published statistics. It is vital therefore when using the 1991 Census migration flows in further work (e.g., sub-national population projections) to use data adjusted for the undercount. See Box 18.1 for Web references to the adjusted data sets on the University of Manchester's MIMAS server.

18.7.4 The Nature of the 2001 Census ODS Migration Tables

The Origin-Destination Statistics (ODS) from the 2001 Census will be provided in three sets: migration tables, journey to work tables, and journey to school tables (Scotland only). Here, we discuss the first set. The ODS Migration Tables will be provided for two area levels. Level 1 tables will be issued for local authority area (i.e., local authority districts in two-tier counties and unitary authorities elsewhere). Level 2 will be produced for electoral wards (as of April 29, 2001) and additionally, information for postcode sectors will be provided in Scotland. The proposed 2001 ODS Migration Table layouts (ONS/GROS/NISRA, 1999a; superseded by GROS/NISRA/ONS, 2000; and subsequently by ONS/GROS/NISRA, 2001f) are reproduced in full in the *CDS Resources CD* in a pdf file containing the document *Census 2001: Origin-Destination Statistics (Final Specifications)* (ONS/GROS/NISRA, 2001f). The tables are extensions and improvements of those published in the 1991 Census SMS. Table 18.6 lists the tables proposed at the two levels and shows how they link back to the 1991 tables. The 2001 Census ODS Migration Tables will improve on the 1991 Census SMS in several ways: (1) they will not be subject to suppression, as disclosure control will be achieved through pre-tabulation record swapping, (2) they will capture migration by students more fully, (3) they will be adjusted for undercount prior to tabulation, and (4) missing origins will be imputed. However, in order to compare 1991 and 2001 Census migration results properly, the user will need to use adjusted 1991 migrant counts (MIGPOP), distribute origin-not-stated migrants in the 1991 Census across known origins, make some allowance for missing student migrants (see Section 18.8), and adjust for 'excess' counts of persons reporting 'no usual address one year ago'.

18.7.5 Access to the SMS and the ODS

The SMS data sets are large and complex in organization. They are held on the MIMAS server for access by registered census users (see Box 18.1). The 1991 Census SMS can be accessed using one of two interface packages: (1) SMSTAB or (2) WICID. Box 18.1 gives detailed references to instructions and Box 18.3 to related file locations.

SMSTAB was written originally by Oliver Duke-Williams as software to read, check, and verify the SMS data sets (Rees and Duke-Williams, 1995a), which was adapted for

Table 18.6 Proposed interarea migration tables for the 2001 Census Origin-Destination Statistics.

Table	Population base	Rows (no. of categories, not including totals or subtotals)	Columns (no. of categories, not including totals or subtotals)	Equivalent table in 1991 SMS
Set SMS1 (level 1, local authority)				
M101	Migrants	Age (25)	Gender (2)	M03
M102	Migrants	Family status of migrant (12)	Gender (2)	M04
M103	Migrants	Ethnic group (6)	Gender (2)	M06
M104	Migrants	Gender by limiting long-term illness (8)	Age (3)	M05
M105	Migrants aged 16–74	Economic activity (11)	Gender (2)	M07
M106	Moving groups of migrants in households	Groups, residents in moving groups (2)	Wholly moving households, other moving groups by persons in groups (6)	M02
M107	Moving groups of migrants in households	Tenure (4)	Wholly moving households, other moving groups by persons in groups (6)	M08
M108	Moving groups of migrants in households	Gender and economic position of head of group (22)	Wholly moving households, other moving groups by persons in groups (6)	M09
M109	Moving groups of migrants in households	Gender by NS-SEC of head of group (22)	Wholly moving households, other moving groups by persons in groups (6)	M09
M110	Migrants resident in Scotland/Wales/Northern Ireland who know Gaelic/Welsh/Irish	Age (10)	Gender (2)	M11

Table 18.6 (*continued*)

Table	Population base	Rows (no. of categories, not including totals or subtotals)	Columns (no. of categories, not including totals or subtotals)	Equivalent table in 1991 SMS
Set SMS2 (level 2, ward/postcode sector)				
M201	Migrants	Age (13)	Gender (2)	M01
M202	Moving groups of migrants in households	Groups, residents in moving groups (2)	Wholly moving households, other moving groups (2)	M02
M203	Migrants	Ethnic group (2)	Gender (2)	M05
M204	Moving groups of migrants in households	NS-SEC of head of group (11)	Wholly moving households, other moving groups (2)	M09

Source: ONS/GROS/ NISRA (2001f).

general use on the MIMAS service of Manchester Computing (details of the software are provided via the *man smstab* command on MIMAS or the Web pages cited in Box 18.1). The user runs SMSTAB by attaching a set of options to the command when issued on-line or accumulates these in a script file. A moderate degree of *Unix* knowledge is needed by the user for success. The easiest way to use SMSTAB is to prepare command files off-line using an editor and then to issue a command such as *smstab–file smsjob.cmd* > *smsresults.txt*, which uses the commands assembled in the file *smsjob.cmd* and sends the output to a file called *smresults.txt*. Several lookup tables have been prepared for common aggregations of wards or sectors or districts that the user can select, or users can develop their own lookup tables. The package is particularly suited to the extraction of migrant-flow matrices for further analysis.

Web-based Interface for Census Interaction Data (WICID) is under current development by Duke-Williams and Stillwell at the University of Leeds. Users access the SMS data via the Web, building up extraction jobs to be run by following on-screen instructions. When all the necessary instructions have been assembled, a query is executed on the SMS database and requested extractions are generated. The WICID uses Web pages to inform users and receive instructions and so has the potential to make available census migration data to a much wider set of users. The software currently provides access to the raw SMS counts, the SMSGAPS counts, and the MIGPOP data from the 1991 Census and will be extended to connect to the 1991 Census Table 100 (see Section 18.8) and the ODS Migration Tables from the 2001 Census.

18.8 THE STUDENT MIGRATION STATISTICS (TABLE 100)

As explained in Section 18.2, the 1991 Census question failed to capture student movement to school or college or university on admission, or student movement from school or college or university on graduation. However, term-time residence of students was recorded and can be compared with usual residence reported (also known as *parental domicile*). Statistics on term-time residence are provided in LBS Table L10 and in SAS Table S10. A special table, Table 100 (see Box 18.3), has been created, which provides a complete matrix for Great Britain (GB) of the districts of usual residence by districts of term-time residence. There are two parts to the table: (a) the first contains resident students for every local authority and (b) the second contains present students resident outside GB for every local authority district. The flows are broken down by sex and age (ages 5–9 years, single years from 10 to 29, ages 30–34 and 35+).

18.9 CONCLUSION

This chapter has reviewed the nature of migration information provided in the 1991 Census of Population. The same question on migration has been included in the 2001 Census but because students are to be recorded at their term-time residence, we should gain better information for this growing and migratory group.

There is an extensive set of tables derived from the 1991 Census describing migration flows at national, regional, and local spatial scales, which have been underutilized. The chapter has presented the structure of these tables and also reviewed the nature of and improvements to the SMS, which hold data on origin to destination flows. From the 2001 Census, an enhanced set of such flow statistics will be produced under the Origin-Destination Statistics umbrella.

Migration data are complex to extract and use because of the double geography attached to flows. Special purpose access software has therefore been developed. Improvements to this software should make accessing and using 2001 Census migration data much easier. In carrying out analysis of these new data, users will want to make comparisons with the 1991 situation; this chapter provides the essentials for such an endeavour.

Box 18.1 Key web links for Chapter 18.

http://www.census.ac.uk/cdu/Datasets/1991_Census_datasets/Area_Stats/Topic_Stats.htm#2	Web pages describing the NMS
http://www.census.ac.uk/cdu/Datasets/1991_Census_datasets/Interaction_Stats/Regional_migration_tables.htm	Web pages describing the RMS
http://www.census.ac.uk/cdu/Datasets/Area_Stats	Area Statistics Web pages (see Chapter 8 for access details)
http://www.geog.leeds.ac.uk/people/o.duke-williams/sms.html	Web pages describing the SMS
http://www.geog.leeds.ac.uk/people/o.duke-williams/smstab/usage.html	Web pages describing how to use SMSTAB, Unix-based software for accessing SMS data sets

http://www.ccg.leeds.ac.uk/wicid/	Web pages describing WICID, Web-based software for accessing SMS data sets
http://www.census.ac.uk/cdu/Datasets/1991_Census_datasets/Area_Stats/Adjusted_data/Undercount_adjusted_census_data/MIGPOP.htm	Web pages describing the MIGPOP version of the SMS. MIGPOP adjusts selected tables of the SMSGAPS version for census undercount
http://www.census.ac.uk/cdu/Datasets/1991_Census_datasets/Interaction_Stats/Table_100.htm	Web pages describing the Table 100 resource (Student Usual and Term-Time Residence)

Box 18.2 Additional CDS resources for Chapter 18.

Chapter22\OriginDestination.pdf	CD Resource with full details of the 2001 Census Origin-Destination Statistics proposals (ONS/GROS/NISRA, 2001f)

Box 18.3 Server resources for Chapter 18.

irwell.mimas.ac.uk/db/census91/topic	Server/directory holding NMS csv files
irwell.mimas.ac.uk /db/census91/regmig	Server/directory holding RMS csv files
irwell.mimas.ac.uk/db/census91/sysfiles	Server/directory holding Area Statistics
irwell.mimas.ac.uk/db/census91/swssms	Server/directory holding ONS version of the SMS
irwell.mimas.ac.uk/db/census91/migpop	Server/directory containing the MIGPOP version of SMS district data adjusted for undercount
irwell.mimas.ac.uk/db/census91/Table 100	Server/directory containing Table 100 files

19

Workplace Data from the Census

Keith Cole, Martin Frost, and Frank Thomas

SYNOPSIS

This chapter outlines the structure of the workplace statistics available from the 1991 Census, describes how these data can be accessed, and reviews selected examples of their use in geographic analyses. These data provide vital information on journeys to work from residence areas to workplaces at a fine geographic level, but have, in the past, been difficult to extract and use. The plans for the provision of enhanced information from the 2001 Census are briefly reviewed.

19.1 INTRODUCTION

The Census of Population is best known as a source of information on the social and demographic characteristics of areas, and it is also an important source of data for measuring the economic characteristics of residents, workplaces, and the patterns of travel that connect the two. This information is derived from a small number of questions on the census form that ask respondents to identify their place of work, the type of job they hold, and the principal mode of transport used for travelling to work. Copies of the questions as they appear on the census form are available in the 1991 Census User's Guide (Dale and Marsh, 1993) and the 1991 Census Definitions (OPCS/GROS, 1992a). From these questions, a data set was developed that offers a range of information directly relevant to many issues of labour market structure and change, much of which is not available from any other source of economic data. Despite the processing being restricted to a 10% sample of forms in all existing censuses, the Workplace Statistics are a highly distinctive, if somewhat underused, data set.

The data are divided into three 'Parts', each of which is split into further tables. Part 'A' lists the characteristics of residents living within specified areas, who hold a job at the time of the census. The detailed composition of the tables is listed in the 1991 Census User Guide 52 (OPCS/GROS, 1992c) with the equivalent for the 1981 Census in 1981 User Guide 156 (OPCS/GROS, 1982). In broad terms, employed residents are classified by the nature of their employment (employee or self-employed), the number of hours they work, their age and family position, the length of their work journey and the mode of transport used, their access to cars within the household, their occupation, and their industry of employment. It is a range of measurements that is unmatched by any comparable national survey. The detail of the occupational and industrial classifications is far greater than that published by the Labour Force Survey (LFS), while the Workplace Statistics have the added advantage of being available for considerably smaller spatial units than the District base of the LFS as a result of their larger sample size.

The profiles of residents contained in Part 'A' are complemented by broadly equivalent details of workplaces in Part 'B'. Once again, these data are a richer source of information about the nature of employment than any alternative national source. They offer an occupational analysis, mode and distance of travel to work, family status and age of worker, and a division between employees and the self-employed—all of which are not found in the Annual Employment Survey (and its predecessors), which is the principal alternative for workplace employment information within small spatial units. The Workplace Statistics form the only record, from the workplace perspective of occupational structure and balance between the employed and the self-employed, for all spatial scales (generally below Local Authority Districts) where it cannot be assumed that the resident-based profiles of the LFS are representative of the structure of workplaces.

This distinctiveness of the Workplace Statistics increases in the case of Part 'C'. The tables within this part catalogue the characteristics of people undertaking a work journey (or staying at home to work), classified by potentially small spatial units. As with Parts 'A' and 'B', individual clients who purchase this 'special' set of tabulations can choose from a variety of zoning schemes in which resident zonings may not necessarily be the same as those applied to workplaces. Of particular academic interest is the ESRC/JISC purchase, which specified census wards as the zoning system for both residents and workplaces in England and Wales, with postcode sectors in Scotland.

The characteristics of travellers recorded in Part 'C' include age, employment status, hours of work, family position, length of journey and mode of travel, car access, and both an occupational and industrial classification. Whenever anyone is recorded as travelling between a pair of zones, all tables are produced, providing counts of travellers falling into the categories specified within the individual tables. In addition, separate sets of tables are produced for each zone (where there is a recorded entry) for all residents who work at home, who do not have a fixed workplace, who do not state their workplace clearly enough on the census form to allow coding, and who work abroad. There is also one small additional group: those that occupy a job within a zone, have completed a census form, but who specify that they live abroad. In 1991, there were just over 17 000 people recorded in this category.

As this outline suggests, Part 'C' is unique in its national scope, level of spatial resolution and classificatory detail. The National Travel Survey records travel behaviour, including work travel, for a sample of residents but cannot approach the spatial detail of journeys recorded in the Workplace Statistics. Individual travel surveys (such as the London Area Transportation Survey of 1991) can sometimes present a more detailed analysis of journey purposes and the personal characteristics of respondents, but none approaches the national coverage available from Part 'C'.

No data set, however, is free from disadvantages. Clearly, the Workplace Statistics share, with all other data based on the Census of Population, the problem that they are collected only rarely (once every ten years) in a setting wherein many other sources of employment and travel information are collected annually. In addition, the Workplace Statistics have a number of distinctive difficulties that arise from their method of collection and from the difficulties inherent is accessing large sets of interaction data. These two issues are considered in the next sections of this chapter.

19.2 THE BASES AND CHARACTER OF THE WORKPLACE STATISTICS

The source of the Workplace Statistics is the part of the census form that asks respondents to provide the address and postcode of their place of work, the type of work they do, the nature of their employer's activity, and the dominant mode of transport they use while travelling to work. As with many apparently simple census questions, it is sometimes difficult to deal flexibly with the complicated reality of many respondents' lives. Although the guidance notes accompanying the census form encourage respondents who move around at work to identify their dominant place of work, this is still within a conceptual framework that assumes that people have one place of residence and one place of work. Exceptions are allowed for respondents who have no clear place of work, such as travelling salesmen who have only periodic contact with any fixed work base, but the forms do not allow the identification of circumstances where people may visit more than one place of work regularly (and may hold more than one job), while doing this from more than one place of residence. Within a society where multiple home ownership is increasing alongside greater fragmentation of people's work functions, this is clearly a disadvantage and accounts, in part, for the small number of unusually long work journeys that are regularly recorded in the travel tables.

Possibly a more significant source of inaccuracy within the data arises from a lack of accuracy in respondents' completion of the census form. The population census is in a relatively strong position when identifying the location of residents through the addresses where census forms are delivered and subsequently retrieved. Residents themselves are likely to have a reasonably accurate memory of their own addresses. The Workplace Statistics, however, are dependent on the clarity with which respondents complete the addresses and postcodes of their place of employment, the nature of their work, and the activity of their employers.

On some forms, the information provided may be of such poor quality that it prevents accurate processing. Postcodes are particularly at risk from missing letters or digits. To minimize this problem in the processing of the 1991 Census, coders used the lists of workplace establishments compiled for the 1989 Census of Employment to complete the faulty addresses of employers whose name and location was clear. Despite these efforts at the processing stage, 3.0% of respondents were classified as having a 'Not Stated' workplace (with a further 0.8% who were members of the Armed Forces, classified as Not Stated). Taking these two groups together with those respondents who had no fixed workplaces (6.2%) means that about 10% of all residents who were known to be working could not be given an identifiable workplace. This amounted to just over 2 364 000 people in Britain in 1991.

It is difficult to 'ignore' 10% of the workforce, however, when calculating the numbers of workplaces within zones. The procedure used by the census is to allocate the workplaces of these people to their known zones of residence. For generally large zones such as local authority districts, such an approach is unlikely to introduce much distortion into the data as many people will live and work within the same zone. At the scale of census wards or small postcode areas, however, it will produce a systematic overestimation of the number of jobs in predominantly residential areas at the expense of employment centres. The overall magnitude of this effect on individual zones can be seen directly from Table B3 (in Part 'B') in which there is a count of Not Stated and Not Fixed workplaces. Profiles of the characteristics of these 'assigned' workers can be obtained by using Part 'C' in which residents falling into these categories are tabulated separately from those with both a known residence and a known place of work.

The problems raised by unusable occupations or employers' activities are less important. In both the occupational and industrial classifications less than 1.0% of respondents, nationally, were classified in the residual not stated or inadequately defined categories.

A final issue arising from the nature of the data relates to the use of postcodes to identify places of work. Postcoded workplaces are allocated to wards, districts, and counties through the Central Postcode Directory (CPD). This is not an exact process. The boundaries of postcode areas do not match those of wards, and there are known to be errors in the Directory that can result in the misplacement of individual postcodes. A review of the accuracy of the CPD undertaken by Office of Population Censuses and Surveys in the mid 1980s found that translation from postcodes to wards produced wards that were accurate in 93% of cases, but also found that only 72% of grid references contained in the CPD were accurate to within 100 m (Raper *et al.*, 1992).

In addition, postcodes are continuously changing, with new codes being created in areas of building and development while existing codes are amended. These effects can be particularly severe in areas of comprehensive development. It has long been claimed informally by the London Docklands Development Corporation that the version of the postcode directory used to process the 1991 Census results omitted some of the newly created codes in areas of dockland development that were current at the time of the census enumeration, producing a systematic underestimation of employment in that area.

Taking all these problems together, it is not surprising that the Workplace Statistics contain a number of apparently anomalous journeys. They contain many examples of people walking implausible distances to work or using underground services where none exist. Some of these problems may be particularly damaging when calculating average distances of travel in which a few very long distance movers can significantly distort such results. In many cases, use of both the median and mean travel distances can be a sensible check on possible distortions.

In conclusion, it should be stressed that no data set is free of difficulties and errors, particularly in the use of postcoded addresses where many surveys share the problem of translating these codes to wards through the CPD. The important conclusion, however, is that the characteristics of the workplace data produced by the census are fundamentally related to the way in which the census is conducted and the type of questions asked on the census form. Although this may seem obvious, it should not come as a surprise that the estimates of numbers of workplaces within zones produced from the Special Workplace Statistics (SWS) can be quite different from those published by the Annual Census of Employment (now the Annual Employment Survey), which pursues a very different operational method.

In general, the Annual Employment Survey will produce larger estimates. In 1991, while the Workplace Statistics estimated there to be 2 967 310 employees in London, the comparable figure from the Annual Census of Employment was 3 254 660, some 9.7% greater. A number of factors contribute to this difference. The Census of Employment measures jobs declared by employers. Some people hold more than one job, while only a single job is recorded in the Workplace Statistics. In addition, students are classified by the census separately from employees, and thus excluded from the Workplace Statistics, but not from the Census of Employment returns by the employers. Finally, as an employment centre such as London is a significant 'importer' of inward commuters, it will suffer a loss of any commuters who are given a 'Not Stated' classification in the processing of the census forms and whose workplaces are allocated to their zone of residence. It is surprisingly difficult to answer unambiguously the simple question—'how many jobs are there in this zone?'

19.3 THE ZONING SYSTEMS IN THE SPECIAL WORKPLACE STATISTICS

Customers can specify either symmetrical or asymmetrical zoning systems for their SWS orders within the following framework. For symmetrical zoning systems, the geographical basis of the zones of residence and workplace must be the same. For England and Wales, residence and workplace zones can be defined in terms of electoral wards, postcode sectors, or local authority districts. For Scotland, residence and workplace zones can be defined in terms of electoral wards, districts, postcode sectors, or Output Areas. For example, a County Council in England might wish to analyse journey to work flows within the county at ward level, but flows into and out of the county at local authority district level for neighbouring counties.

For asymmetrical zoning systems, the geographical basis of the zones of residence and workplace are different. For England and Wales, residence zones can be defined in terms of Enumeration Districts. In Scotland, residence zones can be defined in terms of Output Areas, electoral wards, districts, or postcode sectors. For both England and Wales, and Scotland, workplace zones can be defined in the same terms as used for the residence zones for a symmetrical order (e.g., wards) or unit postcodes. For example, a local authority district in England might wish to analyse journey to work flows within the district from electoral wards to postcode-based workplace zones (e.g., city centre, local airport, or industrial parks), but flows into and out of the district at local authority district level for neighbouring districts.

Enabling users to define residence or workplace zones either in terms of Enumeration Districts and/or unit postcodes does pose a confidentiality risk. By taking the difference between the SWS for two user-defined small zones that only differ marginally, it might be possible to obtain statistics for extremely small areas, such as a single unit postcode that might only contain one household. In order to minimize this problem of 'differencing', the Census Offices have defined a complex set of thresholding and suppression rules for the SWS, which are described below.

19.4 THRESHOLDING AND SUPPRESSION PROCEDURES

Although the SWS are based on the 10% sample, the Census Offices have defined a set of thresholding and suppression rules to prevent the inadvertent disclosure of information about identifiable households or individuals through the differencing of overlapping small areas. These rules vary, depending on whether the zoning system specified by the user is symmetrical or asymmetrical.

For symmetrical zoning systems, the full SWS Sets A, B, and C will be released if the number of employed residents in the 10% sample for each zone (e.g., ward in England and Wales) passes the Small Area Statistics (SAS) thresholds of 50 persons and 16 households.

If the zone fails the SAS thresholds and has greater than five or more persons working in the zone, then SWS Set B is released but SWS Set A is suppressed (except for a count of the total number of employed residents in that zone). If the zone fails the SAS thresholds and has less than five persons working in the zone, the SWS Sets A, B, and C are suppressed (excluding counts of the total number of employed residents and workers in that zone).

For asymmetrical zoning systems, which have been ordered by many local authorities, the SWS Set A and Set C tables will be released for the residence zone (e.g., ED or aggregations of EDs) if that zone passes the SAS thresholds. If the residence zone fails the SAS threshold, the SWS Set A and C is suppressed for that zone (except for a count of the total number of employed residents in that zone).

The SWS Set B and Set C tables will be released for the workplace zone (unit postcode or aggregations of unit postcodes) if that zone has five or more persons working in that zone. If the workplace zone fails this threshold, SWS Set B and C is suppressed for that zone (except for a count of the total number of persons working in each zone).

Suppressed counts in the SWS are not merged into neighbouring areas, as they are in the SAS and LBS. However, the consequences of this are not serious except where very small zones have been specified, for which sampling error is any case likely to make detailed tabulations unreliable.

19.5 MEANS OF ACCESSING THE DATA

The Workplace Statistics can be found in two forms. In a printed form, the Workplace and Transport to Work Topic Report (OPCS/GROS, 1994d) contains information on journey to work flows for the whole of Great Britain, down to local authority district level (including nine city centres and the three special workplace zones of Heathrow, Gatwick, and London Docklands).

Most of the data, at spatial scales smaller than districts, are only accessible in machine-readable form, either as a special order from the Office for National Statistics (ONS) or by using one of the data centres holding all or part of the data. The most comprehensive holding can be found at Manchester Information and Associated Services (MIMAS) (University of Manchester), where the full ESRC/JISC purchase of the Workplace Statistics is available. This covers all of England and Wales at the level of wards, with Scotland being classified at the level of pseudo-postcode sectors. These were the postcode areas current at the time of the census in April 1991, split in some cases to ensure that the areas used in the census could be aggregated into local authority Districts. Via MIMAS, Parts A and B can be accessed using SASPAC software in a way that is broadly similar to the Local Base Statistics (LBS) and SAS tables. For these tables, 1981 equivalents are also available, again in SASPAC format.

Accessing the flow data of Part C is a little more awkward. In the absence of a SASPAC product that will tackle the flow data contained in the census, there are two possible approaches. The software specifically purchased by ESRC for these data (together with the Special Migration Statistics (SMS)) is QUANVERT. The principal advantage in accessing the data in this way is speed. Selecting the flows that link a number of wards or Districts through QUANVERT takes, at most, only a few minutes of execution time. An alternative to QUANVERT is SWSTAB, a program originally developed as part of the ESRC/JISC Census Research Programme at the University of Leeds. This offers users a straightforward way of exploring the numbers and the characteristics of travellers between specified pairs of zones. However, the Fortran code of this program takes longer to read the large data file than QUANVERT. Both approaches are limited to data from the 1991 Census. A new software, WICID, which can be used to access both the 1991 SWS and 1991 SMS using a web interface, will be introduced in 2002 (see Chapter 18 for more details). It will also be possible to access 1981 Census SWS flows between wards converted to 1991 geography.

An alternative site for accessing the Workplace Statistics is NOMIS (the National Online Manpower Information Service), based at the University of Durham. The approach to accessing the data here is integrated within the same procedures used for all of the labour market related data held on this site. The limitation here, however, is that only Parts A and B are available to users (although for both 1981 and 1991); Part C is not available. In addition, NOMIS levies an access cost on all users, whereas data provided via MIMAS are free, at least to users in the academic sector.

19.6 SPATIAL AGGREGATION OF THE DATA

One of the key issues in the use of many small area data sets is how to aggregate small zones, such as census wards, into areas that have meaning in the context of the users' research objectives. In many circumstances, aggregations are determined by the policy issues that underpin the work. For example, analysis of change in the London Docklands is usually associated with selecting wards to produce a 'best fit' approximation to the boundary of the London Docklands Development Corporation. In other situations, aggregations to local authority districts might be the appropriate path. Such aggregations may, at times, be somewhat tedious but pose no more problems than would be experienced using the standard census data sets. There is only one problem for which potential users need to be prepared: in the raw machine-readable form in which the Workplace Statistics are supplied, zones are identified by number rather than the more conventional alphanumeric ward, district, and county codes. Lookup tables linking the numbering sequence on the ESRC/JISC data set at MIMAS (University of Manchester) to the standard alphanumeric sequence are available through the Census Dissemination Unit Web site (see Box 19.1).

In the analysis of the work-travel data contained within Part 'C', however, it is often the case that aggregations based on the administrative hierarchy of wards and districts cut across flows of employees within local labour markets that respond to influences such as distance from city centres, inward or outward direction of travel, and the overall distance of their commuting trips. In essence, these issues require the ability to identify the relative positions of the zones used in the analysis and the use of these positions in any aggregation of zones into larger units.

The most common method used to achieve this is to use centroids as a base for defining the position of zones and straight-line distance measured between them as a way of representing trip distances. It should be noted, however, that if wards are used as the basis for the analysis, the trip distances calculated by this method will not be the same as those recorded in the tables within Parts 'A', 'B', and 'C' of the Workplace Statistics. These distances are based on the postcode of the residence and workplace for each recorded trip. In England and Wales, the calculation uses the National Grid reference of the first address in the postcode contained within the Central Postcode Directory, usually given to the nearest 100 m. In Scotland, the references are to the nearest 10 m and refer to the centroid of the populated part of the postcode.

It should also be noted that, while wards are fairly compact areal units within large urban areas, their size increases greatly in rural areas or small towns where population densities are lower. In these settings, inter-centroid distance becomes an increasingly inaccurate representation of trip distances, particularly when population is unevenly distributed within the wards. This will have the greatest effect on short journeys that involve the crossing of a ward boundary; their apparent distance can be considerably exaggerated.

19.7 APPLICATIONS OF WORKPLACE DATA

In this section, attention turns from the scope and availability of the Workplace Statistics to a selection of applications that best display their distinctiveness. Within the limited space available, these examples concentrate on using the key feature of these data: the ability to identify either work-travel patterns or tabulated work-travel distances.

19.7.1 The Definition of 'Functional' Areas

The definition of functional urban areas and daily urban systems has a long history in Britain, based largely on the use of census data (e.g., see Hall, 1971; Drewett *et al.*, 1976; see also Chapter 5). Various attempts have used the level of commuting into and out from areas to assess their level of employment self-containment. When combined with notions of employment density and size, rules can be developed in which employment centres are identified and 'grown' by progressively attaching those surrounding areas that have relatively high levels of commuting dependence. The process is controlled by an arbitrary set of threshold values that, ultimately, determine the number of functional areas produced.

In part, these approaches have been used in a research setting to produce hierarchical schemes of urban classification in Britain, dividing the urban landscape into cores, rings, and outer rings, and combining these into 280 Local Labour Market Areas covering the whole of Britain (see Coombes *et al.*, 1982; Champion *et al.*, 1987).

The other avenue of development has been in response to the long-standing desire of ONS (and previously, the Department of Employment) to define reasonably self-contained labour market areas for the presentation of unemployment rates. Their underlying logic is that unemployment rates do not accurately represent the difficulty that people may face in obtaining work, unless they are calculated for areas in which residents and jobs are matched together with relatively little crossing of the boundaries by either inward or outward commuters. Defining these so-called *Travel to Work Areas* for the whole country requires a detailed national record of commuting flows, a requirement that can only be met from Part C of the SWS.

Later analyses, using the 1991 Workplace Statistics (ONS and Coombes, 1998) were accompanied by an explanation of the algorithm used to aggregate wards to Travel to Work Areas. Although the procedures are complex in detail, the broad principle was to identify areas that exceeded a minimum size of 3500 working residents and which showed a level of 70% to 75% self-containment of jobs and residents.

19.7.2 Labour Catchments and Associated Measures

The process of how individuals and households match their place of residence with their places of work within both urban and rural labour markets is important to public policies, but poorly understood. Two key areas of concern here are the question of 'balanced communities' in which there is an approximate balance between jobs and residents and the question of how the effects of regeneration policies can be diluted by in-commuters gaining jobs in areas targeted for job creation.

The concept of balanced communities has gained prominence within planning philosophy, partly as a result of the increasing pressure in recent years to plan settlement form in a way that reduces levels of demand for travel, particularly travel to work. Clearly, however, the concept of local self-containment has to be viewed against a background of potentially complex patterns of travel in which individual areas can have high levels of in-commuting and out-commuting simultaneously, even within the same labour force groups. The extent of this complexity has been recently investigated by Breheny, Foot, and Archer (1998) using the information in Parts 'A' and 'B' of the Workplace Statistics. Both parts contain tables showing the numbers of residents (or workers) who cross the boundaries of the zone and the distance they travel. Combining these data has demonstrated that, with present propensities to travel, local balance is elusive, emphasizing the need for public policies to address directly the propensities of residents to travel, possibly by road charging or increases in excise duty on fuel, in addition to physically based planning solutions.

The degree to which the openness of urban labour markets to inward and cross-commuting weakens the link between local residents and local work opportunities raises a related set of issues. While it is clear that urban labour markets are too complex to allow neat exclusive areas to be drawn on maps (Peck, 1997), it is equally clear that it would be misleading to assume that jobs created by regeneration will be filled exclusively by local residents. Clearly, the most desirable data would allow the tracing of individuals within the targeted areas, both before and after policy interventions. Such data are rare, but some knowledge of the commuting fields in which regeneration areas are set can provide indirect evidence of the likelihood of jobs being filled by relatively long-distance inward commuters.

The identification of the full range of distances over which people travel and their individual characteristics requires the use of Part 'C' data to be combined with the grid coordinates of zones to allow the estimates of travel distances to be made. The product of one such analysis is shown in Figure 19.1. This shows the profiles of distances travelled by men into jobs within the London Docklands. It shows clearly the much closer 'local' links between residents and jobs in manual employment in contrast to the other socio-economic categories in which work-travel trips have a significantly higher average length. In this example, travel distance is approximated by the straight-line distance between the centroids of origin and destination wards.

The flow data, however, also allow the shape of catchment areas to be examined as well as the data include the unique identification of both the ward of origin and the ward of destination for all work journeys. A view of the catchment area linked to jobs in the Docklands is shown in Figure 19.2. In this map, all districts that contribute at least 1.0% of Docklands employment are shaded. The strength of the link between Docklands and districts in the east, particularly in Essex, is strikingly clear.

19.7.3 The Environmental Implications of Worktravel

In an era of increasing sensitivity to issues of traffic congestion, environmental sustainability of travel, and the impacts of household growth on urban form, a new use of the

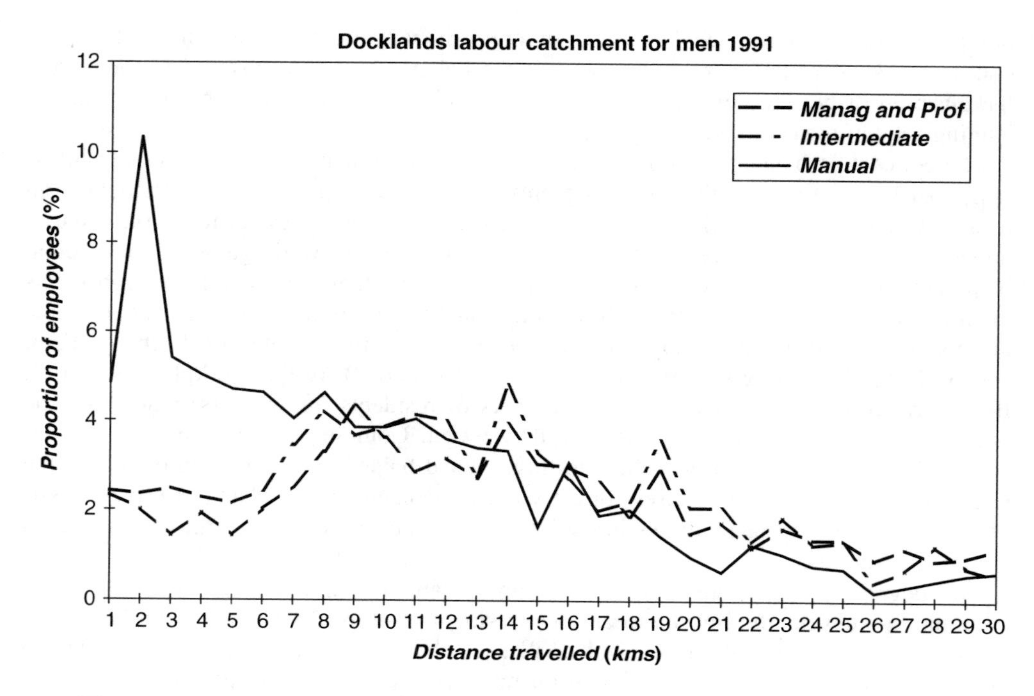

Figure 19.1 Distances travelled to work by men to jobs in London's Dockland.

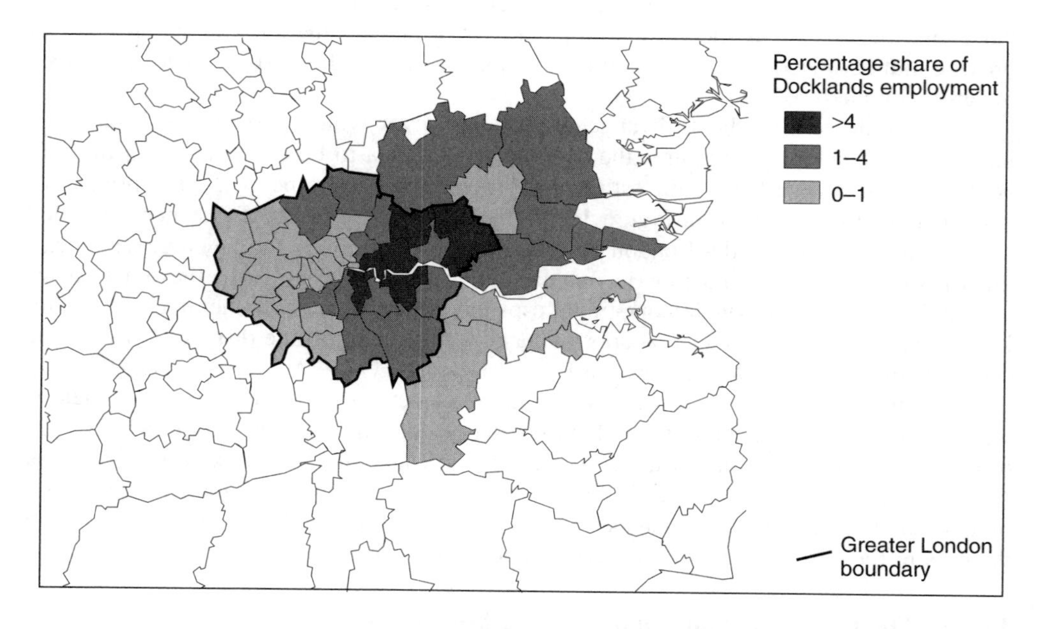

Figure 19.2 The catchment area for employment in London's Dockland.

Workplace Statistics has been in the assessment of the environmental impacts of work journeys. These have been given particular prominence in debates on urban form as they are, in aggregate, partially determined by the relative locations of employment centres and homes, both of which can be influenced to some degree through planning decisions.

The environmental impacts have been assessed using both the 'static' travel intensities of zones of residence and zones of employment (Banister and Banister, 1994) recorded in Parts 'A' and 'B' of the data, and the full travel data from the journey matrices recorded in Part 'C' (Frost *et al.*, 1997), using information from both 1981 and 1991 Censuses. Travel distances can be linked to energy through a range of energy consumption levels per kilometre for each mode of transport (for sources see, Commission of the European Communities, 1992; Department of Energy, 1990). There are some benefits to be gained from using the full travel data in Part 'C' of the Workplace Statistics as these data allow the direct measurement of both the distances travelled and modes of transport used. Use of Parts 'A' or 'B' in this setting requires an estimation procedure that combines evidence from the two separate tables that catalogue these characteristics.

An example of the application of the Workplace Statistics to the estimation of energy consumption is shown in Tables 19.1 and 19.2 (Frost, Linneker, and Spence, 1997). The relatively intense use of energy in both inward and outward commuting and its rapid rate of increase between 1981 and 1991 is clear from these results. It is possible, in a full analysis of these data, to go further and identify the contributions made by individual modes. If this is done, it can be seen that London's relatively favourable profile, given that it receives some of the country's longest commuting journeys, can be directly attributed to the place of rail transport in the capital's commuting structure.

Table 19.1 Energy use per capita, 1991 (million joules/per person trip).

City	Within	Out of	Into
London	9.9	73.3	63.2
Birmingham	11.7	90.9	70.0
Manchester	13.0	95.0	83.1

Source: Frost, Linneker and Spence, 1997, from SWS, 1991 Census; ESRC/JISC purchase.

Table 19.2 Change in total energy use, 1981 to 1991 (%).

City	Within	Out of	Into
London	1.6	65.7	16.8
Birmingham	17.0	55.7	36.8
Manchester	20.0	41.7	45.2

Source: Frost, Linneker, and Spence, 1997 from Special Workplace Statistics, 1991 Census; ESRC/JISC purchase.

19.8 PROPOSALS FOR SPECIAL WORKPLACE AND TRAVEL STATISTICS FROM THE 2001 CENSUS

As a result of extensive consultation with users and research by the General Register Office Scotland, an enhanced set of commuting statistics will be delivered by the Census Offices, based on the 2001 Census. Here, we précis the summary of proposals contained in ONS/GROS/NISRA (2001f: p. 4–5), which provides full specifications and is available on the ONS or GROS Web sites (Box 19.1) and the enclosed *CDS Resources* CD.

The *Special Workplace Statistics* (SWS) will be based on journeys to work and produced for England and Wales and Northern Ireland at three area levels. The SWS1 will house flows between local authorities, SWS2 flows between 2001 electoral wards, and SWS3 flows between Output Areas. The *Special Travel Statistics* STS will report on journeys to work or study for Scotland for the same area levels as SWS, except that STS2 will also be available for postcode sectors. Part of each STS table will correspond to the SWS for elsewhere in the United Kingdom and could be requested separately. The SWS and STS for the 2001 Census will contain the flow elements of the 1991 SWS, with outputs for workers by zone of destination having been moved into the Standard Area Statistics.

Compared with 1991, the 2001 SWS will be based on 100% of the forms collected in the census, which was the case only for Northern Ireland in 1991. The SWS together with the SWS part of the STS will be a single UK product for each area level. The geography will be more straightforward with symmetric zones only. The full postcode will be imputed for workplaces (United Kingdom) and places of study (Scotland), so that the problem of dealing with 'address not stated' categories will be avoided. If the standard products do not meet the user's precise needs, it will be possible to request tailored outputs for non-standard zones or containing non-standard tables, provided the user can pay the cost and the request is judged not to risk disclosure of information about identifiable individuals.

Box 19.1 Key web links for Chapter 19.

http://www.census.ac.uk/cdu/Software/	SWSTAB extraction software

Box 19.2 Additional CDS resources for Chapter 19.

Origin Destination.pdf	Final version of proposals for the Origin-Destination Statistics, including detailed specification of the SWS and the Special Travel Statistics (STS) (ONS/GROS/NISRA, 2001f)

Part VI

Planning for 2001 Census Outputs

20

New Questions for the 2001 Census

John Dixie and Daniel Dorling

SYNOPSIS

This chapter introduces the new and revised questions chosen by the Government for inclusion in the 2001 Census. It shows what form the questions take and discusses some of the key issues concerning those questions and highlights how they might be used in future census analysis. Although finally excluded from the 2001 Census, the case for and against an income question is also reviewed.

20.1 INTRODUCTION

The choice of census questions always involves a difficult balance between users' needs for additional information and the requirement to keep the burden on the public within reasonable bounds. This time the public consultation has been unprecedented in its scope and depth. However, there will remain issues of contention over what was asked in the census and what should have been asked (Rees, 1998b). Here, with the exception of income we concentrate only on those questions that were new or revised from 1991. We do not consider those questions that were put forward but excluded from further consideration early on in the consultation process (see HM Government, 1999 or Rees, 1999 for further information).

The final form of the census was decided by Parliament early in the year 2000. In short, the significant changes in census content from 1991 are new questions on:

general health;
provision of unpaid care;
time since last paid employment;
size of employer's organization; and
a voluntary question on religion;

and major revisions to the questions on:

qualifications;
relationship within the household;
ethnic group; and
accommodation.

A strong case has been made for information about income and a question was trialled in the census rehearsal. However, uncertainties about the impact upon response rates led the Government to abandon a question on income for the 2001 Census. Instead, the Government Statistical Service is undertaking research on whether needs could be met by alternative sources of data.

20.2 QUESTIONNAIRE DESIGN

The style of the census questionnaire changed between 1991 and 2001. The objective was to make the form as clear and easy to understand as possible by minimizing the instructions and the burden on respondents. The new census questionnaire was designed particularly with a view to increasing coverage (the proportion of households that returns a form). The changes also enabled the bulk of the responses to be captured by scanning and automatic mark and character recognition.

As the form had to be as compact and legible as possible, space on the form was at a premium. The complexity of the questions and any explanatory text had to be limited, and this constrained the topics that could be included.

> 'Topics shown in testing to require substantial explanation, such as proficiency in English, have not been proposed for inclusion.' (Moss, 1999: 29)

20.3 CHANGES IN POPULATION DEFINITIONS

In 2001, people were recorded where they were *usually resident*. One major change from the 1991 definition of 'usually resident' is that students and schoolchildren were enumerated as usually resident at their term-time address. In addition, basic demographic data (age, sex, marital status, and relationship) were collected at their parental address. This enables all children and students to be included in statistics on family composition.

20.4 NEW QUESTIONS

The criteria for accepting any topic (new or old) for inclusion in the census are that:

- there is a demonstrated need for the information;
- users' requirements cannot be adequately met by information from other sources;
- the topic should be shown, in tests, to have no significant adverse effect on the census as a whole, particularly the level of public response; and
- practicable questions can be devised to collect data that is sufficiently accurate to meet users' requirements. (HM Government, 1999).

New questions increase the cost and complexity of taking the census, and so their introduction must always be strongly justified.

Five new questions were proposed for inclusion in the 2001 Census. One of these, a question on religion, was voluntary. It is important that readers realize that these five questions have been scrutinized in great detail. A large number of other possible new questions were examined and excluded during the preliminary user consultation that began in April 1995 and continued up to June 1998.

The programme of testing possible questions for the 2001 Census has been much more extensive than for any previous census, and has covered both the possible new questions and those asked in previous censuses. It has included cognitive research into how people understand question wording, small-scale quantitative tests, and two major census Tests in 1997 and 1999. The objective has been to ensure that *all* questions proposed for the census meet the criteria of practicality and reliability.

In addition, the Office for National Statistics (ONS) has carried out a sample survey of more than 2000 households after the census rehearsal in April 1999 to measure the quality of responses to the proposed questions. This survey provided a further check on the effectiveness of the questions, and on the public's views on the topics asked, before the final decision on census content was taken by Parliament.

20.4.1 Income

There has been a long and sustained call for a new question to be asked on income in the 2001 Census (Dorling, 1999).

> 'A question on income has not previously been included in a census in the United Kingdom. However, consultation with users throughout 1995-98 clearly identified a widespread requirement for information on this topic. Income data is seen as a more discriminating variable than occupation or housing condition for the purposes of identifying areas of affluence or deprivation. Users consider that the information is critical in responding to Government initiatives on inequality, social exclusion and deprivation. They emphasize the importance of the availability of information on income for small geographical areas, and the ability to use this data in conjunction with other socio-economic information collected in the census. The main requirement is for a measure of household income, particularly in relation to households at the lower end of the income scale.' (Moss, 1999: 34)

If there had been an income question, it would have been asked of individuals, as this would have been much less intrusive than asking household income. The responses could have been summed up to give an approximate value for household income. The question would not have attempted to discern sources of income. It would also have contained a small number of income bands as the prime requirement of most users is for information about the less well-off, because people on higher incomes appear to object more strongly to stating their income, and because testing has shown that the width of the bands affects acceptability. The possible form of the question considered but not asked is shown in Figure 20.1.

For an idea of the uses to which the income question could have been put to if it were included in the census, one may look at the analyses of the US Censuses of population over the last few decades in which, for example, income information from the census is regularly used as an indicator of deprivation.

Table 20.1 shows the distribution of annual individual income for Britain in 1997, calculated from the British Household Panel Survey (BHPS), which had a sample of just over 9000 people answering this question. It illustrates the kind of information that could be produced for every ward in Britain if statistics were available at small area level.

At a regional level, the Labour Force Survey and the Family Resources Survey are useful sources of information. However, no official survey includes enough people in ethnic minority groups to be able to calculate reliably the degree of income inequality between different ethnic groups in Britain, let alone allowing for differences in age structure or other relevant variables.

What is your total current gross income from all sources?

- *Do not deduct* Tax, National Insurance, Superannuation or Health

Insurance payments

- Tick the box that covers your income

Count all income, including

Earnings

Pensions

Benefits

Interest from savings or investments

Rent from property

Other (for example maintenance payments, grants)

Per week		**or** **Per year (approximately)**
Nil	☐	Nil
Less than £60	☐	Less than £3000
£60 to £119	☐	£3000 to £5999
£120 to £199	☐	£6000 to £9999
£200 to £299	☐	£10 000 to £14 999
£300 to £479	☐	£15 000 to £24 999
£480 or more	☐	£25 000 or more

Figure 20.1 The proposed question about individual income.

It has been suggested that the census may not be the best way to provide this information at a small area level. It is claimed that there is evidence, both from the 1997 Census Test and from the more recent census rehearsal in 1999, that the question is not acceptable to some members of the public.

> 'Analysis of response rates in the 1997 Census Test suggested that the inclusion of an income question lowered overall response in terms of the proportion of census forms returned, from 57.4 per cent to 54.6 per cent.' (Moss, 1999: 34)

In addition, the programme of testing possible questions has shown that, even after allowing the banding of the response categories, the question may not be answered with accuracy.

The Census Offices decided that it would be of no overall benefit to the census users if information of limited precision about income were to be obtained at the cost of

Table 20.1 Individual annual income in Britain in 1997.

Income	Men (%)	Women (%)	Ratio (men:women)
Under £3000	10.4	23.8	0.4
£3000–5999	12.2	26.9	0.5
£6000–9999	19.4	22.7	0.9
£10 000–14 999	20.5	14.0	1.5
£15 000–25 000	24.2	9.6	2.5
Over £25 000	13.2	2.9	0.5
	100.0	100.0	1.0

Source: BHPS wave 7 (weighted) annual income by sex (1.9.96–1.9.97) (sample N = 9064).

introducing unknown bias into the whole data set because of increased under-enumeration. For this reason, as the White Paper says, '. . . the preferred approach is to identify possible alternative means of securing relevant information.' It is true that if the information were to be produced from administrative sources, an additional benefit might be that it could be published more frequently than once a decade, for example, annually or quarterly. However, administrative data cannot be linked to census records easily, thus denying the opportunity to investigate the relationship between income and other personal and area characteristics recorded in the census.

20.4.2 General Health

The 2001 Census included a new question on the general health of the population. It is not possible, in a short census question, to go into precise definitions and so the question has to depend upon self-assessment.

> 'The new question on general health asks respondents to assess their own health over the preceding 12 months as either 'Good', 'Fairly good', or 'Not good'. This information has been shown to be a good predictor for the use of health services. Testing revealed that there were a range of interpretations and references used by respondents in answering the question. However, there was broad agreement between responses and levels of medical attention sought during the past year. It was determined that, although subjective, the question meets the main requirements of users.' (Moss, 1999: 32)

The Limiting Long-Term Illness question, introduced in the 1991 Census, was retained for 2001. It has been used for allocating resources to the health service and predicting the use of medical services. In essence, the reason for asking the two questions is that the wording of the Limiting Long-Term Illness question implies permanent disability, whereas the general health question will collect information about a greater range of conditions.

20.4.3 Provision of Unpaid Care

This new question distinguishes between people providing up to 19 hours, 20 to 49 hours, and 50 or more hours of unpaid care to someone who is ill or disabled.

There are a number of aspects of caring that are of concern to census users and there was a vigorous debate in the user community about what would be the most valuable focus for this question. The topic is not a simple one, and the question wording required careful testing to find an effective question.

> 'In recognition of the increasing amount of voluntary help provided, there was strong support for information on the provision of unpaid care. Users were initially concerned with obtaining information on whether care was provided to someone inside or outside the respondent's household. The question has been refocused to obtain a measure of the amount of time spent providing care, to help provide information to support the Government's Carers Strategy. Testing indicated that the definition of care was not well understood and an explanatory note giving examples of care has been included in the final question in order to improve the quality of response.' (Moss, 1999: 33)

Because of the change in focus, the testing of this question has been extended, using the ONS Omnibus Survey.

The break point of 20 hours is important for government policy on the support of carers, but some users may wish for a finer breakdown of the 1 to 19 hours category. It may be that cross-tabulating against the individual's other usual activities such as 'hours worked', and perhaps an allowance for commuting time, will aid interpretation of the breakdown of the census statistics.

20.4.4 Size of Work Organization

This question asks people aged between 16 and 74 who have ever worked whether their employer (or last employer) had 1 to 9, 10 to 24, 25 to 499, or 500 or more people working at the place where they work (or worked). The self-employed would answer for the number of people they employ (or employed).

This question is needed for the new official socio-economic classification.

> 'Consideration has also been given to the collection of information to enable the derivation of the new national statistics socio-economic classification. Following evaluation of alternative questions including size of workplace and size of organization, and questions covering supervisor/manager responsibilities, it was agreed that the 2001 Census would include questions on size of workplace and supervisory responsibilities. While size of organization would provide the ideal information for differentiating social classes based on occupation, workplace details will be adequate for all but a minority of workers.' (Moss, 1999: 33)

For researchers interested in the size of organizations as such, the Census of Employment provides a convenient source.

20.4.5 Religion

In Great Britain, a new question: 'What is your religion?' was included for the first census of the new millennium. The main user requirement was apparently to determine the respondent's religious affiliation to supplement the question about ethnic group. Testing indicated that most respondents found the topic acceptable. In addition, some respondents saw the question as an opportunity to further define their cultural identity, or considered that religion was a better indicator of their ethnicity and culture than ethnic group. In response to parliamentary concerns about civil liberties, this question alone, out of all census questions, is voluntary.

What is your religion?

- *This question is voluntary*
- *Tick one box only*

! None

! Christian (including Church of England, Catholic, Protestant and all other Christian denominations)

! Buddhist

! Hindu

! Jewish

! Muslim

! Sikh

! Any other religion, *please write in*

....

Figure 20.2 The new question about religion in England and Wales.

The form of the question posed in England and Wales is shown in Figure 20.2. In Scotland, two voluntary questions were asked, distinguishing between current religion and religion of upbringing. The format of the Scottish questions is similar to that used for England and Wales, but with some distinctions made between Christian denominations (Church of Scotland, Roman Catholic, Other). The census in Northern Ireland has included a question about religion since 1861, concentrating on differences in Christian denominations at the expense of other religions, a tradition continued in 2001. As in Scotland, for 2001 the Northern Ireland Census used two questions to distinguish between current religion and religion of upbringing.

As with the ethnic group question in the 1991 Census, the question on religion is expected to provide a wealth of opportunities for social and cultural analyses. There is, at present, very little reliable information available about the religious make up of the country. Census users, however, would have preferred a question on income.

20.5 MAJOR QUESTIONS REVISED

Four questions were revised from their 1991 format for the 2001 Census as a result of consultation with census users. In general, these revisions increase the amount of information collected.

20.5.1 Qualifications

The 1971, 1981, and 1991 Censuses asked a question about degrees and professional and vocational qualifications such as teaching and nursing, but did not cover school leaving qualifications. It thus had limited meaning for the majority of the population. Before 1971, age at the time of leaving school was asked.

For the next census, users have stated a requirement for information about a much broader range of qualifications. In view of the many changes to the educational system in recent decades, and the very large number of qualifications that a person may have, not to mention variations in Scotland and Northern Ireland, the development of a new question was a considerable challenge. In addition, it was required to have only tick-box responses, in order to contain the cost of processing.

> 'For the 2001 Census, information will be collected on broad levels of qualifications for all people aged 16 years and over, using pre-coded tick boxes. In England and Wales, the question will also collect information on whether people have teaching, medical, nursing and/or dental qualifications. As there is less of a requirement in Scotland and Northern Ireland it is not proposed to collect information on specific professional qualifications there.' (Moss, 1999: 33)

The extended qualification's question offers a great step forward for social analysts, primarily because it includes a category for 'no qualifications' and covers the most common qualifications such as GCSE, CSE, and GCE O and A levels. It allows research to be undertaken on issues such as race discrimination in employment practices, by area. The inclusion of vocational qualifications also enables learning in later life to be measured. Because tick-box responses were used, 100% of the responses will be processed, which is another welcome improvement.

As with any revised question, there would be some reduction in comparability over time. It should be possible to estimate changes in the levels of qualifications measured in the 1971 to 1991 Censuses using just census statistics. For the new levels of qualifications to be covered in 2001, the census will need to be used in conjunction with other sources to estimate changes.

20.5.2 Accommodation

A useful revision of the accommodation question allows the accommodation of families with small children or infirm or elderly people to be differentiated by floor level.

> 'A question is proposed on the lowest floor level of accommodation, providing a measure of households and people living in potentially unsuitable accommodation. This question is new to the censuses for England and Wales and Northern Ireland, and is based on one previously asked in Scotland. In Northern Ireland there will also be a new question on the number of floor levels in the accommodation, however there was less of a requirement for this information in the rest of the United Kingdom.' (Moss, 1999: 34)

20.5.3 Relationship Within the Household

In 1991, respondents were asked to list the 'head of household' first on the census form. The relationship of the other household members to this person was the only information collected. This was clearly unsatisfactory in households with complex structures. Even in simpler households, to be fully effective the question relied on the concept of 'head of household', which by 1991 was becoming a contentious term. It also relied on the respondent following the instruction to put the 'head of household' as 'person number one' on the form. The 1991 question on family structure was clerically coded, which meant that only 10% of replies could be processed. This limited the range of reliable statistics that could be produced at small area level.

In 2001, the question has been asked in the form of a tick-box matrix so that the relationship of each member of the household to every other member will be recorded, and all the responses can be processed. To limit the burden on the form-filler, the relationship was only asked one way. Thus, person 2 was asked his/her relationship to person 1; person 3 was asked his/her relationship to person 1 and person 2; person 4 was asked his/her relationship to persons 1, 2, and 3, and so on.

This matrix approach requires an additional response from each successive member of the household. Thus, 'person 18' in an 18-person household would need to respond no less than 17 times! The question was therefore further limited so that in the larger households, from person 6 onwards, information has been collected only about a person's relationships to 'person 1' and the 2 preceding people on the form.

Again, this was a challenging question to devise and was extensively tested. The Census Offices were somewhat surprised, but heartened, to find that it was possible to frame an effective question in this format.

There are many ways in which complex relationships could be tabulated, and this is an important topic for the current consultations about outputs. But it is good that a meaningful question has been asked. The revision of this question will have important implications for a great deal of research. In particular, it could alter the methodology by which estimates of future housing need in Britain are calculated.

20.5.4 Ethnic Group

The 1991 Census question on ethnic group was a major innovation and has helped shape the subsequent collection of statistics about ethnicity in many administrative systems. However, measured against current requirements, it has several drawbacks.

> 'In England and Wales, users' requirements for additional information on people of mixed ethnic origin, and demands to sub-divide the 'White' population, in particular the 'Irish', have been met by an expanded *ethnic group* question. New response categories provide optimum comparability with information from the 1991 Census, while at the same time meeting the needs of those who prefer to describe themselves as 'Black British' or 'Asian British'. Test results indicated that the topic of ethnicity was acceptable to the public.' (Moss, 1999: 32)

The new ethnicity question posed in the 2001 Census is shown in Figure 20.3.

In Scotland, the users' requirements for information about ethnicity were in line with the 1991 Census question, but with the addition of the 'mixed ethnic group' and 'Irish' categories. In Northern Ireland, where the question was proposed for the first time in a census, the requirement was similar, but with the addition of a category for 'Irish Traveller'. Simpler extensions of the 1991 question have been implemented for these countries.

Although the responses throughout the United Kingdom will be broadly comparable, some users may wish that the same question was asked everywhere. It is important to remember that balance must always be struck between having a uniform approach and having questions that reflect local conditions and needs for information. Perhaps more surprising is the use of the same question in both rural and large metropolitan areas.

What is your ethnic group?

• Choose one section from (a) to (e) then tick the appropriate box to indicate your cultural background

(a) White

! British

! Irish

! Any other White background

Please write in below

....

(b) Mixed

! White and Black Caribbean

! White and Black African

! White and Asian

! Any other mixed background

Please write in below

....

(c) Asian or Asian British

! Indian

! Pakistani

! Bangladeshi

! Any other Asian background

Please write in below

....

(d) Black or Black British

! Caribbean

! African

! Any other Black background

Please write in below

....

(e) Chinese or Other ethnic group

! Chinese

! Any other

Please write in below

....

Figure 20.3 The revised question about ethnic group in England and Wales.

20.5.5 Employment Status

In the 1971, 1981, and 1991 Censuses, respondents only had to tick one box to say that they were unemployed or seeking work. In 2001, the 1991 multi-tick question on employment status was replaced with a series of five separate questions. To identify themselves as unemployed, respondents had to return all of: 'no' on question 18 (Were you doing any work last week?), 'yes' on question 19 (Were you actively looking for paid work during the last four weeks?), 'yes' on question 20 (If a job had been available, could you have started it within two weeks?), 'no' on question 21 (Last week, were you waiting a job already obtained?), and 'none of the above' on question 22 (Last week were you retired, student, looking after home/family, permanently sick/disabled, none of the above?). The questions of the 2001 Census, particularly question 20, imply a more stringent definition of unemployment than that used in previous censuses, but one more in line with that used in allocating welfare benefits. In practical terms, it is at present unclear whether or not these changes will have any significant impact upon reported levels of unemployment.

20.6 CONCLUSION

Five new questions were introduced and five revised within the auspices of the Census Act. All these new questions and revisions will be extremely useful, whether they are to aid comparison across the United Kingdom, to increase the detail of the information collected, or to add new topics to the census data set.

Such is the importance of the national census that some users will no doubt feel that, for their purposes, it would have been better to have excluded one topic in order to make room for some other. But the content of the census must be decided taking into account the needs of all user sectors, and be acceptable to the public. Hopefully, all users will feel that the many improvements to the census content proposed for 2001 far outweigh any disadvantages from their perspective.

Box 20.1 Key web links for Chapter 20.

http://www.statistics.gov.uk/census2001/pdfs/whitepap.pdf	On-line copy of 2001 Census white paper

21

A One Number Census

Ian Diamond, Marie Cruddas, and Jennet Woolford

SYNOPSIS

The 1991 Census is estimated to have 'missed' 2% of the population it was attempting to count, causing the Office for National Statistics (ONS) to issue two sets of revised census population counts. To avoid a repeat of this situation, in the 2001 Census, further steps are being taken to maximize response rates. These are outlined briefly. A new and far more significant departure is the undertaking of a post-census Census Coverage Survey (CCS). The statistical justification for a CCS is presented, along with an outline of its implementation and use.

21.1 INTRODUCTION

One of the major uses of the decennial UK Census is in providing figures on which to rebase the annual population estimates. This base needs to take into account the level of under-enumeration in the census, which has traditionally been measured from data collected in a post-enumeration survey (PES) and (at the national level) through comparison with the estimate of the population based on the previous census. In the 1991 Census, although the level of under-enumeration was not high (estimated at 2.2%), it did not occur uniformly across all socio-demographic groups and parts of the country. There was also a significant difference between the survey-based estimate and that rolled forward from the previous census. Further investigation showed that the PES had failed to measure the level of under-enumeration and its degree of variability adequately.

For the 2001 Census, maximizing coverage was a priority. A number of initiatives were introduced to help achieve this, for example:

- the census forms were redesigned to make them easier to complete;
- population definitions for the census were reviewed;
- post-back of census forms was allowed for the first time; and
- resources were concentrated in areas where response rates were the lowest.

Despite efforts to maximize coverage in the 2001 Census, it is only realistic to expect there will be some degree of under-enumeration. The One Number Census (ONC) project aims to measure this under-enumeration, provide a clear link between the census counts

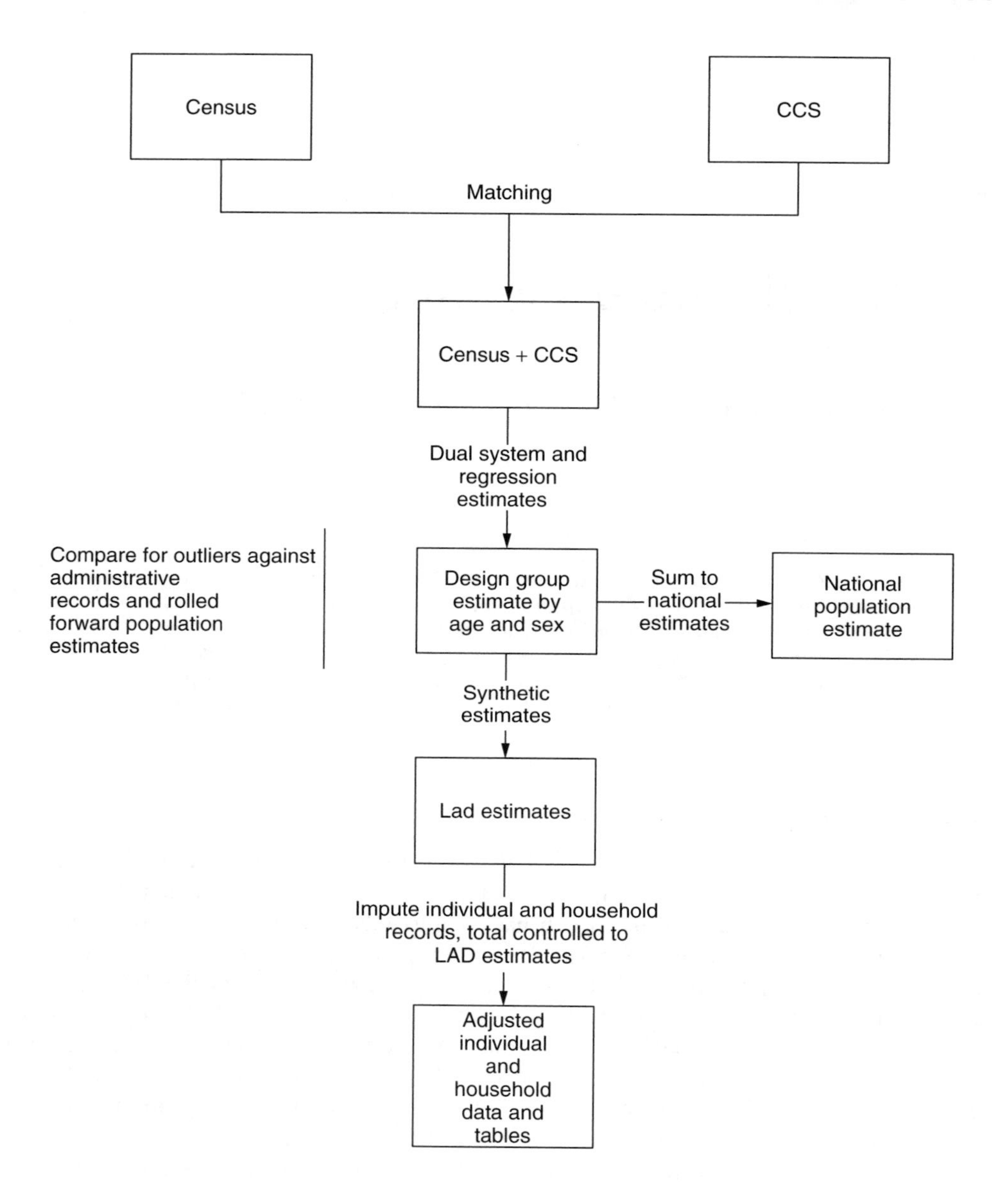

Figure 21.1 A schematic overview of the One Number Census process.

and the population estimates, and adjust all census counts (which means the individual level database itself) for under-enumeration.

Figure 21.1 illustrates the ONC process, which comprises six stages:

1. A CCS re-enumerated a sample of postcodes (geographical units of around 15 households). The survey collected data on a small number of key variables central to measuring under-enumeration.

2. The CCS data will be matched, using a probability-based matching procedure, against individual census records.
3. Combined regression and dual system estimation (DSE) will be used to produce estimates of the population based on the census and CCS, by age and sex, for each area of a broad regional stratification of the United Kingdom. These regions, each with a population of around 0.5 million, are referred to as *Design Groups* and are large Local Authority Districts (LADs) or groups of smaller LADs. The size of the Design Groups was selected to ensure a high efficiency of the design, on the basis of a simulation study. The LADs are important units of resource allocation in the United Kingdom. There are more than 400 LADs of varying population sizes.
4. LAD estimates will be derived from the Design Group estimates using synthetic estimation.
5. National, Design Group, and LAD estimates will be compared with a set of 1991-based estimates to assess their plausibility. In the event that any estimate is implausible, a contingency strategy will be used.
6. Individual and household level records will be imputed for those estimated to have been missed by the census.

21.2 THE DESIGN OF THE CENSUS COVERAGE SURVEY

Following the 1991 Census, a Census Validation Survey (CVS) was carried out in England, Scotland, and Wales. This survey aimed to estimate net under-enumeration and to validate the quality of census data (Heady *et al.*, 1994). The second of these aims required a complete re-interview of a sample of households that had previously been enumerated in the census. This requirement was costly because of the time required to fill out the complete census form, resulting in a small sample size. It also meant that the ability of the CVS to find missed households was compromised, as no independent listing of households was carried out.

An alternative strategy is required for 2001. Administrative records were found not to be accurate enough to measure census quality to the required precision. It was therefore concluded that a PES was needed with a clear objective and different design. The CCS (as the PES is known in 2001) addresses coverage exclusively. Focusing on coverage allows for a shorter doorstep questionnaire. Savings in time can be translated into a larger sample size. Information on question response error in the census data will be obtained from other sources, particularly the question testing programme, the 1997 Census Test, and through a separate quality survey carried out in 1999.

The CCS is a postcode-unit based survey, re-enumerating a sample of postcode units rather than households. It is technically feasible to design a household-based CCS by sampling delivery points on the UK Postal Address File, but the incomplete coverage of this sample frame makes it unsuitable for checking coverage in the census. Consequently, an area-based sampling design was chosen for the CCS, with census Enumeration Districts (EDs) as primary sampling units and postcodes within EDs as secondary sampling units. Sub-sampling of households within postcodes was not considered because coverage data from all households in a sampled postcode is necessary for the estimation of small area effects in the multilevel models proposed for stage six of the ONC.

Subject to resource constraints, the CCS sample design will be optimized to produce population estimates of acceptable accuracy for the 24 age–sex groups defined by sex (male/female) and 12 age classes: 0 to 4, 5 to 9, 10 to 14, 15 to 19, 20 to 24, 25 to 29, 30 to 34, 35 to 39, 40 to 44, 45 to 79, 80 to 84, 85+. The ages 45 to 79 have been combined because there was no evidence of any marked under-enumeration in this group in 1991. This age grouping will be reviewed before finalizing the CCS design (not known at the time of writing).

Under-enumeration in the 2001 Census is expected to be higher in areas with particular characteristics. For example, people in dwellings occupied by more than one household (multi-occupancy) have a relatively high probability of not being enumerated in a census. In order to control for the differentials, EDs within each Design Group are classified by a 'Hard to Count' (HtC) score. This score was chosen to represent social, economic, and demographic characteristics that were found to be important determinants of under-enumeration by the ONS and the Estimating With Confidence Project (Simpson *et al.*, 1997a). The variables making up the HtC score were still to be finalized at the time of writing. The prototype HtC score used in the CCS Rehearsal, which was undertaken as part of the 1999 Census Rehearsal, was based on the following variables from the 1991 Census:

- percentage of young people who migrated into the Enumeration District in the last year;
- percentage of households in multiple-occupied buildings; and
- percentage of households that were privately rented.

For sample design purposes, the HtC score has been converted to a five-point HtC index, with each quintile assigned an index value from 1 (easiest to count) to 5 (hardest to count). At the design stage, the role of the HtC index is to ensure that all types of EDs are sampled. Further size stratification of EDs, within each level of the HtC index, based on the 1991 Census counts improves efficiency by reducing within-stratum variance.

The second stage of the CCS design consists of the random selection of a fixed number of postcodes within each selected ED. The practicalities of conducting the CCS are described by Dixie (1999).

21.3 MATCHING THE CENSUS COVERAGE SURVEY AND CENSUS RECORDS

The estimation strategy requires the identification of the number of individuals and households observed in both the census and the CCS and those observed only once. Under-enumeration of around 2 to 3% nationally means that, although absolute numbers may be large, percentages are small. Thus, the ONC process requires an accurate matching methodology.

The independent enumeration methodologies employed by the census and the CCS mean that simple matching using a unique identifier common to both lists is not possible. Furthermore, simple exact matching on the variables collected in common by both methods is out of the question as there will be errors in both sets of data caused by incorrect recording, misunderstandings, the time gap, and errors introduced during processing. The size of the CCS also means that hand matching is not feasible. Thus, a largely automated process involving probability matching is necessary.

Probability matching entails assigning a probability weight to a pair of records on the basis of the level of agreement between them. The probability weights reflect the likelihood that the two records correspond to the same individual. A blocking variable, for example, postcode, is used to reduce the number of comparisons required by an initial grouping of the records. Probability matching is only undertaken within blocks as defined by the blocking variables.

Matching variables such as name, type of accommodation, and month of birth are compared for each pair of records within a block. Provided the variables being compared are independent of each other, the probability weights associated with each variable can be summed to give an overall probability weight for the two records. Records are matched for the census record that most closely resembles the CCS record in question if the likelihood of them relating to the same household or individual exceeds an agreed threshold.

The CCS data will be used for two purposes: to enable the data to be matched against the census and to identify the characteristics of under-enumeration via the modelling process, so that adjustments can be applied to the whole population. In order that the second part is not biased by the first, the matching and modelling variables should be as independent as possible.

The initial probability weights used in 2001 will have been calculated from the data collected during the 1999 Census Rehearsal. These weights will be refined as the 2001 matching process progresses.

As the data are structured both geographically and by individuals within households, we use this structure within the matching strategy.

The key stages of the matching are as follows:

1. Use blocking variables to reduce the number of comparisons made
2. Match households
3. Match individuals within matched households
4. Clerically check any CCS forms left unmatched.

21.4 ESTIMATION OF DESIGN GROUP AGE–SEX POPULATIONS

There are two stages to the estimation of Design Group age–sex populations. First, a DSE method is used to estimate the number of people in different age–sex groups missed by both the census and the CCS within each CCS postcode. Second, the postcode level population counts obtained from these DSEs are used in regression estimation to obtain final counts for the Design Group as a whole.

To start, we describe the DSE component of this methodology. It is unlikely that the union count (i.e., the total of those counted in the census and/or CCS) for an area will constitute a complete count. DSE assumes that:

1. the census and the CCS counts are independent and
2. the probability of 'capture' by one or both of these counts is the same for all individuals in the area of interest.

When these assumptions hold, DSE gives an unbiased estimate of the total population. Hogan (1993) describes the implementation of DSE for the 1990 US Census. In this case, assumption (i) was approximated through the operational independence of the census and PES data capture processes, and assumption (ii) was approximated by forming post-strata on the basis of characteristics believed to be related to heterogeneity in the capture probabilities.

In the context of the ONC, DSE will be used with the census and the CCS data as a method of improving the population count for a sampled postcode, rather than as a method of estimation in itself. That is, given matched census and CCS data for a CCS postcode, DSE is used to define a new count, which is the union count and an adjustment for people missed by both the census and the CCS in that postcode. This DSE count for the sampled postcode is subsequently used as the dependent variable in a regression model, which links this count with the census count for that postcode.

The regression model is based on the assumption that the 2001 Census count and the dual system adjusted count within each postcode satisfy a linear regression relationship with a zero intercept (a simple ratio model). However, for some age–sex groups, there is the possibility of a non-zero intercept as in some postcodes, the census can miss all the people. For such age–sex groups, an intercept term α_d will be added to the ratio model described in the following text. This issue is currently being researched.

It is known from the 1991 Census that undercount varies by age and sex and by local characteristics; therefore, a separate regression model within each age–sex group for each HtC category within each Design Group is used. Let Y_{id} denote the DSE count for a particular age–sex group in postcode i in HtC group d in a particular Design Group, with X_{id} denoting the corresponding 2001 Census count. Estimation will be based on the simple zero intercept regression model:

$$\begin{aligned} E\{Y_{id}|X_{id}\} &= \beta_d X_{id} \\ \text{Var}\{Y_{id}|X_{id}\} &= \sigma_d^2 X_{id} \\ \text{Cov}\{Y_{id}, Y_{jf}|X_{id}, X_{jf}\} &= 0 \quad \text{for all } i \neq j;\, d, f = 1, \ldots, 5 \end{aligned} \tag{21.1}$$

Substituting the Ordinary Least Squares (OLS) estimator for β_d into Equation (21.1), it is straightforward to show (Royall, 1970) that under this model the Best Linear Unbiased Predictor (BLUP) for the total count T of the age–sex group in the Design Group is the stratified ratio estimator of this total given by:

$$\hat{T} = \sum_{d=1}^{5} \left\{ T_{Sd} + \sum_{i \in R_d} (\hat{\beta}_d X_{id}) \right\} = \sum_{d=1}^{5} \hat{T}_d \tag{21.2}$$

where T_{Sd} is the total DSE count for the age–sex group for CCS-sampled postcodes in category d of the HtC index in the Design Group and R_d is the set of non-sampled postcodes in category d of the HtC index in the Design Group. Strictly speaking, the model specified by Equation (21.1) is known to be wrong because the zero covariance assumption ignores correlation between postcode counts within an enumeration district (ED). However, the simple OLS estimator (21.2) remains unbiased under this mis-specification, and the OLS estimator is only marginally inefficient under a non-zero covariance structure (Scott and Holt, 1982).

The variance of $\hat{T} - T$, the estimation error associated with Equation (21.2), can be estimated using model (21.1). Unlike Equation (21.2), this is sensitive to mis-specification of the variance structure (Royall and Cumberland, 1978). Consequently, as the postcodes are clustered within EDs, the conservative ultimate cluster variance estimator will be used.

This is given by:

$$\hat{V}(\hat{T} - T) = \sum_{d=1}^{5} \frac{1}{m_d(m_d - 1)} \sum_{e=1}^{m_d} (\hat{T}_d^{(e)} - \hat{T}_d)^2 \tag{21.3}$$

where $\hat{T}_d^{(e)}$ denotes the BLUP for the population total of category d of the HtC index based only on the sample data from ED e and m_d is the number of EDs in HtC group d.

The aforementioned estimation strategy represents a regression generalization of the Horvitz–Thompson DSE estimator proposed in Alho (1994). As a postcode is a small population in a generally small geographic area, and with the counts split by age and sex, the DSE homogeneity assumption should not be seriously violated. In the situation in which people missed by the census have a higher chance of being missed by the CCS than those counted by the census, one would expect the regression estimator based on the DSE count to underestimate, but to a lesser extent than the regression estimator based on the union count. When the reverse happens, and the CCS is very good at finding the missed people (the requirement for getting unbiased estimates when using the union count in the regression estimator), one would expect the DSE count regression estimator to overestimate. However, unless these dependencies are extremely high, one would not expect a gross error.

21.5 LOCAL AUTHORITY DISTRICT ESTIMATION

Direct estimation using the CCS produces only estimates by age and sex for each Design Group. In the case of an LAD with a population of approximately 500 000 or above, this will give a direct estimate of the LAD population by age and sex. However, for the smaller LADs clustered to form Design Groups, this will not be the case, although all LADs will be sampled in the CCS. For these LADs, it will be necessary to carry out further estimation and allocate the Design Group estimate to the constituent LADs.

Standard small area synthetic estimation techniques are used for this purpose. These techniques are based on the idea that a statistical model fitted to the data from a large area (in our case the CCS Design Group) can be applied to a smaller area to produce a synthetic estimate for that area. The problem with this approach is that, although the estimators based on the large area model have small variance, they are usually biased for any particular small area. A compromise, introduced in the 1980s, involves the introduction of random effects for the small areas into the large area model. These allow the estimates for each small area to vary around the synthetic estimates for those areas. This helps reduce the bias in the estimate for a small area at the cost of a slight increase in its variance (Ghosh and Rao, 1994).

As described in the previous section, direct Design Group estimation is based on the linear regression model (21.1) linking the 2001 Census count for each postcode with the DSE-adjusted CCS count for the postcode. This model can be extended to allow for the multiple LADs within a Design Group by writing it in the form

$$Y_{idl} = \beta_d X_{idl} + \delta_{dl} + \varepsilon_{idl}$$

where the extra index $l = 1, \ldots, L$ denotes the LADs in a Design Group, δ_{dl} represents an LAD 'effect' common to all postcodes with HtC index d, and ε_{idl} represents a postcode-specific error term. The addition of the δ_d term represents differences between LADs that have been grouped to form a Design Group.

This regression model can be fitted to the CCS data for a Design Group and the LAD effects δ_{dl} estimated. For consistency, LAD population totals obtained in this way will be adjusted so that they sum up to the original CCS Design Group totals, and are always at least as large as the 2001 Census counts for the LAD.

21.6 IMPUTATION OF MISSED HOUSEHOLDS AND INDIVIDUALS

This final stage of the ONC process starts by modelling the probability of being counted in the census in terms of the characteristics of individuals and households. This is possible in CCS areas where there are two independent counts of the population. These models are applied to all individuals and households counted by the census to calculate their coverage weights. The coverage weights are calibrated to agree with the total population estimates by age–sex group and by household size for each LAD.

The imputation procedure is based on the fact that there are two processes that cause individuals to be missed by the census. First, when there is no form received from the household and therefore all household members are missed. Second, when contact with the household fails to enumerate all household members and therefore some individuals are omitted from the form. These two processes are treated separately by the methodology.

21.6.1 Creating Household Coverage Weights

After the census and the CCS, it can be assumed that all households within CCS areas fit into one of the following categories:

1) counted in the census, but missed by the CCS;
2) counted in the CCS, but missed by the census; and
3) counted in both the census and the CCS.

Underlying this is the assumption that no household is missed by both. Although this is an unrealistic assumption, the calibration process accounts for such households. The final imputed database is constrained to the population estimates at the Design Group level. Categories (1) to (3) in the preceding text define a multinomial outcome variable that can be modelled for each LAD using a logistic specification. On the basis of this model, the probability $\theta_{jidl}^{(t)}$ that household j in postcode i in HtC group d in LAD l has outcome t can be estimated. For outcomes $t = 1$ and $t = 3$, this estimated probability will be a function of the characteristics of the household as measured by the census. This model can therefore be extrapolated to non-CCS areas to obtain estimated coverage probabilities for all households. Consequently, for each household j counted in the census, a household (h/h) coverage weight

$$w_{jidl}^{h/h} = \frac{1}{\theta_{jidl}^{(1)} + \theta_{jidl}^{(3)}}$$

can be calculated. In general, the weighted sums of households of different sizes computed using these weights will not agree with the corresponding estimates for the LAD.

Consequently, these weights are calibrated (via an iterative scaling procedure) and hence these constraints are satisfied.

21.6.2 Creating Individual Coverage Weights

Coverage weights for individuals counted by the census are obtained using similar assumptions to those described for households. It is assumed that if a household is counted by the census only, no individuals from that household are missed by the census. Similarly, if the household is counted by the CCS only, it is assumed that no individuals from that household are missed by the CCS. Although this assumption is violated in practice, the extra people are accounted for by constraining to estimated totals at the LAD level. Using these assumptions, it is only necessary to consider individuals in households counted by both the census and the CCS. In this case, the possible categories are:

1. counted in the census, but missed in the CCS;
2. counted in the CCS, but missed by the census;
3. counted in both the census and the CCS.

Matched census and CCS data and an assumed multinomial logistic model are used to estimate the probability $\pi_{kjidl}^{(r)}$ that individual k in household j in postcode i in HtC group d in LAD l has outcome r. As with the household model, the individual probabilities for outcomes $r = 1$ and $r = 3$ depend on individual and household characteristics as measured in the census. Therefore, they can be extended to allow computation of coverage probabilities for all individuals counted by the census within households also counted by the census. For each such individual (ind), therefore, a coverage weight

$$w_{kjidl}^{\text{ind}} = \frac{1}{\pi_{kjidl}^{(1)} + \pi_{kjidl}^{(3)}}$$

can be calculated.

21.6.3 Donor Imputation for Missed Households

The next stage of the imputation process involves imputing missed households. Households are split into impute classes defined by similar household characteristics and processed sequentially in order of increasing coverage weight. When the cumulated weighted count of the households gets more than 0.5 ahead of the cumulated unweighted count, a new household is imputed. The donor household is defined by the characteristics of the impute class as well as those households with the current weight and not only donates the household characteristics but all the individuals within the household as well. This process ensures that the total number of households after imputation matches the estimated LAD total. It will also correspond to totals defined by any other variables to which the household weights have been calibrated.

21.6.4 Donor Imputation for Missed Individuals

This is the most complex stage of the imputation as adding individuals to households changes the structure of the recipient household. This stage is best thought of in two parts.

The first identifies how many individuals need to be imputed and obtains the appropriate donors. Individuals are processed sequentially, in order of coverage weight within impute class. When the cumulated weighted count exceeds the cumulated unweighted count by more than 0.5, an individual needs to be imputed. The impute class and weight define the basic characteristics of that person. A donor household is then found that contains a person of the required type. Second, the person is imputed into a 'nearby' recipient household. The recipient household is the household nearest to the donor household in both space and household structure. The imputed person is added into the recipient household. The recipient household is subsequently subject to census edit checks to ensure internal consistency.

21.6.5 Pruning and Grafting of Individuals

The preceding stages of imputation add individuals to the census database, either as part of an imputed household or as an addition to a counted household. Typically, this results in an excess of synthetic individuals on the database. The final stage of the imputation process therefore is to make sure that the totals of individuals match LAD totals by age and sex and that the resulting household size distribution is correct. A process of 'pruning off' and 'grafting on' imputed individuals from the database is then carried out until these key LAD totals are achieved.

Eventually, an individual level database will be created that will represent the best estimate of what would have been collected had the 2001 Census not been subject to under-enumeration. Tabulations derived from this database will automatically include compensation for under-enumeration and therefore all add to the 'One Number'.

Box 21.1 Key web links for Chapter 21.

http://www.statistics.gov.uk/nsbase/census2001/IntroOneNumber.asp	Up-to-date information on the ONC programme

22

An Output Strategy for the 2001 Census

Chris Denham and Philip Rees

SYNOPSIS

This chapter has been prepared by Phil Rees, based in part on material prepared by Chris Denham and his colleagues in the Census Organizations. The chapter presents plans for producing output from the 2001 Census. The principles governing output design are first set out, building on the strategy outlined in the 1999 White Paper on the 2001 Census. The topics to be covered are then reviewed. In principle, it will be possible for census users to order whatever tables they wish from the census database, subject to confidentiality considerations. The chapter then describes the data sets and products being planned from the 2001 Census, as proposed at time of writing, which will be more detailed and useful than at any previous census. Arrangements for dissemination, which are the subject of vigorous activity, development, and negotiation in the 2001 to 2002 period are then discussed. The arrangements may not, in subsequent years, take exactly the shape predicted but the advance knowledge of intentions will help users both shape their own plans and also contribute to the evolution of dissemination practice; maximum ease of extraction and minimum barriers to use are the goals of both census providers and user organizations.

22.1 INTRODUCTION

The 2001 Census is the first UK Census of Population of the twenty-first century. It incorporates many essential features that have characterized censuses in the twentieth century: near-complete enumeration of the population, presentation of timely reports to the Parliament, and provision of more detailed statistics in computer-readable form. The outputs will incorporate a number of important innovations, such as integration of statistical output with linked geographic data, imputation of virtually all missing items in household records, and the addition of missing households and individuals, provision of quicker and cheaper customized outputs, harmonized provision of core outputs across the separate parts of the United Kingdom (England, Wales, Scotland, and Northern Ireland), and Web access to a core set of data products. As a convenient shorthand, we refer to the three Census Offices collectively in this chapter. The Office for National Statistics (ONS), based in London and Titchfield, has responsibility for the census in England and Wales;

the General Register Office Scotland (GROS), based in Edinburgh, has responsibility for the census in Scotland; the Northern Ireland Statistics and Research Agency (NISRA), based in Belfast, has responsibility for the census in Northern Ireland. In planning for the 2001 Census, all three Census Offices worked closely together; major decisions were taken by the UK Census Committee in which all three agencies participated. Figure 1.1 provides contact details for each Census Office, from which readers can obtain the latest information on 2001 Census outputs.

The proposals for outputs from the 2001 Census were outlined in successive documents in 1999 and refined in 2000 and 2001:

- a White Paper on the 2001 Census was published in March 1999 (HM Government, 1999), outlining the views of Her Majesty's Government on the form and conduct of the 2001 Census;
- a consultative paper on census outputs was published for the major round of consultation meetings held by the Census Offices in March and April 1999 (ONS, GROS, NISRA 1999b); and
- supplements to the consultative paper were produced that gave details of proposals for particular data products (e.g., the Standard Tables). Of particular use to users planning future data analyses was a document on proposed data classifications (ONS/GROS/NISRA, 1999c).

Users in central government, local government, the health service, planning bodies, business organizations, and the academic community were asked to comment on the proposals and to state their requirements. During the remainder of 1999, 2000, and in 2001, the Census Offices revised output proposals in response to user comments, carrying out additional consultations where there was a conflict between the requirements of different users. The current state of output proposals is provided on the ONS Web site (see Box 22.2).

The publication of output proposals had been preceded by two years of consultation between the Census Offices and users. The consultation took place through meetings of an Outputs Working Group (OWG), organized by ONS, to which user sectors sent representatives. The OWG user representatives provided feedback to the Census Offices on output proposals and made proposals of their own (e.g., about output geography). Within user sectors, there were meetings to gather views about output data needs. The Economic and Social Research Council (ESRC) and Joint Information Services Committee (JISC) funded a series of four Workshops Planning for the 2001 Census (Rees, 1998c), the results of which were published as Working Papers (Rees, 1997d; 1998b; 1998d; 1998e).

Census data are produced under the authority of a series of Census Acts: 1920 for Great Britain, 1969 for Northern Ireland (Marsh, 1993c). There are two types of output authorized by the Acts:

- statistical reports laid before Parliament and
- statistical abstracts requested by users, either prepared in *standard form* to meet the common requirements of a large number of customers, or prepared in *special form* to meet their own specific requirements.

In the 1991 Census, the Census Offices prepared a new form of abstract, the Samples of Anonymised Records (SAR) (see Chapter 14) at the request of the academic community,

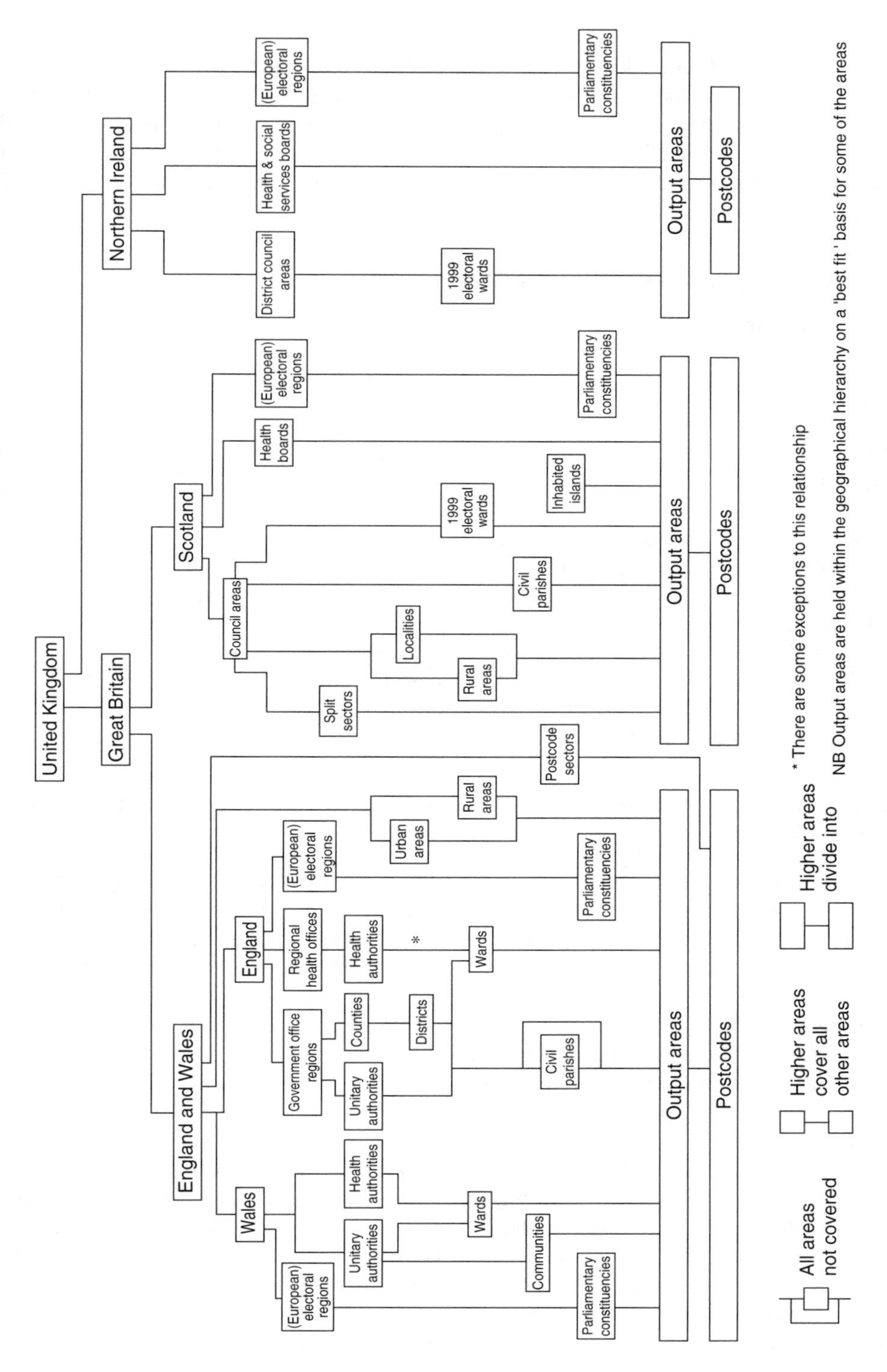

Figure 22.1 The geographical hierarchy of output from the 2001 Census.

and, to help offset the full additional marginal cost to ESRC (with the support of JISC), ESRC was granted an exclusive licence for both academic and commercial purposes.

In the rest of this chapter, we review the general principles to be adopted by the Census Offices in preparing and delivering outputs (Section 22.2), the topics to be covered in the 2001 Census (Section 22.3), the data products proposed by the Census Offices and by user sectors (Section 22.4), and the proposals by the Census Offices and user sectors for the dissemination of census data (Section 22.5).

22.2 OBJECTIVES OF CENSUS OUTPUT DESIGN

The Census Offices recognize as part of their mission that

> 'the effort and cost of taking a census is only worthwhile when results meet needs and are delivered effectively' (ONS 1999e, p.5).

The overall aim of the Census Offices is to deliver high-quality output. To achieve this aim, output needs to be:

- comprehensive, comparable over areas and over time;
- flexible enough to meet *ad hoc* needs;
- accessible for use on PCs through intuitive and up-to-date products;
- delivered on time and error-free;
- good value for money, so that costs do not constrain use; and
- subject to innovations that enhance the fitness for purpose of census data.

These objectives are discussed in turn.

22.2.1 Harmonization of Standard Census Outputs and Production of Territory Specific Outputs

In producing 2001 Census outputs, ONS, GROS, and NISRA will harmonize data products across the United Kingdom as far as possible. Although the emphasis here is on products harmonized across the United Kingdom, there will be both the need and the opportunity to produce territory-, region-, and area-specific output. A small number of questions are unique to England, Wales, Scotland, and Northern Ireland and reflect different requirements. There are also differences in needs, for example, between urban and rural areas. In some urban areas, there is a need for detailed tables on ethnic groups but in rural areas the ethnic minority numbers may be too small for such tables to be published, given confidentiality concerns. Two solutions have been proposed. The first is to produce sets of univariate counts for the full classification of a variable such as ethnic group. These do not disclose any information about individuals. The second is that a detailed set of ethnic group tables will be produced for the local authorities and their wards in which there are large ethnic minority populations, as an 'extension' to the Standard Tables.

Harmonization over time has in previous censuses not been given the attention that it deserved, but this omission has been repaired by the development of lookup tables such as the enumeration district to postcode directory (ONS), the All Fields Postcode Directory (ONS), higher order indexes (GROS and NISRA), and lookup tables derived by several academics (see Chapter 5).

22.2.2 Flexible Census Outputs

Inevitably, standard outputs are designed as compromises to meet the general needs of a large number of users. Individual users will have more specific needs. Traditionally, these have been met by production of special tabulations by the Census Offices. The costs have been high and processing queues build up because of a shortage of skilled database programmers. With faster and easier to use tabulation software (the SUPERSTAR suite from Space-Time Pty), this bottleneck should be overcome once the 2001 Census master database has been created and signed off.

22.2.3 A Variety of Data Formats and Delivery Methods

The Census Offices will be producing standard and bespoke outputs in the computer formats most convenient to users. As well as traditional flat, comma-separated variable files in text format (for which there will still be a demand for advanced analysis by user's own software), the Census Offices will produce outputs in standard spreadsheet, database, and Geographic Information System (GIS) formats for analysis on users' desktops. There is also a high demand for access to data products over the World Wide Web. The academic community has developed Web interfaces, funded through a series of ESRC/JISC Development Programmes, for access to the Census Area Statistics (CAS) (see Chapter 8), to the Migration and Workplace Statistics (Chapters 18 and 19), to Linked Censuses over Time (see CD-ROM), and to Geographical Conversion software (developed by Simpson and Yu, 2000, Yu and Simpson, 2000). These are being linked to mapping and visualization software (e.g., Chapter 7).

22.2.4 Timetable for Census Outputs

At the time of writing, the pre-announced timetable mentioned in the Census White Paper (HM Government 1999) for the results has not been published, but the target for the first report that will provide key statistics for areas throughout the country concurrently is December 2002, with further standard output becoming available in tranches in the first half of 2003. The timetable will not be crystallized until it is clear that the processing of the census is proceeding to plan. There will, of course, be pressure from users for early release, but this should not be at the expense of data quality. The Census Offices plan to involve key user organizations in testing at the pre-release stage in 2002 and 2003 to help ensure a near error-free first release.

22.2.5 Charges for Census Outputs

The initial objective of the Census Offices was to deliver 2001 Census outputs at no greater real charges than 1991 Census outputs, but with enhanced functionality, using better information systems technology to hold down costs. A major new development is that standard output will be free in effect to all end users. The ONS have funds, under the Treasury Invest and Save programme with contributions from user organizations, for a *Census Access Project* (*CAP*) to cover the costs that must be recovered under the Census Act 1920 and to develop better ways of delivering outputs. Similar developments apply in Scotland and Northern Ireland. See Section 22.5 for more details.

22.2.6 Innovations in Census Outputs

Several decisions have been taken by the Census Offices which will further enhance the utility of 2001 Census outputs compared with previous censuses.

The *first* innovation is the decision that all responses on private household and communal establishment forms will be processed. The Census Offices have outsourced the data capture process to an external information services company, Lockheed-Martin. The company is able to apply its expertise in scanning, optical character recognition, and neural net classification to achieve a high hit rate for the automatic classification of write-in answers on difficult-to-code topics.

The *second* innovation is the decision to link all data products with machine-readable Metadata. Information on data product specifications, variable definitions, and variable interpretations will be made available in computer readable files for simultaneous release with the corresponding statistics. These Metadata will be produced in two forms: as help (documents to be read by users on-line or after printing) or as searchable databases (see Chapter 23). Considerable experience has been built up in developing data dictionaries and glossaries for the 1991 Census, both within the Census Offices and in user organizations. Ideally, as is pointed out in Chapter 23, the user should be able to discover where the information is located on a particular topic, at what geographic scale, and in what form.

A *third* innovation to be introduced in the outputs from the 2001 Census is the integration of geographic and statistical data. This will be achieved in different ways, depending on the data set. For example, each set of output area (OA) statistics will be accompanied by a matching digital file of OA boundaries in the form of topologically consistent output area polygons. The digital boundaries will be provided in a variety of common GIS and mapping formats either in stand-alone form or integrated into combined statistics–mapping databases. Consistent and matching labels and accurate centroids and reference (seed) points will be supplied. Topographical background maps in raster form will also be included in the service where required to aid location of OAs, replacing 1991 hard copy reference maps, although 2001 maps will be printed on request.

A *fourth* innovation is the assignment of a unique georeference to each enumerated household and communal establishment. This georeference will be the Ordnance Survey Address-Point™ product in England and Wales, the Ordnance Survey Northern Ireland COMPAS™ product in Northern Ireland, and a reference from the GROS GIS database in Scotland. Such georeferencing will open up a wealth of applications for census data or make them much more precise, removing the 'noise' of having to assign households the georeference of their postcode. It is very encouraging that the main players (the Mapping Agencies, the Royal Mail, and the Census Offices) are collaborating in a project called GRIDLINK to ensure a pooled georeference database of the highest accuracy and currency.

Attention is now focused on the specific data products to be produced from the 2001 Census. At the time of writing, the detailed specifications were still being finalized. Details may have changed by the time one reads this chapter.

22.3 TOPICS COVERED FOR THE 2001 CENSUS

The 2001 Census contained six more questions than the 1991 Census. The nature of these questions and the needs that they meet are discussed in Chapter 20. Table 22.1 sets out the

Table 22.1 Topics in the 2001 Census.

No.	Topics
For all properties occupied by households and all unoccupied household accommodation:	
1	The address, including postcode
2	Type of accommodation
3	Names of all residents Names and usual addresses of visitors on census night (optional)
4	Tenure of accommodation
5	*Whether rented accommodation is furnished or unfurnished (in Scotland only)*
6	Type of landlord (for households in rented accommodation)[a]
7	Number of rooms
8	Availability of bath and toilet
9	Self-containment of accommodation
10	**Lowest floor level of accommodation**
11	***Number of floor levels in the accommodation (in Northern Ireland only)***
12	Availability of central heating
13	Number of cars and vans owned and available
For residents:	
14	Name, sex and, date of birth
15	Marital status
16	Relationship to others in household
17	Student status
18	Whether or not students live at enumerated address during term-time
19	Usual address one year ago
20	Country of birth
21	*Knowledge of Gaelic (Scotland only), Welsh (Wales only), and Irish (Northern Ireland only)*[a]
22	Ethnic group[a]
23	**Religion**[a]
24	***Religion of upbringing (Scotland and Northern Ireland)***
25	**General health**
26	Long-term illness
27	**Provision of unpaid personal care**
28	Educational and vocational qualifications
29	Economic activity in the week before the census
30	**Time since last employment**
31	Employment status
32	**Supervisor status**
33	Job title and description of occupation
34	***Professional qualifications (England)***
35	**Size of workforce of employing organization at place of work**
36	Nature of employer's business at place of work (industry)
37	Hours usually worked weekly in main job
38	Name of employer

(continued overleaf)

Table 22.1 (*continued*)

No.	Topics
39	Address of place of work[a]
40	Means of travel to work[a]
41	***Address of place of study (in Scotland only)***
42	***Means of travel to place of study (in Scotland only)***

Source: H.M. Government (1999); Census Factsheet 9 (*http://www.statistics.ac.uk/*).
Note: **Bold** indicates a new question (compared with the 1991 Census); *Italics* indicates question to be used in only one part of the United Kingdom.
[a]Response categories vary among parts of the United Kingdom.

topics covered in the 2001 Census. A core set of 34 topics was included in the censuses of England, Wales, Scotland, and Northern Ireland. Three of these core topics, questions on ethnicity, address of place of work, and means of travel to work, vary in pre-defined categories used among the different parts of the United Kingdom. An additional question on religion was asked in England, Wales, and Northern Ireland. An additional question on Welsh, Gaelic, and Irish was asked in Wales, Scotland, and Northern Ireland, respectively. In Scotland, information was also gathered on the address of the place of study and means of travel to the place of study. It should be stressed that names of residents and visitors are collected for internal Census Office verification only and will not be used in any form of output for 100 years.

22.4 DATA SETS AND DATA PRODUCTS FROM THE 2001 CENSUS

22.4.1 Overview

The term *data sets* here refers to the general category of output produced, while data products are the specific outputs within data sets. We first describe the range of data sets to be produced and then consider the products in detail. Table 22.2 lists the *Standard Area Statistics* planned as output from the 2001 Census. The term *Standard Area Statistics* describes all the data sets to be released that include tables of count and derived statistics for geographical areas, no matter what their spatial scale may be. Table 22.2 has three columns. The first column identifies the *data product*, the second column defines the 'universe' or geographical *coverage* for which the product will be produced, the third column specifies the spatial *scale* for which statistics will be reported. So, for example, the majority of the Standard Tables will be produced for wards (local electoral areas) within the whole of the United Kingdom and a minority of the Standard Tables will be specific to each constituent part of the United Kingdom.

The list of Standard Area Statistics data sets includes Topic Statistics, Standard Tables, CAS, Key Statistics, which are all 'single-geography' data sets. The Origin–Destination Statistics, comprising the migration and journey to work statistics (United Kingdom), and the journey to school statistics (Scotland) have a double-geography classification (see Chapters 18 and 19 for details). Note that terminology for these data sets has changed, compared with the 1991 census. The following equivalences hold:

Table 22.2 Standard Area Statistics planned for the 2001 Census.

Data products	Geographical coverage	Spatial scale
Census geography		
Resident population and household counts	UK	Postcodes
Key statistics	UK, England, Wales, Scotland, Northern Ireland	OAs[a], wards[b], LGAs[c], GORs[d] (England), UK parts[e], health authorities, urban and rural areas
CAS[f]	UK, England, Wales, Scotland, Northern Ireland	OAs, wards[b], LGAs[c], GORs[d] (England), UK parts[e]
Standard Tables	UK, England, Wales, Scotland, Northern Ireland	Wards[b], LGAs[c], GORs[d] (England), UK parts[e]
Postal geography		
CAS	UK, England, Wales, Scotland, Northern Ireland	Postal sectors, postal districts, postal areas
Electoral geography		
CAS	UK, England, Wales, Scotland, Northern Ireland	Parliamentary constituencies, European constituencies,
CAS	Wales	Welsh Assembly constituencies
CAS	Scotland	Scottish Parliament constituencies
CAS	Northern Ireland	Northern Ireland Assembly constituencies
Civil parishes and communities		
CAS	England	Civil Parishes
CAS	Wales	Communities
CAS	Scotland	Civil Parishes
Grid geography[g]		
CAS	Northern Ireland	Grid squares

Note: Release of data products for the areas listed will be subject to areas being above threshold. For example, it will be necessary to amalgamate a number of parishes and communities and ward with neighbouring areas so that populations of the new zones pass confidentiality thresholds.

[a]OAs: Output Areas—small areas defined as groups of unit postcode areas.

[b]Wards: electoral wards for local government elections.

[c]LGA: Local Government Areas. In England, LGAs are London Boroughs, former metropolitan districts, unitary authorities, districts within shire counties; in Wales, LGAs are unitary authorities; in Scotland, LGAs are council areas; in Northern Ireland, LGAs are Local Government Districts.

[d]GOR: Government Office Regions.

[e]UK parts: England, Wales, Scotland, Northern Ireland.

[f]CAS: Census Area Statistics.

[g]Grid squares: Only proposed as a standard data set by NISRA for Northern Ireland.

1991 Census	**2001 Census**
National Statistics	
Topic Reports	To be determined (probably in the form of census topic reports and 'multi-source' topic reports including census output)
Area Statistics	
	Standard Area Statistics (SAS) = ST + CAS + KS
Local Base Statistics (LBS)	Standard Tables (ST)
SAS	Census Area Statistics (CAS)
Monitor Statistics	Key Statistics (KS)
Interaction or Flow Statistics	
Special Workplace Statistics (SWS)	Origin–Destination Statistics (ODS)
Special Migration Statistics (SMS)	(with SMS, SWS, and STS components)

Associated directly with the area and origin-destination statistics will be two geographical products. First, digital boundary data will be produced for output areas, wards, local government areas, and for postcode sectors. Second, lookup tables will be produced that link the two building bricks of the 2001 Census, the OA and its unit postcode or part postcode component, with other areas at the time of the census, with other areas that are created after the census in the following decade as they are revised, and with areas used for previous censuses. Using these lookup tables, which will be comprehensive and rapidly available once the 2001 Census outputs have been released users can create estimates of change.

Microdata products are planned. The Government (in the White Paper) has indicated a willingness to produce again SAR released in safe form for users to analyse, and ESRC/JISC has plans to request their production. Details of the SAR data sets will be discussed among the parties in 2001 to 2002 (see Chapter 14 for further details).

The ONS has approved plans for the linkage of 2001 Census records to the Longitudinal Study (LS) for England and Wales, while a project funded by the Scottish Higher Education Council is constructing a Scotland LS linking sample 1981 and 1991 Census records, in preparation for adding 2001 Census records. Users, whose projects have been approved, will be able to request vetted tables or analyses from the LS (see Chapter 16 for further details).

22.4.2 Census Reports Laid before Parliament

At the time of writing, the exact form of output to be laid in reports before Parliament, corresponding to the Topic Statistics produced in previous censuses, had not been determined. It is likely that a set of summary tables will be produced for the United Kingdom as a whole, for its parts, and for the regions of England (i.e., the area statistics described later), together with a CD-ROM containing the machine-readable versions of the printed tables in common computer formats (spreadsheet or database), also to be made available on the Census Offices Web sites (see Figure 1.1). In a second tranche, a commentary on the statistics illustrated with graphs and maps may be produced, placing the information in a temporal context (comparisons with 1991 Census statistics) and in a spatial context (comparisons with European Union statistics).

22.4.3 Standard Area Statistics

We summarize here the proposals for the Standard Area Statistics as current in September 2001. The proposals for Key Statistics and Standard Tables were in their final form but those for the CAS were subject to one further round of consultation in 2001. Full details are provided in the relevant consultation papers published by ONS/GROS/NISRA (see Box 22.1).

Postcode Statistics

For unit postcode building bricks, only counts of resident population (by gender) and households will be provided, which will be mainly useful in re-aggregating more detailed census statistics to user-defined areas.

Key Statistics (KS)

These statistics will provide key counts and percentages for all geographical levels from the Output Area to the United Kingdom and will be released as the first results from the 2001 Census in paper format as Reports to the UK Parliament, the Scottish Parliament, the Welsh Assembly, and the Northern Ireland Assembly, with a CD containing all versions at a UK level. They will be available as computer-readable files from National to OA levels via Census Organization Web sites (Figure 1.1) and on CD. Table 22.3 lists the tables into which the counts and percentages will be arranged. ONS/GROS/NISRA (2001b) provides full details (see Box 22.1).

Census Area Statistics (CAS)

These will be provided for Output Areas and aggregations of OAs. Table 22.4 summarizes their content. They will consist of a set of 68 univariate tables (giving information for a single variable), a main set of 118 CAS tables, cross-classifying variables in a similar way to the 1991 Census SAS as shrunken versions of the parallel Standard Table, two UK theme tables on dependent children and older persons, and two tables for England and Wales using only the 16-way classification of ethnicity to deliver counts for the ethnic groups. Together, this set of statistics contains some 6000 unique counts for each area together with univariate tables proving 700 unique counts. The 2001 CAS will differ from the 1991 SAS in a significant respect. To avoid even the perception of disclosure, counts in tables will not only be subject to imputation and record swapping, but will also be randomly perturbed and rounded to the nearest three. This means that OA counts are unlikely to sum to the corresponding ward, local government area, government office region and UK part counts, although this dependent upon implementation details yet to be decided.

Standard Tables

These will be output for wards (postal sectors in Scotland) and larger areas. The Census Offices consulted users in 1999 and 2000 on the form and content of these tables.

Table 22.3 Summary of proposed Key Statistics for the 2001 Census.

Table	Title	Table	Title
1	Usual resident population	18a	Occupation groups—all people aged 16–74 in employment
2	Age structure	18b	Occupation groups—males aged 16–74 in employment
3	Living arrangements	18c	Occupation groups—females aged 16–74 in employment
4	Marital status	19	Qualifications and students (England and Wales)
5	Migration	20	Qualifications and students (Scotland)
6	Country of birth	21	Qualifications and students (Northern Ireland)
7	Ethnic group (England)	22a	National statistics—socio-economic classification—all people aged 16–74
8	Ethnic group and language (Wales)	22b	National statistics—socio-economic classification—males aged 16–74
9	Ethnic group and language (Scotland)	22c	National statistics—socio-economic classification—females aged 16–74
10	Ethnic group and language (Northern Ireland)	23	Travel to work (England, Wales, and Northern Ireland)
11	Religion (England and Wales)	24	Travel to work and place of study (Scotland)
12	Religion (Scotland)	25	Dwellings, household spaces, and accommodation type
13	Religion (Northern Ireland)	26	Cars or vans
14	Health and provision of care	27	Tenure (England, Wales, and Northern Ireland)
15a	Economic activity—all people aged 16–74	28	Tenure (Scotland)
15b	Economic activity—males aged 16–74	29	Rooms, amenities, central heating, and floor level
15c	Economic activity—females aged 16–74	30	Household composition

Table 22.3 *(continued)*

Table	Title	Table	Title
16	Hours worked	31	Households with dependent children
17a	Industry of employment—all people aged 16–74 in employment	32	Lone parent households with dependent children
17b	Industry of employment—males aged 16–74 in employment	33	Communal establishment residents (England, Wales, and Scotland)
17c	Industry of employment—females aged 16–74 in employment	34	Communal establishment residents (Northern Ireland)

Note: Full details of the Key Statistics are given in ONS/GROS/NISRA (2001b). See Box 22.1 for CD resource.

Table 22.4 Summary of proposed Census Area Statistics for the 2001 Census.

Table set	Description
Univariate tables	UV01, total population to UV68 all people in communal establishments (Northern Ireland)
CAS theme tables	CAST01, dependent children in households (UK), CAST02, older people (UK), CAST03, ethnic group—people (England and Wales), CAST04, ethnic group of household reference person—households (England and Wales)
UK tables	CAS01, age by gender and resident type to CAS67 age of Family Reference Person (FRP) and dependent children by approximated social grade
England and Wales tables	CAS68 to CAS96: the numbering sequence parallels that of the Standard Tables but for many numbers there is no CAS equivalent of the Standard Table
Scotland tables	CAS97 to CAS143: the numbering sequence parallels that of the Standard Tables but for many numbers there is no CAS equivalent of the Standard Table
Northern Ireland tables	CAS144 to CAS171: the numbering sequence parallels that of the Standard Tables but for many numbers there is no CAS equivalent of the Standard Table

Note: Full details of the CAS are given in ONS/GROS/NISRA (2001d). See Box 22.1 for CD resource.

Table 22.5 Summary of proposed Standard Tables for the 2001 Census.

Table set	Description
Theme tables	T01, dependent children to T23, Irish language—Northern Ireland
UK tables	S01, age by gender and resident type to S65, gender and age by general health and limiting long-term illness (no S66, S67)
England and Wales only tables (some available for either Northern Ireland or Scotland)	S68, gender and age by ethnic group to S95, residents (non-staff) in communal establishments and gender and limiting long-term illness and type of communal establishment by age
Wales only table	S96, gender and age by country of birth and knowledge of Welsh
Scotland only tables	S97, gender and age by ethnic group to S143 age and highest level of qualification by ethnic group
Northern Ireland only tables	S144, gender and age by ethnic group to S171, residents (non-staff) in communal establishments and gender and limiting long-term illness and type of communal establishment by age
Armed Forces tables	AF1, residents in employment in Armed Forces to AF4, UK and foreign Armed Forces by residence and workplace

Note: Full details of the Standard Tables are given in ONS/GROS/NISRA (2001d). Details of the Theme tables are given in ONS/GROS/NISRA (2001g). See Box 22.1 for CD resource.

Table 22.5 lists the tables proposed (see ONS/GROS/NISRA, 2001d and Box 22.1 for details). There will be 23 Theme tables and 65 tables for all parts of the UK, 28 tables specific to England and Wales (with some available for either Northern Ireland or Scotland), 1 special to Wales, 47 specific to Scotland, and 28 specific to Northern Ireland. Finally, there will be a set of 4 Armed Forces tables for areas with high numbers of military. The Theme tables take a particular sub-group of the population and cross-classify this population against a series of related variables. The Main tables are traditional cross-classifications of multiple variables in single tables. The Standard Tables achieve delivery of a large core of common statistics across the UK and also provide tables for different parts tailored to the characteristics of the populations living there.

Geographies

Standard Area Statistics will be produced for a variety of geographical areas as indicated in Table 22.2. The census geography consists of a hierarchy of areas, as constituted on 29 April 2001: smaller areas nest exactly or nearly exactly into larger areas. Figure 22.1 shows the proposed geographical hierarchy for all parts of the UK. Postcodes, as defined at the time of the 2001 Census, will constitute the basic building bricks, although they will be split where they cross administrative boundaries, that is, where Address Points and additional residential addresses found in the Census fall in two or more administrative areas. Otherwise unit postcode polygons are fitted to administrative boundaries. Postcodes will be aggregated to Output Areas (see Chapter 3 for details of their design) for publication of statistics. Output Areas (OAs) will then be aggregated to local government wards on an exact basis throughout the UK. OAs also aggregate to communities (Wales) and civil parishes (England, Scotland), which are important for historical comparisons. OAs can be summed to form Parliamentary Constituencies. From wards can be built local government areas in all parts of the UK and Government Office Regions in England. From either wards or OAs can be built Health Authorities (Wales, England), Health Boards (Scotland), Health and Social Service Boards (Northern Ireland).

Postal geography intersects with these administrative hierarchies and output for postcode sectors will probably be built independently in England and Wales. In Scotland, output areas will aggregate directly to 'split sectors', which are entire postcode sectors (e.g., EH12 7) when they fall within council areas and part postcode sectors when they straddle council area boundaries.

It will be possible for users to build their own reporting zones from Output Areas (Chapter 4 gives advice on methods of zone design) and to request, subject to confidentiality constraints, statistics for other areas built from postcodes. The Census Offices also plan to establish georeferenced address bases as part of the census operation so that, in principle, again subject to confidentiality constraints, users can request statistics for any geographical network of areas. After the 2001 Census this database will be updated for any changes in postal, local government, health or other statutory areas, as the need arises, so that statistics for revised geographical hierarchies can be generated.

22.4.4 Origin–Destination Statistics

A suite of improvements to origin-destination statistics will be effected in the 2001 Census outputs. Details are discussed in Chapter 18 and ONS/GROS/NISRA (2001f) (see Box 22.1). (1) The local authority flow statistics will not be subject to suppression. (2) The journey to work statistics will be based on all household returns, not just 10% of them. (3) The statistics will cover the whole of the UK, filling in holes left in origin-destination matrices in the 1991 products. (4) Information on the journey to study location by school children and students will be made available in Scotland. (5) Moving groups besides wholly moving households will be identified. (6) The classifications used in the tables will be updated and improved. (7) Missing origins in the migration statistics and missing workplaces in the journey to work statistics will be imputed. As in 1991, flow statistics from the 2001 Census will be produced at several different spatial scales: local authority to local authority, ward/postcode sector to ward/postcode sector, and OA to OA (just for the journey to work and school statistics). The tables attached to each flow have the greatest detail for local authority to local authority flows. Rather simple statistics are to be provided for ward/postcode sector to ward/postcode sector flows. The OA to OA commuting tables will contain just classifications by method of travel.

22.4.5 Digital Boundary Data and Lookup Tables

The Census Offices propose to make radical changes in the range of geographical products supported. They will: produce geographic reference guides, geographic directories, digital boundaries, geographic centroids, reference maps generated from the GIS used in enumeration district and OA planning and derived information such as area measurements, densities; and will either use an urban/rural classification and thus a definition of urban areas prepared for wider Government purposes or, if such a classification does not become available, one developed specifically for the Census. All products will be made available in electronic format, and on paper if required.

Table 22.6 lists the main digital boundary products and geographical lookup tables. In theory, users can use OA boundaries and build, using a GIS, higher area boundaries from these. However, experience with the 1991 digital boundary data suggests that, for a majority of users, the data volumes involved are overwhelming. So, there will be a demand for simplified digital boundaries suitable for thematic mapping, which may be supplied by the Census Offices, by the academic community or by other agencies. The lookup tables will contain two key sets of information: the composition of OAs in terms of unit postcodes and the link between output areas and higher areas. These products will be maintained over time from census date 2001 to reflect changes in unit postcodes and other geographies.

Table 22.6 Digital boundary data and geographical lookup tables for the 2001 Census.

Data products	Coverage	Scale
Digital boundary data		
Digital boundaries	England	OAs, wards, LGAs; postcode sectors
Digital boundaries	Wales	OAs, wards, LGAs; postcode sectors
Digital boundaries	Scotland	OAs, wards, LGAs; postcode sectors
Digital boundaries	Northern Ireland	OAs, wards, LGAs; postcode sectors
Lookup tables		
OA directory	England	OAs to other areas, postcodes to OAs and other areas
All fields postcode directory		
OA directory	Wales	OAs to other areas, postcodes to OAs and other areas
All fields postcode directory		
OA directory	Scotland	OAs to other areas, postcodes to OAs and other areas
Higher areas index		
OA directory	Northern Ireland	OAs to other areas, postcodes to OAs and other areas
Postcode directory		

22.4.6 The Samples of Anonymised Records and Small Area Microdata

The SARs were a product commissioned and purchased from the Census Offices by ESRC/JISC for the 1991 Census of Population. Details of the data sets and plans for the 2001 Census are outlined in Chapter 14. Table 22.7 summarizes the proposals put forward by the Census Microdata Unit (University of Manchester) after extensive research into confidentiality considerations and user requirements. It is proposed that a 2001 Census Household SAR be generated largely on the same basis as in 1991. This is a confidentially sensitive data set and fine geographical breakdowns are inappropriate. For the Individual SAR, two proposals have been made: an increase in sample size from 2 to 3% to make possible reduction in the population thresholds for which samples can be produced from 120 000 to around 70 000 people. The lower threshold would permit the production of samples for a majority of local authorities. The third proposal is for a new data set, the feasibility of which has been demonstrated (Dale and Elliot, 1998). The Small Area Microdata (SAM) would be a 5% sample produced for low population thresholds

Table 22.7 Microdata samples for the 2001 Census.

Data products	Coverage	Scale
Household Sample Anonymised Records		
1% Household SAR	UK	GORs[a] (England), Wales, Scotland, Northern Ireland
1% Household SAR	England	GORs
1% Household SAR	Wales	Wales
1% Household SAR	Scotland	Scotland
1% Household SAR	Northern Ireland	Northern Ireland
Individual Sample Anonymised Records		
3% Individual SAR	UK	LGAs[b] above threshold or combinations
3% Individual SAR	England	LGAs above threshold or combinations
3% Individual SAR	Wales	LGAs above threshold or combinations
3% Individual SAR	Scotland	LGAs above threshold or combinations
3% Individual SAR	Northern Ireland	LGAs above threshold or combinations
Small Area Microdata		
5% SAM	UK	Wards above threshold or combinations
5% SAM	England	Wards above threshold or combinations
5% SAM	Wales	Wards above threshold or combinations
5% SAM	Scotland	Wards above threshold or combinations
5% SAM	Northern Ireland	Wards above threshold or combinations

Note: These proposals are subject to approval by the Census Offices and to detailed negotiation concerning exact specification.
[a]GORs: Government Office Regions.
[b]LGAs: Local Government Authorities.

(between 7000 and 30 000 people), with very broad coding of variables. The Census Offices are evaluating proposals for SARs and SAM derived from the 2001 Census, and a full specification should be agreed in 2002.

22.4.7 The Longitudinal Study

No anonymised individual data are released from the LS—the data set that links individuals across censuses—but user requested tables and statistical analyses are provided, subject to confidentiality protection. Chapter 16 provides details of the Longitudinal Study and the proposed addition of 2001 Census data for study members. The LS covers only the population of England and Wales. However, Scottish academics (Boyle) have been funded to create an LS for Scotland by the Scotland Higher Education Funding Council with the cooperation of the GROS.

22.5 DATA DISSEMINATION

Once the Census Offices have generated the 2001 Census data products, they will be disseminated to users. Arrangements for doing this are currently evolving rapidly. The account here states the position as of mid-2001. We first outline plans for Web dissemination and subsequently describe the cooperative project that will fund the means of dissemination. The new arrangements for reuse and publication of census data are then outlined, followed by a brief account of plans for the ESRC/JISC-funded programme of support for census data use within the HE/FE community over the 2001 to 2006 period.

22.5.1 Web Access to Census Data

Web access to area statistics from the 1991 Census was provided to registered academic users via CASWEB from 1998 (see Chapter 8). In 2001 ONS launched Neighbourhood Statistics, a comprehensive service to deliver small area derived from administrative, survey and census sources. A small amount of 1991 Census information is made freely available to all users via the National Statistics Web site *www.statistics.gov.uk*. The plan is to provide access to the whole of the 2001 Census area statistics from OA scale upwards from the via Neighbourhood Statistics on the Web, although 'bulk' delivery would be via free-standing media in current technical circumstances.

The 1991 Census data sets made available via ESRC/JISC Data Support Units to registered users within the academic community are likely to be made available to all users via the Web under arrangements for licensing described in Section 22.5.4 in parallel with their release on Census Offices' Web sites.

Special arrangements will be needed for making available 2001 Census Data for which ESRC/JISC had an exclusive licence in 1991 (the Samples of Anonymised Records). Such data sets will be free at the point of use to academic users but there are probably charges for non-academic users.

Arrangements for data sets purchased by user organizations such as ESRC/JISC but not placed on the National Statistics Web site, such as the Origin–Destination Statistics, still need to be negotiated at the time of writing.

22.5.2 The Census Access Project

ONS have won some £2.1 million Invest and Save ISB programme funds from the Treasury to support the Census Access Project (CAP) over the 2000 to 2003 period. This has the twin aims of making census data freely exchangeable and usable across partners without need of special arrangements and of making census data much more accessible and intelligible through the development of on-line documentation, metadata and help systems and through the development of point-and-click user interfaces. An article in *Census News* No 45 (April 2001, pp. 5–6) gives fuller information on the CAP. Census News is available from *census.customerservices@ons.gov.uk* or on-line via the National Statistics Web site at *www.statistics.gov.uk/census2001/cennews.asp*.

CAP Partners are four in number: Department of Local Government, Transport, and the Regions on behalf of central government departments, the Department of Health for the National Health Service, the Local Government Association on behalf of local government, and ESRC/JISC on behalf of Higher Education and Research and Further Education. CAP Partners are committed to providing £175k each in addition to the ISB funding to cover the CAP for England and Wales, whereas GROS and NISRA have made funding arrangements that will cover Scotland and Northern Ireland, respectively. For this investment, partners and the users they represent will get free in effect access to an integrated set of statistics, geography and metadata. This set would include: the Key Statistics, the CAS, the Standard Tables, the boundaries of census areas including output areas, wards, local government areas, and all the zones that could be built from these units, together with the lookup Tables/Directories linking OAs, postcodes and other areas.

The relationship between the CAP arrangements and the Census Offices' Web strategy is being worked out. Without the CAP there would be no customer-funded products to place on the Web. The public sector investment by CAP partners will make census data available to all private sector users, public sector users, and individuals within the UK, free at the point of use. Users will be able access the data via ONS or GROS or NISRA Web sites, or via ESRC/JISC Census Programme Web sites or via their organization's intranet or via personal CDs.

22.5.3 Licensing Use of Census Data in Other Products

On the 3rd April 2001, Cabinet Office Minister Ian McCartney announced the introduction of the 'Click-Use' on-line licence, which went live on Her Majesty's Stationery Office (HMSO) Web site, as of 1st April 2001.

> 'Under the new plans, businesses will be able to utilize and communicate Government information, providing new and exciting products for their customers. Such information includes data about health, house prices, school performance tables, crime, census results and guidance issued by departments. Most of this material can reused free of charge. Libraries, local government and members of the public will also be able to take out licences. ... Applying for a licence is simplicity itself. All that users have to do is visit HMSO's Web site, complete a simple application form and hit the submit button. The licence then takes immediate effect.' (Source: *http://www.hmso.gov.uk/docs/mccartney.htm*).

The new arrangements are spelt out in detail on the HMSO Web site under Licensing and Copyright (*www.hmso.gov.uk/copyhome.htm*). There is a Weblink from there to Click Use Licence Application and from the Census pages of the ONS Web site. If

Table 22.8 The ESRC/JISC 2001 to 2006 Census programme: timetable, milestones, and deliverables, 1999 to 2006.

Month, year	Milestone	Deliverable
August 1999 to July 2000		
August 1999	External assessor evaluates census programme	Batty (1999)
October 1999–February 2000	Coordinator prepares census programme proposal	Three drafts census programme proposal
May 2000	ESRC approves programme	
July 2000	ESRC and JISC agree revised programme	£3.8 million budget: 57% ESRC, 43% JISC
August 2000	Coordinator prepares revised programme	Fourth draft of census programme proposal
August 2000 to July 2001		
October 2000	JISC approves programme	
October 2000–November 2000	Call for outline proposals for data support and registration	Six proposals shortlisted December 2000.
February 2001	Full proposals submitted	Proposals for refereeing.
April 2001–July 2001	Commissioning of data units	Interviews. Five data units chosen. Budgets agreed.
May 2001–June 2001	Call for full proposals for Longitudinal Study support unit	Proposals for refereeing
August 2001 to July 2002		
August 2001	New data units start	Enhanced data and user support.
August 2001–October 2001	Commissioning of LS support unit	Interviews. One data unit chosen. Budget agreed.
November 2001	New LS support unit starts	Enhanced user project support
December 2001–February 2002	Call for coordinator proposals	Proposals for refereeing
March 2002	Interviews of candidates	Coordinator selected
July 2002	New coordinator starts	New inspiration and ideas. Liaison with Census Offices for speediest delivery of data products.
August 2002 to July 2006		
Programme disseminates census data to HE and FE communities for use in research and teaching.		

government produced information is made available under click-use arrangements, then a user can register for a licence very simply on-line and does not have to pay for reuse of the information or its inclusion in another product. An article in Census News 45 (p. 6) entitled 'New terms for the reuse of 1991 Census output' interprets the licensing arrangements.

There are several different types of census information which will need licensing, some via the 'click-use' licence and others via tailored licence agreements. Tailored licence agreements will need to be reached through negotiation involving HMSO, the Census Offices (ONS/GROS/NISRA), ESRC/JISC and other user organizations.

The Census News 45 article indicates

> 'that material from the 1991 Census reports, LBS and Small Area Statistics (SAS) can now be reused and published without extra cost. . . . Existing electronic versions of material in Census reports and any other existing Census output wholly or substantially produced in electronic forms for Government purposes will be available for reuse and publication through the new scheme'.

22.5.4 The ESRC/JISC 2001 to 2006 Census Programme

Chapter 1 outlined the Census Data System, which has operated for the 1991 to 2001 decade to deliver data products to academic census users. From 2001, a new ESRC/JISC Programme will operate that aims to deliver to all members of Higher and Further Education Institutions and Institutes recognized for receipt of research awards the data sets licensed for *academic purposes*. By academic purpose is meant any use, the results of which are placed in the public domain. This is normally research funded by the Higher Education Funding Councils or Research Councils. It also covers use of the data by students in their taught course or dissertation research.

Table 22.8 looks five years into the future and sketches out the proposed timetable for delivering to academic census users the data products of the 2001 Census and suggests how research activity might be organized with the ESRC/JISC programme. It should be stressed that the proposal is subject to revision, approval and implementation. Equivalent timetables will apply to Central Government, Local Government, Health Authority, and Business Users. Even in the post-socialist world in which we live, flexible five-year plans have value in helping organizing research activity.

22.6 CONCLUSION

The outputs strategy described in this chapter is an ambitious one. It involves a number of important innovations that exploit progress in data processing technology, geographical information systems technology, and information networking technology. It also reflects increasing sophistication in the needs of census users and a sustained drive to meet those needs. The strategy has built on the past experience of the Census Offices but has also benefited from a set of detailed consultations with census users. Bodies representing user sectors are putting in place administrative arrangements and software that will mean speedy delivery of the 2001 Census to their users. We feel confident that the output strategy, successfully implemented, will renew and enhance the Census Data System described in this book.

Box 22.1 Additional CDS resources for Chapter 22.

UK Discussion Paper.pdf	ONS/GROS/NISRA (2001a)
Key Statistics.pdf	ONS/GROS/NISRA (2001b)
CAS.pdf	ONS/GROS/NISRA (2001c)
Standard Tables.pdf	ONS/GROS/NISRA (2001d)
Master Index.pdf	ONS/GROS/NISRA (2001e)
Origin Destination.pdf	ONS/GROS/NISRA (2001f)
Theme Tables.pdf	ONS/GROS/NISRA (2001g)
Classifications.pdf	ONS/GROS/NISRA (2001h)

Box 22.2 Key web links for Chapter 22.

www.statistics.gov.uk/census 2001/cennews.asp	Web site containing Census Newsletters, which provide the latest information about census output plans and products
census.customerservices@ons. gov.uk	E-mail address of ONS Census Customer Services: contact for paper copy of Census Newsletters
www.hmso.gov.uk/copyhome. htm	Web site of Her Majesty's Stationery Office where information about obtaining 'click-use' licences for Crown Copyright material including Census Core Statistics is provided

23

Metadata for the 2001 UK Census: Recommendations

Paul Williamson and Neil Lander-Brinkley

SYNOPSIS

The sheer volume of planned outputs from the 2001 Census means that many users will require a guide just to navigate their way around them. In addition, the apparent simplicity of the published results will mask an underlying complexity concerning such matters as population and geographic base, adjustments for under-enumeration, definitions of terms, and historical comparability. This chapter outlines the information about census outputs (metadata) that will be required by different types of user, ranging from the novice to the expert, to make the most out of the census. A key emphasis is upon the provision of machine-readable (and searchable) metadata integrating information from a wide range of traditionally separate sources.

23.1 INTRODUCTION

To the casual user, census outputs are simple tabulations of a 100% sample survey, requiring little or no background knowledge to interpret. In fact, a range of potential hazards exists to catch the unwary. These include lack of awareness of the full range of census outputs, selection of a suitable population base, and the meanings of certain terms. For example, is the user interested in the population present on census night, all residents or private household residents only, and what is the difference between a dependent and a dependant? At ward level and below, further problems arise, including 'special' enumeration district (ED)s, thresholding, barnardization, undercount, and the statistical reliability of 10% counts. All of this points to the need for metadata—information and guidance on the interpretation of census data. This chapter outlines the possibilities that exist for creating metadata about 2001 Census outputs, drawing upon the lessons to be learnt from the 1991 Census.

Metadata have been defined as 'data about data'. In fact, the deliberately broad, almost all-embracing nature of this definition hints at one of the biggest problems with metadata—knowing where to draw the line. This problem manifests itself in a number of ways. First, the breadth of census metadata provision can range from a simple definition of terms and an index of published census counts through to information on statistical reliability and the reasoning behind the selection, exclusion, and wording of various census questions. Second, the depth of metadata provision can vary. For example, if information is provided on census question selection, should this be in brief summary form or should it comprise the minutes from every relevant Census Advisory and Planning group meeting held during the intercensal period? One guide might well be user relevance, but information needs vary considerably between potential users. Third, there is sometimes difficulty in distinguishing the fine line between metadata and the data they describe. For example, are imputation rates data on non-response or metadata concerning census data quality?

In Sections 23.2 and 23.3, the range of outputs and the history of metadata provision for the 1991 Census are briefly reviewed (in part summarizing and updating Williamson *et al.*, 1995). Drawing upon the lessons to be learnt, the final section of the chapter makes a series of recommendations regarding metadata provision for the 2001 Census. Although the chapter does draw upon published Census Office output strategy documents where appropriate, it does not directly reflect the official view of the Census Offices of the United Kingdom. As a result, the recommendations are best viewed as a 'wish-list' of the types of information that could be included in a 2001 Census metadata product by National Statistics (NS), given sufficient time, resources, and user demand.

23.2 1991 CENSUS OUTPUTS

To the end-user, one of the most important improvements in the 1991 Census compared with previous censuses was the greatly increased range, quantity, and detail of census output. Two samples of anonymized records were released for the first time, the various Special Topic volumes all improved in coverage upon their 1981 counterparts, and additional Special Topic volumes, not previously compiled, were also released. However, perhaps the single most important change was the increase in the number of statistics available for small geographic areas.

In 1981, the 53 published tables of census Small Area Statistics (SAS) contained 5500 counts. For the 1991 Census, the number of SAS tables was increased by just over half to 87, whilst the number of counts trebled to over 15 500. However, whereas for the 1981 Census the SAS were the only source of information at ward level, for 1991 both the SAS and Local Base Statistics (LBS) provide ward level data. Taking the LBS into account, there was a greater than ninefold increase in the number of table counts available between censuses for small areas. As a result, the task of remembering in detail the full wealth of information available from the census passed beyond the average user's reach.

Allied to the explosion in data provision, the 1991 Census also saw a marked increase in the complexity of published census output. When designing table layout for the SAS and the LBS, the NS (then the Office for Population Censuses and Surveys) had to strike a compromise between table 'legibility' and efficiency of page usage. This led to the concatenation (joining) of cross-tabulations sharing common elements into larger, single tables. Although the need for table concatenation is understandable, it does lead to increased table complexity, as illustrated by the layout of SAS Table 46 (see Figure 23.1). There are numerous ways in which this table may be split into individual cross-tabulations. Figure 23.1 presents a solution compatible with the table classification approach adopted in the construction of METAC91, an unofficial 1991 Census metadata product (described later).

Table S46 Households with dependent children: housing; Households with dependent children; residents in such households												
		Persons per room		Amenities			Tenure					
Household composition	Total house holds	Over 1 and up to1.5 persons per room	Over 1.5 persons per room	Lacking or sharing use of bath/shower and/or inside WC	No central heating	Not self-contained accomm-odation	Owner occupied	Rented privately	Rented from a housing association	Rented from a local authority or new town	No car	Total persons in house holds
All households with dependent children **Households of 1 adult with 1 or more dependent children** Dependent child(ren) aged 0–4 only Dependent child(ren) aged 5 and over only Dependent child(ren) aged 0–4 and 5 and over	**1**	**2**		**3**		**4**	**5**				**6**	**7**
Dependent children in households of 1 adult with 1 or more dependent children All dependent children Dependent children aged 0–4 Dependent children aged 5–15 Dependent children aged 0–17	**8**	**9**		**10**		**11**	**12**				**13**	**14**
Other households with dependent children Dependent child(ren) aged 0–4 only Dependent child(ren) aged 5 and over only Dependent children aged 0–4 and 5 and over	**15**	**16**		**17**		**18**	**19**				**20**	**21**
Persons in other households with dependent Children All adults All dependent children Dependent children aged 0–4 Dependent children aged 5–15 Dependent children aged 0–17	**22**	**23**		**24**		**25**	**26**				**27**	**28**
Households with 3 or more dependent children Households with 3 or more persons aged 0–15	**29**	**30**		**31**		**32**	**33**				**34**	**35**
Households with 4 or more dependent children Households with 4 or more persons aged 0–15	**36**	**37**		**38**		**39**	**40**				**41**	**42**

Figure 23.1 The layout of 1991 Census SAS Table 4.6.

23.3 1991 CENSUS METADATA

23.3.1 'Official' OPCS Census Metadata

To support use of the 1991 Census, the Census Offices have produced a variety of metadata. These include a 'Definitions' volume (OPCS/GROS, 1992a), published in advance of the first census results, describing the meaning and derivation of census variables such as social class. The various published County and Special Topic volumes also contain appendices in which key census terms (e.g., 'economically active') are explained to aid table interpretation. Perhaps as importantly include indexes of table contents, although typically restricted to no more than the identification of two-way tabulations. Thus, for example, it is possible to look for a tabulation on long-term illness by age, but not for a tabulation on long-term illness by age by sex. Various joint Office for Population Census and Surveys and General Register Office Scotland (OPCS/GROS) user guides also contained information on aspects of table layout and content, but suffer from similar limitations (OPCS/GROS, 1992e; 1992f).

In-house, OPCS-held information on the 1991 Census table contents in the form of a glossary of all census variables, a series of TAU (programming) statements, and a Table and Information Monitoring System (TIMS). The glossary contains information on the definition and derivation of each census variable used in output tables, but lacks any link between variable records and the census table(s) with which they were associated. The TAU (programming) statements were used by OPCS to generate the derived variables required for census output. They contain the same information as the Glossary in a less immediately accessible form, but with the addition, in some cases, of references to associated census tables. The TIMS was an on-line database that theoretically defined all tables in terms of their component variables. However, for a variety of reasons, this ideal has never been fully achieved.

Some two years or more after the release of the initial census data, OPCS also published a General Report (OPCS/GROS, 1995c). This report dealt with a wide variety of issues, including the history of question selection and testing, census delivery and publicity strategies, and edit and imputation methodologies. Perhaps, even more importantly, the report addressed a range of data quality issues, providing estimates of census undercount and question response error rates, as well as information on overall item imputation rates and the sampling errors associated with 10% coded data. Data quality issues were also considered further in two volumes reporting on results from the 1991 Census Validation Survey. Heady *et al.* (1994) addresses coverage issues, while Heady *et al.* (1996) addresses question response error rates in greater detail. Although of great interest, a univariate approach was taken to all of these data quality issues. As a result, the error and imputation rates for population sub-groups defined by two or more variables in combination are not considered. Finally, a set of historical tables (OPCS/GROS, 1993e) provides a time series of population counts for present census areas, disaggregated by age and sex. These counts may be viewed as metadata because they aid comparison of census counts from consecutive censuses, although for rather coarse geographical areas.

The main problems with the official 1991 Census metadata relate to timeliness, comprehensiveness, and the division of metadata into a number of overlapping products. But perhaps the biggest weakness in the official 1991 Census metadata provision is data on census geographies. Paper copies of ward and ED maps have been made available to users, but at a price per sheet that has generally made the purchase of sufficient maps to cover a city, let alone the whole country, prohibitive. In addition, most users required this information in machine-readable rather than printed form. As a result, users have had to resort to alternative, if officially sanctioned, commercial products. On the other hand, the NS has made available lookup tables that locate postcodes within statutory authority areas and enumeration districts within postcodes, although again at prices somewhat prohibitive to the casual user. Further details may be found in the guide to official census products published as Appendix B in OPCS/GROS (1995c).

23.3.2 Other Sources of 1991 Census Metadata

Since the release of the 1991 Census, a wide range of semi-official and unofficial products have been created to supplement the official census metadata. Perhaps one of the earliest and best examples is The 1991 Census User's Guide (Dale and Marsh, 1993), which contains a great variety of background information on the 1991 Census. Chapters in the book cover such diverse topics as the history behind the choice of census content and coverage, census-data editing and imputation procedures, guides to census geography, the sample of anonymized records, and census output. It is perhaps best viewed as a 'semi-official' guide to the census, as the Census Offices supplied most of the information upon which the book is based, and vetted the contents of many chapters. However, the book in fact had its genesis in academia, including the original inspiration, editorial control, the great majority of the writing and the funding of production costs as part of a joint Economic and Social Research Council (ESRC) and Joint Information Services Committee (JISC) Initiative on the 1991 Census. Although covering much of the same ground as the OPCS's own General Report, the User's Guide was both considerably more timely and generally more in-depth, with the exception of only a lack of hard information on data quality. The later Census Users' Handbook (Openshaw, 1995a) is in contrast a determinedly 'unofficial' guide to the census, mainly focused on offering advice on how to analyse census data. Whether this 'training material' should also be viewed as census metadata is left for the reader to decide. Other census metadata provided by sources outside of the Census Office include a range of *ad hoc* reports on sample variability and data quality or reliability (e.g., Cole, 1994; Campbell *et al.*, 1996; Sandhu, 1993; Simpson and Middleton, 1997).

In addition to publishing census statistics for areas the size of local authority districts and above, the Census Offices released machine-readable census statistics for small (sub-district) areas. A number of user communities, for example, the academic sector, centralized the purchasing of these data to negotiate price discounts. In some cases, this has led to the creation of centralized metadata. Perhaps the best example is the Census Dissemination Unit (CDU), funded by the ESRC/JISC Census Initiative to support academic access to census data. The CDU has drawn together, in one location, a range of metadata, including a description of available census outputs, lists of unique ward and ED identifiers, and lookup tables to help aggregate 1991 EDs into other contemporary and historical geographies. This information has been supplied mainly in machine-readable form over the Internet, ensuring that it is always as up-to-date as possible. The CDU also provides 'metadata' on how to run the data extraction software SASPAC91 (MCC, 1994). The more casual user requiring access to unpublished sub-district level census data either has to rely upon published reports, produced by local authorities and others on an *ad hoc* basis, or on access to one of a number of commercial software packages that bundle together census data and data extraction software. These packages typically include at least limited metadata, but often include a full set of census geographies.

As mentioned earlier, the Census Offices entered into commercial agreements with two organizations, the ED-LINE consortium and Geographical Data Capture Ltd, to create and sell digitized census boundary data. The relatively high costs and complexity of these products also led to centralized purchasing and provision strategies. For example, in the academic sector machine-readable boundary data are supplied via the ESRC/JISC-funded UKBORDERS. However, integrating these map and census data still requires considerable technical expertise. Less expert users are forced to rely upon commercial packages that provide easy-to-use mapping facilities, but typically incorporate a relatively limited range of census data (e.g., SCAMP-2 from Claymore Services Ltd.). For those without access to such software, the final resort is once again often a printed map from an *ad hoc* council or other reports published locally. CASWEB (see Chapter 8) offers a solution to these problems, at least for the academic sector, directly integrating census geography with census data access in a highly user-friendly manner.

One final source of unofficial census metadata is METAC91, a metadatabase created in response to the need for a more comprehensive index of the 1991 LBS/SAS Census output (Williamson *et al.*, 1995). METAC91 boasts of a number of features including comprehensive description of all LBS/SAS output and support of multiple search types. It is freely disseminable and is based on fast and user friendly software. A copy of this software is provided as part of the accompanying *CDS Resources*. However, although addressing a number of the problems with pre-existing census indexes outlined earlier, METAC91 is not without its own problems. These include no graphical representation of table layouts, no guarantee of total accuracy, non-standard naming of census variables, release over a year after publication of the first LBS/SAS data, no coverage of special topic reports, and lack of ongoing updates and revisions (hence, no coverage of LBS Table 100).

23.4 METADATA FOR THE 2001 CENSUS

Reviewing the lessons to be learnt from the 1991 Census reveals the importance of timely and integrated metadata as part of overall census outputs and highlights the possibilities offered by computer- rather than paper-based metadata products. It is reassuring, therefore, that in a recent consultation document on proposed outputs for the 2001 Census, the following statement of intent was included:

> 'The Census Offices will employ technology current at the time of delivery to enable the means of dissemination to be as flexible as possible, with *greater emphasis on electronic media* as a means of delivering statistical information *backed by co-ordinated and comprehensive metadata*.' (ONS/GROS/NISRA, 1999a: para 1.5, authors' emphasis)

More specifically, the document declares that:

> '. . .comprehensive information about the data ('metadata') will be produced on 2001 Census definitions and concepts, output classifications, geography, data quality and coverage. As far as possible, metadata will be available before or with statistical output, and will be accessible in a number of formats, including via the Internet. To assist customers using Census data, training material will be developed if there is sufficient interest.' (ONS/GROS/NISRA, 1999a: para 3.35)

This amounts to an ambitious and laudable set of targets at which to aim. What remain to be determined are the precise details concerning the content and delivery of the proposed metadata product(s). The following two sub-sections consider each of these aspects in turn.

23.4.1 Content

Despite the wealth of information made available concerning the 1991 Census, room for improvement remains. Figure 23.2 presents one view of the possible range of information that could be included in a 2001 Census metadata product. Key elements of such a product are arranged in concentric rings around the core census product of the data themselves, with items closest to the centre of greatest importance. Table 23.1 indicates the 1991 Census equivalent for each item mentioned in Figure 23.2, highlighting any known shortcomings.

Central to any census metadata must be a detailed geography of census output. Put bluntly the provision of areal data without supporting geographical boundaries is next to useless. However, the provision of map boundary data immediately raises issues of cost and copyright, especially given the current intention to use Ordnance Survey data as the basis from which to create output area boundaries. There are also problems concerning the delivery of boundary data in a user-friendly form that novices can use. Should the Census Offices' metadata include rudimentary mapping software or should provision of

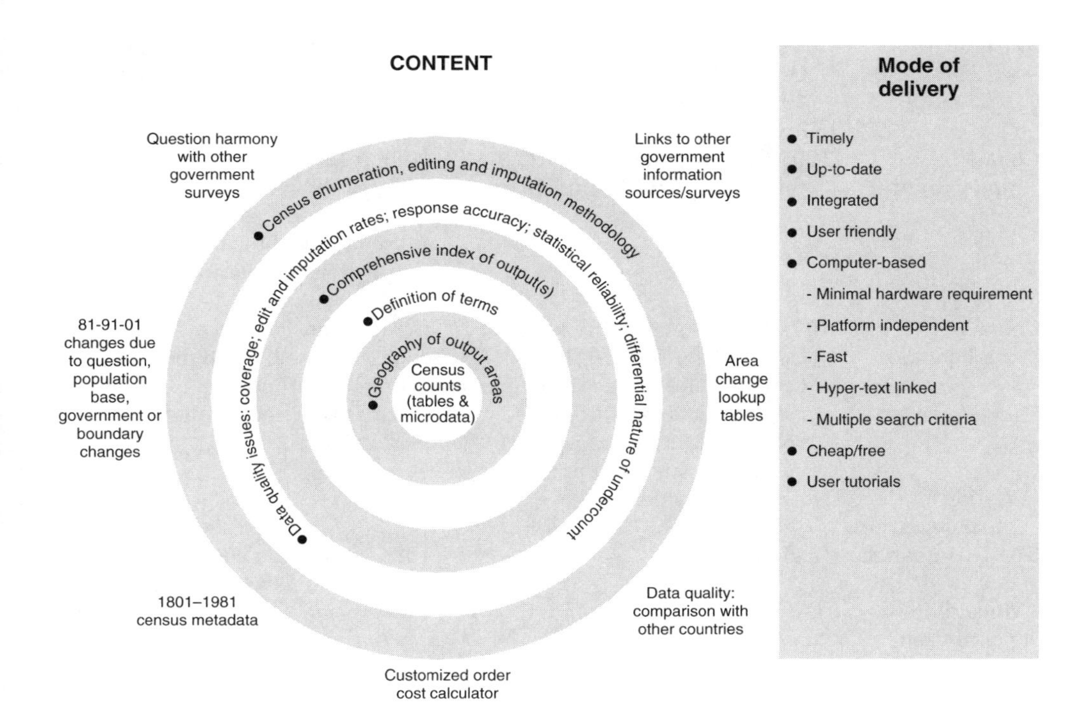

Figure 23.2 The ideal 2001 Census metadata.

Table 23.1 1991 Equivalents of the recommended 2001 Census metadata.

Metadata	1991 Equivalent	Shortcomings in 1991
Definition of terms	1991 Census definitions volume	Paper-based; not hypertext linked; not integrated directly with other census metadata
	Appendices to county volume and special topic reports	Partial coverage
	The 1991 Census user's guide	Paper-based; not hyper-text linked; not integrated directly with other census metadata
Index of outputs	Indexes to county volume and special topic reports, OPCS census user guides	Not fully comprehensive (bivariate tabulations only), with separate indexes for each product
Coverage	1991 Census validation survey: coverage report, 1991 Census general report	No disaggregation of estimated undercount other than by age, sex, and broad geographical area
Edit and imputation rates	1991 Census general report	Univariate analyses only; no disaggregation by population sub-group or geographical area
Response Accuracy	1991 Census validation survey: quality report	Univariate analyses only; no disaggregation by population sub-group or geographical area
Statistical reliability	1991 Census general report, OPCS census user guides, The 1991 Census user's guide, *ad hoc* academic reports	Users left to calculate reliability of specific counts using published formulae
Census undercount	1991 Census general report, estimating with Confidence 'Gold' Standard	Undercount disaggregated by age, sex, and broad geographical area, greater geographical breakdown but undercount still only disaggregated by age and sex
Census methodology	1991 Census general report, 1991 Census user's guide, 1991 Census definitions, 1991 Census county and special topic volumes, 1991 Census validation survey reports	Comprehensive but piecemeal; paper-based; not hyper-text linked; not integrated directly with other census metadata
Question harmonization	No 1991 equivalent	
81-91-01 changes due to question, population base or area changes	LBS Table 01	1991 population count given for 1971 and 1981 population base definitions; no 1971 and 1981 population counts for 1991 definition

Table 23.1 *(continued)*

Metadata	1991 Equivalent	Shortcomings in 1991
	Historical Tables' special topic volume	Previous census counts disaggregated only by age and sex for each current census area; no alternative series for historical census areas or for other variables of interest
Area change lookup tables	Machine-readable lookup tables	Many geographies covered, but prohibitively expensive and difficult to use for casual users; released after census data
Data quality: comparison with other countries	1991 Census validation survey coverage report	None
1801–1981 Census metadata	Census reports 1801–1981	Piecemeal and paper-based
Links to other relevant data sources	CSO Guide to official statistics 1990	Paper-based; univariate; no reference to variable classification (e.g., quinary or single year of age)
Customized output cost calculator	Census Office guidelines	Paper-based guidelines

this functionality be left to commercial providers as for the last census? This is clearly a complex issue, but one that needs to be resolved and at a price end-users can afford. Of almost equal importance are definitions of the terms used in census output, such as 'population base', 'social class', and 'imputation'. In both 1981 and 1991, the requirement for such information was admirably answered by the relevant census 'Definitions' volume. The challenge for the next census is to deliver the same information in machine-readable form, preferably well integrated with other census metadata while at the same time satisfying the continuing (although diminishing) paper-based market.

Having established the geography and meaning of census output, perhaps the next most useful metadata element would be a comprehensive index of census outputs, enabling users to navigate their way around the various census outputs available. As the construction of METAC91 demonstrated for the 1991 Census, this amounts to more than just a detailed index of standard table and census area statistics contents. For example, a user should not only be able to search the index to locate specific multiway tabulations of variables, but also specific aggregations of variables (e.g., single or quinary year age groups) and to check for availability of data at different geographic scales. This index would also, ideally, include pointers to other data sets when appropriate. For example, the 1991 SARs offer the flexibility to estimate more detailed multiway tabulations at local authority district than those published in the LBS.

Of less interest to the casual user, but essential to any serious census data analyst, is information on census data quality. This ideally includes an indication of the reliability of answers given, item, household, and individual imputation rates, in each case disaggregated by geographical area and type of household or individual. Information on the estimated statistical reliability of each and every census count released would also be welcome—information that may well arise naturally as an end product of the One Number Census process. A number of users will have considerable interest in the various methodologies underlying census enumeration, the Census Coverage Survey, data editing, and imputation. These could usefully be gathered together in one metadata product, rather than being scattered in various official and semi-official sources, as was effectively the case for the previous census.

Peripheral to these core items of metadata, a wide range of other information could usefully be made available to end-users (see Figure 23.2). These include the degree of compatibility in variable definitions among different government data sources (data harmonization), comparison of data quality and methodology issues with findings and practices from other national census agencies, the reasoning behind the inclusion or exclusion and wording of the various census questions, and time-series of key data, allowing users to assess the impact of changes over time in geographical boundaries, variable definitions, and population bases. For time-series, of course, the distinction between data and metadata is perhaps becoming dangerously blurred.

At present, the Census Offices are planning to produce a Census Quality Report that will address many of the issues raised earlier, including coverage, comparability with data from other sources, timeliness, completeness, and the quality of data processing (ONS, 1999f). The Census Quality Report is also planned to include population definitions and full details of the entire One Number Census methodology, including census data processing (editing and imputation) and the design and implementation of a supplementary Census Coverage Survey. Further plans include the reporting of achieved imputation rates and response quality problems identified both by the Census Quality Survey, conducted in 1999 as part of the Census Rehearsal, and by comparison with previous United Kingdom and international census experience. The Census Offices are also considering the provision of computer-based training material. This would effectively comprise selected census metadata, combined with a specially written front-end designed to guide novice users through the intricacies of using census data.

23.4.2 Mode of Delivery

To maximize functionality and cross-linkages, the 2001 Census metadata are likely to be computer-based, and ideally should be delivered as an integral part of the data they are intended to support. It almost goes without saying that the software used to access the metadata should be user-friendly. In today's context, this is taken to mean a quick-response 'point-and-click' hypertext-linked product embedded in a Microsoft Windows–like interface. Ideally, this product would also be platform independent and impose minimal hardware requirements. It is likely, however, that user demand (and possibly statutory obligation) will require the production of at least some census metadata in a paper-based format. For cost and other reasons, the latter will effectively provide only a subset of the full electronic version, the precise nature and content of which is yet to be decided.

At present the two preferred modes of electronic metadata delivery are via CD-ROM and via the Internet. The main disadvantage of a CD-ROM based product is that once released it is frozen in time. As a result, errors cannot be corrected, nor can additional clarification be added in the light of user feedback. A further problem is that elements of the overall metadata will be available at different times. For example, the background on the reasoning underlying the selection and wording of each census question could be made available now that the relevant census legislation has been passed. Evaluation of results from the Census Quality Survey, giving an indication of probable response reliability and patterns of question non-response, could also theoretically have been available on or before Census Day, 2001. On the other hand, information on edit and imputation rates is unlikely to be available until after the release of census data in around mid-2003. As ensuring forward compatibility of released metadata with future products is probably an impossible challenge, this implies that later releases of the metadata will have to include revised versions of all earlier metadata products. This gives rise to concerns over the final cost to the end-user of purchasing the entire set of metadata.

As an alternative mode of delivery, supply of metadata over the Internet using a Web browser looks attractive, but this approach also suffers from problems. If interactive on-line access to the metadata is provided, then heavy end-users could end up with high connection charges. As the size of the various metadata elements is likely to be large, downloading the latest version to a local PC could prove equally expensive. On the other hand, provision via the Internet might well suit casual or infrequent users, as access costs are likely to be relatively low, and access to the latest version of the metadata could be guaranteed.

23.4.3 Progress to Date

Taken together, the range of consultation documents issued as part of the 1999 Census Output Roadshows themselves form an impressive range of metadata, including provisional definitions of terms, table layouts, and a history of 2001 Census developments to date. Updated, these documents are likely to form a key part of the proposed 2001 Census metadata. The metadata proposals published to date also suggest that the Census Offices have indeed learnt key lessons from the 1991 Census. Equally importantly, these proposals appear to enjoy widespread and enthusiastic user support. However, at the time of writing, the Census Offices still have to resolve a number of key issues. These include:

1. the viability of providing supporting geographic (boundary) data
2. identification of a suitable software platform for the proposed electronic metadata product
3. ensuring that metadata elements released at different times integrate seamlessly
4. final cost to end-users

23.5 CONCLUSION

Only time will tell what final form the 2001 Census metadata will take. It is highly probable that time and resource constraints will make it impossible to fulfil the entire ‘wish-list’ set out earlier. Key decisions also depend in part upon the technological and software developments that occur between now and April 2001, when the first metadata are likely to be released. However, users can perhaps look forward to a marked improvement on the metadata provision made for the 1991 Census. For further news on the latest metadata developments, readers are referred to the Census Office Web site listed in Box 23.1.

Box 23.1 Key web links for Chapter 23.

http://www.statistics.gov.uk/2001census/	The latest news on 2001 Census developments, including the latest information on the planned 2001 Census metadata
http://census.ac.uk	A selection of on-line 1981 and 1991 Census metadata, created under the aegis of the Census Advisory Committee to the ESRC and maintained by the academic Census units established under the ESRC/JISC 1991 Census initiative

Box 23.2 Additional CDS resources for Chapter 23.

MetaC91	Self-extracting user-friendly software indexing 1991 Census LBS and SAS outputs plus user guide

24

Testing User-Requested Geographies

Oliver Duke-Williams and Philip Rees

SYNOPSIS

Users require census outputs for a wide range of geographies, covering administrative, postal, and grid-based areas. This demand for multiple geographies is further exacerbated by changes in boundaries over time. However, the Census Offices are reluctant to release counts for multiple geographies. They impose additional output production costs and threaten respondent confidentiality through areal 'differencing' (subtraction of the counts for one geography from another). Advances in computing power are making the issue of cost increasingly redundant. There remains the issue of confidentiality. This chapter outlines a method by which the Census Offices could quantify the risk of disclosive differencing when considering the release of census data for additional user-requested geographies.

24.1 INTRODUCTION

24.1.1 Geography in the Census

For a census to be taken, and for the results to be useful, it is necessary to divide the country into a number of component areas. From the point of view of planning and conducting a census, this subdivision is necessary for the efficient collection of data; the country is divided into a number of enumeration districts (EDs), which are designed such that each will provide a reasonable workload for a single enumerator. In contrast, virtually all users of the census, whether public or private sector, and regardless of their line of work, want to know about the population of specific places that are rarely coincident with the areal extent of single EDs or even aggregations of EDs.

The primary reason for the census to be taken is to provide the Government with the information that it requires for its own planning needs, both at national and local scales. The 2001 Census White Paper (HM Government, 1999) recognizes, however, that it is not just the Government itself that requires such information but a variety of users:

> 'Government, local and health authorities, commercial businesses and the professions need reliable information on the number and characteristics of people and households if they are to conduct many of their activities effectively' (HM Government, 1999: para. 5)

A problem that arises from this shared requirement for information is that users have different preferences for the geography that should be used to subdivide the country when reporting the results of the census, including local government areas, postal districts, and TV regions, each of which is an example of alternative *geographies*. Discussions with various user groups (Rees, 1997a) has indicated three main types of geographies in which users are interested:

- *Administrative geographies:* a common method of subdivision is to use the boundaries defined for local and national governmental administrative purposes. These units are typically wards and districts, and larger areas such as counties and regions. The reporting of data for these areas fulfils the basic requirements of the census as a method of providing information needed for planning.
- *Postal geographies:* the next most obvious subdivision of the country is by some postcode-based geography. Generally speaking, health authorities and business users use these geographies, as their databases of patients and customers record residential locations using postcodes.
- *Grid type geographies:* an alternative to the subdivision of the country using irregular polygons (such as local government wards or postal sectors) is to use a regular lattice or grid. This is typically defined using the National Grids (the grids are defined separately for Great Britain and Northern Ireland). Environmental scientists often prefer grid geographies, especially when they are being used in conjunction with large-scale data sets that have been collected by remote sensing.

All of these have strengths and weaknesses as output geographies. Administrative geographies are useful to planners who have to allocate resources for the same areas, but they are not always meaningful to other people. Members of the public will not necessarily know which ward they live in, and are certainly not in a position to know which ED they live in. This problem is overcome by the use of postcodes, as most people know either the full unit postcode of their house, or at least the postal sector. In Scotland, postcodes have been used in the 1981 and 1991 Censuses as the base for the output geography, creating a more meaningful set of small area units. However, postcodes are not always easy to work with. Firstly, their definition is ambiguous: rather than being defined as a particular polygon, a unit postcode is simply a set of specific delivery points. This is significant, because researchers often wish to use polygons to process data. This might be because they want to use the area of the postcode as a denominator when producing some index, or because they wish to draw a thematic map in which polygons will be shaded according to some variable. In most cases the set of delivery points that form a unit postcode are simply a set of residential households in a particular street, and in such cases it is a straightforward task to draw a polygon around these points (using a Theissen or Voronoi polygon). There are, however, many other cases in which it is less easy to construct such polygons, because the individual delivery points are not neatly clustered in physical space.

A second problem with postcodes as a geographic base is that they are not static, but rather change over time, with new postcodes being introduced and others being changed (Raper *et al.*, 1992). The problem with postcodes changing over time is also a feature of administrative geographies; indeed, one of the most important uses of census data is as an input to the re-drawing of electoral boundaries, to ensure equal representation for equal population. This tendency for areas to change over time is problematic for many academic users of the census, who often use the data to study change over time, and consequently would prefer to use a temporally consistent geography.

24.1.2 The Desire for Multiple Geographies

In short, there are a number of alternative geographies for which users of the census would like to see data reported. The actual number of geographies that people would like to use is larger than might be expected, because of the way in which geographies change over time. Researchers wishing to study the way in which an area has changed may want to get hold of current data that has been re-aggregated using the boundaries that were in force in 1971, 1981, or 1991, or any year in between. As we look towards the 2001 Census, this issue seems to be particularly thorny: the local geography of Britain has been subject to a series of changes (Wilson and Rees, 1999) that have taken a sufficiently long time to implement that data for each calendar year in the late 1990s seems to be based on a different geography!

The need for data for alternative geographic schemes has been reflected in the past by the publication of data for multiple geographies. The actual geographies for which data were published from the 1991 Census varied across the United Kingdom. In England and Wales the Small Area Statistics (SAS) were originally published for both EDs and wards. In Scotland, the data were originally published for output areas based on postal sectors. In Northern Ireland, a subset of counts was initially created for a grid-based geography, the full Northern Ireland SAS only subsequently being created for EDs. With the exception of Scotland, therefore, data were made available using more than one geography. (Although data published subsequently for other geographies, such as postal sectors and parishes, have in all cases been approximated through aggregations of the basic building blocks identified earlier.) There has been particular concern about this practice because of the potential that people may be able to compare two sets of counts and derive data for a new area through subtraction of one set of counts from another. This process has been called *differencing* (Denham, 1993) and is often cited as a reason why researchers cannot be provided with data for every geography that they desire.

There is another important reason the Census Offices are reluctant to make data available for multiple geographies, and that is quite simply the cost of creating the data. In the past, generation of SAS type data sets has required a considerable amount of resources (both human and computing). However, because of the continuing rapid advances in technology (as predicted by Moore's Law), handling census data is a process that has become somewhat easier. Databases that were considered enormous around the time of the last census are now considered commonplace, and where once (but quite recently!) expensive specialized computer hardware would be required to process census data, it is now feasible to process the data on standard and comparatively cheap equipment.

This leads us to a situation in which the financial argument against the production of data for multiple geographies will no longer be seen as a convincing one. An additional hurdle in the past has been that there may not be sufficient geographic detail captured in census records to allow us to re-aggregate the data in whatever way we like. To be able to have full flexibility of re-aggregation, it is necessary that all records must be georeferenced with as much accuracy as possible. In previous censuses, data was stored for each household, although the location of that household might not have been known to a greater degree of accuracy that the unit postcode (typically one unit postcode refers to some 20 to 50 households). No accurate subdivision of data below this level would be possible. However, since the last census, databases such as ADDRESS-POINT and COMPAS have been introduced, which give the locations of mail delivery points to an accuracy of 1 m or better in Great Britain and Northern Ireland, respectively. If these are used (as is expected) in the collection of data for the 2001 Census, then we will be in the situation in which arbitrary spatial aggregation of data is both technically and financially feasible, although there will still be practical problems with the small proportion of households that are not captured accurately in these databases.

24.1.3 The Desire for User-Requested Geographies

It is impossible for the Census Offices to provide a set of data that will satisfy all users, even if they do so for more than two geographies, because they cannot predict the geographies that will be required in the future. Future demand will include both data for new sets of administrative boundaries, which will themselves be influenced by the data collected in the 2001 Census, and also new interests based on questions that researchers have not yet thought of asking. It is inevitable that users will decide that they want data for geographies other than those that have already been published, and will want to ask the Census Offices to provide data for these geographies. This is all the more likely if it is apparent that the technical and fiscal hurdles to producing such data no longer exist. These future geographies may be termed *user-requested geographies*.

Whereas it will now be technically possible for the Census Offices to fulfil users' requests, the question of whether they can do so and retain the confidentiality of the data remains. In the rest of this chapter a closer look is taken at the problem of differencing, and at the related arguments for and against the release of data for multiple geographies.

24.2 THE PROBLEM OF DIFFERENCING

Figures 24.1 and 24.2 show an example of a case in which there is potential threat from differencing. Figure 24.1 shows two geographies—EDs (solid lines) and postal sectors

Figure 24.1 Multiple geographies for part of Leeds.

Figure 24.2 The problem of differencing for postal sector 'LS7 1'.

(dotted lines)—overlain on each other. EDs are generally much smaller in extent than postal sectors, and it can be seen that there are many areas where one or more EDs fit wholly inside a particular postal sector. One postal sector where this occurs (LS7 1) is picked out in grey.

Figure 24.2 shows in greater detail this postal sector together with the boundaries of all the EDs that overlap the sector. The postal sector is shown here in dark grey. There are a number of EDs that fit wholly inside the postal sector, and these are shown in light grey. The remaining EDs are the ones that overlap the sector, but are not wholly contained within it. The boundaries of these EDs are shown in the figure, and the parts of them that fall outside the boundary of the postal sector are stippled. There are a number of operations that we could perform on these areas to reveal new data. First, we could add up the values for a particular variable (or set of variables) for the EDs shaded light grey, and subsequently subtract them from the same set of variables published for the postal sector. This would give us a new count, which referred only to the area of Figure 24.2 that is shaded dark grey. Secondly, we could add up the values of a variable for all the EDs (both the overlapping set that are part stippled, part shaded dark grey, and the included set that are shaded light grey), and subsequently subtract from that total the value of the same variable for the postal sector. This would reveal a second new count, which referred to only the part of Figure 24.2 that is stippled. A terminology has been introduced for these regions involved in the differencing process. The wholly included areas that can be subtracted from their container (i.e., the light grey EDs in Figure 24.2) are termed a *nugget*. The region that is revealed through this subtraction (the dark grey remnant of the postal sector in the example) is termed the *inner halo*, whereas the outlying parts of overlapping areas (i.e., the stippled parts of EDs) are called the *outer halo*.

On the face of it, this indicates that there is a problem; we have successfully produced counts for new areas, which the Census Offices had not explicitly published. However, the data are protected in a number of ways before they are released, which serve to reduce the apparent threat from differencing to respondent confidentiality. The SAS data are blurred, before they are published, through a process in which an unknown proportion of cells in any cross-tabulation are modified by adding or subtracting 1 from their value (Marsh, 1993a). When it comes to differencing, this is highly significant, because the error will be additive: if we have to say that any cell in the counts for a particular area might have an error of +1 or −1, then the counts produced by the subtraction of one area from another would lead to a possible error in the range +2 to −2 in the resulting differenced data. In practice, as in the earlier example, we are generally adding several small areas together to be able to carry out a differencing 'attack', and this means that the potential error is increased still further. In Figure 24.2, we are using a total of five light grey EDs to difference the postal sector, and the total produced will therefore have a maximum error for any variable within the range +5 to −5.

An additional form of protection comes from the use of thresholds. For data to be released for any area from the 1991 Census, they were required to pass certain thresholds. In England and Wales and in Northern Ireland, for data to be released for an ED (or any other area), it was required to contain at least 50 persons and 16 households; for any area that failed to meet these criteria only three basic counts were published. The counts for the suppressed area were then merged into those of a neighbour. In Scotland, output areas were specifically designed so that all Output Areas were above such thresholds. The effect of thresholding serves to complicate the differencing procedure. If one of the areas to be used was suppressed, then it is necessary to find the neighbouring area that the counts were merged into, and to make sure that that area is included in the set of areas being used in the differencing analysis. This is both time consuming and is likely to lead to larger numbers in the halo regions.

A final form of protection comes from the legal undertakings that census users make when they acquire the data (or a suitable computer account to use the data). Among other things, users agree not to reveal information, or claim to be able to do so, about individuals. It is to this promise that differencing relates. If it is possible to use differencing to reveal new halo regions that contain very small numbers of people, then it might possible to 'identify' people with certain sets of characteristics in the resulting cross-tabulations. Although, this would not reveal information about named individuals without a considerable degree of a priori knowledge, it would certainly be considered dangerous by the Census Offices. In practice such a claim would be hampered by the uncertainties introduced to the data through the blurring procedure, but it might not stop a malicious party from claiming that it is possible to identify individuals within the census, and thus reducing public confidence in the census and the Census Offices.

24.3 DESIGN OF SOFTWARE FOR TESTING DIFFERENCING

As described previously, there are a variety of protection measures that reduce the threat from differencing. One of the key protection devices is the legal one. The front line defence, however, is statistical. Before data for a new geography can be released, the Census Offices have to assess the risk that is posed by differencing and decide whether many (or any) differenced areas could potentially contain such a small number of people that concerns about the ability to identify individuals (that is, to indicate that one can determine that an individual has certain sets of characteristics), would be justifiable.

At present, no systematic way of assessing this risk is available to the Census Offices. Consequently, a prototype software system has been developed at Leeds University (Duke-Williams and Rees, 1998) to assess the threat from differencing posed by different geographies, using outputs from the 1991 Census. Such a system could be used by various statistical agencies to help them assess the degree to which both new standard geographies (i.e., those for which data are produced and made widely available) and proposed bespoke geographies (i.e., those requested directly by specific users) presented a risk, given the geographies that had already been produced. Such a tool would obviously be increasingly desirable if users were to demand the sort of flexible control over geography that is outlined in Section 24.1.3.

The prototype system was used to conduct a number of tests to assess the practical risks from differencing. To carry out such experiments, a method is needed to assess the level of risk of disclosure in any given set of outputs. There has been considerable debate in the literature over methods of risk assessment and over the question of what exactly constitutes 'risk'. One of the problems that arises is that the variables reported in the census have differing degrees of sensitivity, but that in all cases this sensitivity is dependent on a subjective view of the data. A relatively simple practical measure of risk was used for experimentation: whether or not data could be discovered for new areas, and if so, whether the numbers of residents and households in the new area were above or below the thresholds used for safe publication in the 1991 Census. On the one hand, this test is coarse, as we know that the thresholds are a coarse measure: the level of risk of disclosure is a function of both total numbers and the composition of a population, but the use of a threshold only considers the former criterion. On the other hand, however, the test is a useful one, as it is easy to perform and mirrors past procedure.

The software used a number of components loosely joined together, including ESRI's GIS package ARC/INFO and a number of custom-written programs. These programs were used to search for areas that are wholly contained within areas in an alternative geography, and then to carry out the various subtractions of data required to calculate new values for the halo areas created. These new counts were subsequently compared with the 1991 publication thresholds.

In any attempt to difference two geographies, three sets of data are required:

- accurate population counts for areas in the first geography
- accurate population counts for areas in the second geography
- a list of intersections between areas in the two geographies.

These data can be gathered in a number of ways from different sources, although it is important to maintain consistency across all data sets. The experiments conducted with the prototype system used both published digital boundaries for the Yorkshire and Humberside regions, and a separate list of intersections, taken from the Postcode to Enumeration District (PC/ED) directory. In all cases, 1991 boundaries were used, as this was the period for which the most data were available in a consistent form.

An initial task was the derivation of an accurate set of population counts for both geographies being compared. Although some counts are available directly from the census, this is not always the case, especially when arbitrarily defined boundaries are being used. An accurately geo-coded base population data set is therefore required, which can be used with any set of boundaries to build aggregates of the numbers of persons and households in each area. In our case, we used a synthetic population, because we did not have access to a true data set of this kind. This synthetic population was developed by randomly generating household location coordinates in a controlled manner, such that each 1991 ED contained the correct number of households. Household records from the 1991 Sample of Anonymized Records were then randomly assigned to these location points to create a synthetic population of around 5 million people living in 2 million households. This resulted in a population with a reasonable degree of realistic spatial distribution, but no spatial demographic patterns. However, because our tests only compared total numbers of persons and households, this was not a problem. For more sophisticated tests, either the original master data would be required or a procedure such as that outlined by Williamson *et al.* (1998) would be needed to generate a more realistic synthetic population.

A table of intersections was constructed using the ARC/INFO command *identity*, which was used to overlay the two geographies and create a new coverage consisting of all intersections, with all polygons in the new coverage having two labels, one from each source geography. These label pairs were written to an external file for subsequent processing. An alternative approach is to generate a table of intersections based solely on the presence of households in overlapping areas—if two sets of boundaries overlap on a map, but there are no households in the overlapping area, then no intersection would be recorded with this method. An example of the latter approach is the published Postcode to ED lookup table. This method offers greater accuracy and control, but is dependent on the availability of an accurate map of household locations.

For experiments with grid-based geographies, new sets of digital boundaries had to be created for input to ARC/INFO in order for data to be tabulated. In the case of regular grids, the task of creating a grid of cells of the required size was trivial. However, for variable-sized grids a rather more sophisticated program was required, which used a quadtree structure (Finkel and Bentley, 1974) to build an index for the data it handles. This index was then used to create a set of cells that was the smallest possible decomposition that contained a minimum count of people. This led to a set of variable-sized grid cells, with large cells in sparsely populated areas and much smaller cells in densely populated urban centres. An example of such a grid is shown in Figure 24.3.

To search for areas at risk of differencing, the table of intersections for the two geographies was processed and two re-ordered tables were created—Table 'A' lists each area in the first geography in turn, giving a list of overlapping areas in the other geography, and Table 'B' lists each area in the second geography, showing all overlapping members of the first geography. The tables were then examined in turn. For any area in Table A,—say A1—we can take the list of overlapping areas—perhaps B23, B47, and B64—and look up each one in Table B. In most cases, the respective entries in Table B will show that in addition to the area currently being studied in Table A (in this example, area A1), there are also intersections with other components of Table A. However, it is also possible that Table B will show that a particular area only has a single entry; in this example, area A1. If this is the case, then it means that the area in Table B must be wholly contained with the current focus area in Table A, as it is the only area in A it overlaps with. When such areas are found, it is possible to carry out a differencing procedure and reveal data for a new area. For each area, we developed lists of all wholly contained areas and all other overlapping areas. If there was one or more wholly contained area, then their counts were summed and subtracted from the total for the focus area.

24.4 RESULTS OF EXPERIMENTS

Three main sets of experiments were carried out using the prototype system; these were comparisons of EDs with postal sectors, EDs with regular grid cells, and EDs with

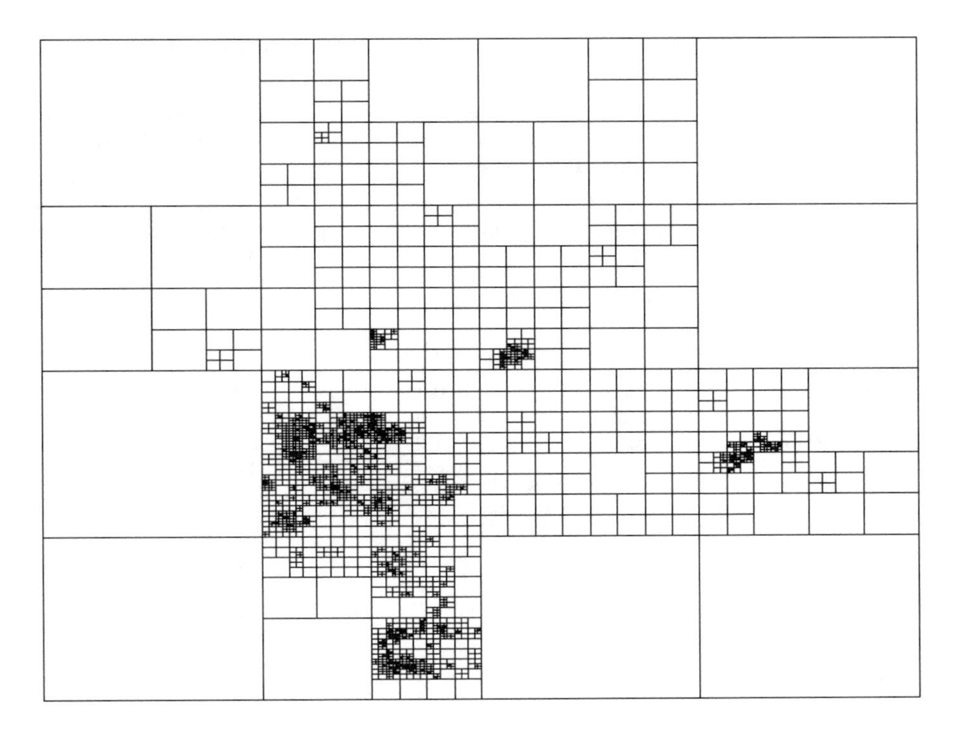

Figure 24.3 A variable-sized grid geography for Yorkshire and Humberside.

variable-sized grid cells. The results from these experiments are shown in Tables 24.1 to 24.3, and summarized in Table 24.4. In Tables 24.1 to 24.3, there are three sets of statistics. The first part shows how many areas could be investigated using the differencing test, how many could not be because they did not wholly contain any areas in the other geography, and how many were rejected because they were close to the edge of the region and thus had intersecting polygons which overlapped with the background. These cases were problematic and could not be fully handled by the prototype system. The second part of each table shows the distribution of person counts in the halo regions revealed, whilst the third part shows the distribution of household counts in the halo regions. The total number of haloes in the second and third panels (i.e., for persons and households) should be equal to twice the number of areas for which differencing was possible, because each successful test reveals information about both haloes. The class limits used to group the counts are the standard 1991 ED threshold (50 persons, 16 households), twice these thresholds, and subsequently three times the threshold (for households), or four times the threshold (for persons).

Table 24.1 shows, in columns 2 and 3, the results for the comparison of postal sectors and EDs, using the digital boundaries and the synthetic population. Column 2 in Table 24.1 shows that the majority of postal sectors can be differenced using this method, but all of this majority produce large haloes that contain significantly larger-than-threshold populations. Column 3 in Table 24.1 reveals that only two EDs can be differenced. The small frequency for EDs which can be differenced is as expected, because the average size

Table 24.1 Results of a comparison of EDs and postal sectors.

Category	**Using synthetic populations and digital boundaries**		**Using census counts and PC/ED directory**	
(1)	**EDs inside postal sectors (2)**	**Postal sectors inside EDs (3)**	**EDs inside postal sectors (4)**	**Postal sectors inside EDs (5)**
Areas	PSs	EDs	PSs	EDs
Differencing not possible				
Not contained	50	10421	49	10685
At edge of region	212	0	0	0
Differencing possible	511	2	698	13
Total	773	10423	747	10698
Haloes				
Above threshold				
≥200 persons	1022	3	1366	16
100–199 persons	0	0	9	2
50–99 persons	0	1	3	
Below threshold				
<50 persons	0	0	6	0
Total	1022	4	1384	18
Haloes				
Above threshold				
≥48 households	1022	4	1372	4
32–47 households	0	0	3	0
16–31 households	0	0	3	0
Below threshold				
<16 households	0	0	5	0
Total	1022	4	1381	4

of postal sectors is much greater than the average size of EDs. Thus one is unlikely to find cases of a postal sector being contained within an ED. Both of the cases found are in city centres, where the density of unit postcodes (mainly business addresses) is much higher than in the residential areas outside city centres. Table 24.1 also shows, in columns 4 and 5 respectively, results for the comparison of postal sectors and EDs using the relationships between postcode units and EDs using the PC/ED directory to provide a list of intersections. As with the previous comparison, most of the postal sectors can be differenced, but few EDs can be differenced. A larger number of haloes could be revealed using the second method, and this will always be the case when published lookup tables are used to generate a list of intersections between geographies as these lookup tables are much 'cleaner' than the lists of intersections created by comparing digital boundaries. When intersection sets are generated by overlaying two boundary sets from different sources, slight differences in the digital representation of lines will lead to large numbers of sliver polygons being generated. These sliver polygons thus make the resulting intersection set less 'clean' than sets generated by other methods.

Table 24.2 Results of a comparison of EDs and grid cells.

Category	10-km grid		5-km grid		1-km grid	
	EDs inside cells	**Grid Cells inside EDs**	**EDs inside cells**	**Grid Cells inside EDs**	**EDs inside cells**	**Grid Cells inside EDs**
(1)	**(2)**	**(3)**	**(4)**	**(5)**	**(6)**	**(7)**
Areas	Cells	EDs	Cells	EDs	Cells	EDs
Differencing not possible						
Not contained	10	10422	225	10422	13780	9826
At edge of region	122	0	260	0	1343	0
Differencing possible	65	0	226	0	863	596
Total	197	10422	711	10422	15986	10422
Haloes						
Above threshold						
≥200 persons	129	0	445	0	1711	424
100–199 persons	1	0	6	0	8	379
50–99 persons	0	0	0	0	2	238
Below threshold						
<50 persons	0	0	1	0	5	151
Total	130	0	452	0	1726	1192
Haloes						
Above threshold						
≥48 households	130	0	451	0	1720	938
32–47 households	0	0	0	0	1	112
16–31 households	0	0	1	0	5	92
Below threshold						
<16 households	0	0	0	0	0	50
Total	130	0	452	0	1726	1192

One feature of both columns 4 and 5 in Table 24.1 that is immediately apparent is that the total numbers of haloes is inconsistent. Column 4 shows that 698 postal sectors can be differenced, potentially creating 1396 haloes, but both sets of halo frequencies have totals slightly below this figure. The reason for this disparity is the way in which postal sectors that overlap with the background are handled; the system used for the experiments cannot always process such cases, and some areas have to be ignored.

Table 24.2 reports the results for three sets of regular grids that were generated. Columns 2 and 3 give the results for a grid with cells with a side of 10 km, columns 4 and 5 give the results for a grid with cells of side 5 km, and columns 6 and 7 give the results for tables with a side of 1 km. Both of the larger grids (10 and 5 km) use cells that are considerably bigger than the average ED, and therefore it is not surprising that the results show that no EDs could be differenced using these grids (columns 3 and 5). However, although many of the grid cells, especially for the larger grids, could be differenced, this tended to generate large haloes containing populations of greater-than-threshold size (columns 2, 4, and 6). In contrast, the smaller grid has an average size which is smaller than an ED. Given that the size of EDs varies considerably, whereas the grid has cells of a fixed

Table 24.3 Results of a comparison of EDs and variable-size grid cells.

Category	**Minimum number of households per cell**							
(1)	**256 EDs inside grid cells (2)**	**Grid cells inside EDs (3)**	**128 EDs inside grid cells (4)**	**Grid cells inside EDs (5)**	**64 EDs inside grid cells (6)**	**Grid cells inside EDs (7)**	**32 EDs inside grid cells (8)**	**Grid cells inside EDs (9)**
Areas	Cells	EDs	Cells	EDs	Cells	EDs	Cells	EDs
Differencing not possible								
Not contained	715	10422	2574	10422	6055	10417	14643	10107
At edge of region	101	0	122	0	173	0	200	0
Differencing possible	574	0	476	0	280	5	149	315
Total	1390	10422	3172	10422	6508	10422	14992	10422
Haloes								
Above threshold								
≥200 persons	1148	0	944	0	545	8	276	580
100–199 persons	0	0	8	0	14	2	20	49
50–99 persons	0	0	0	0	1	0	1	1
Below threshold								
<50 persons	0	0	0	0	0	0	1	0
Total	1148	0	952	0	560	10	298	630
Haloes								
Above threshold								
≥48 households	1148	0	952	0	559	10	297	630
32–47 households	0	0	0	0	1	0	1	0
16–31 households	0	0	0	0	0	0	0	0
Below threshold								
<16 households	0	0	0	0	0	0	0	0
Total	1148	0	952	0	560	10	298	630

size, this means that in practice some EDs are larger than the grid cells and some are smaller. Hence both EDs and grid cells can be differenced using the 1-km grid, as shown in columns 6 and 7.

As the size of regular grid cells becomes smaller, it is increasingly easy to find sets of grid cells that can be aggregated to give a close approximation of an irregular area (such as an ED). If one can achieve a good approximation of the shape of the containing area, the halo regions revealed will be physically small, and thus might be expected to contain small populations. As the grid size increases, more cells become large enough to contain at least one ED, and so more cells will permit differencing. However, EDs are generally irregularly shaped, and will be unlikely to achieve a close approximation of a square. The result of this is that the halo regions will tend to be large and, therefore, one would expect them to contain relatively large populations.

Table 24.3 reports the results for four sets of grids constructed with grid cells of variable size. The grids were built using minimum thresholds of 256, 128, 64, and 32 households (i.e., multiples of the 1991 household threshold for EDs). The last of these four grids

Table 24.4 Summary of results of comparing EDs with other geographies.

Type of geography	Number of areas	Average area $(km)^2$	Test for EDs inside other geography			Test for other geography inside EDs		
Description			% Areas differencing possible	% Differenced haloes sub-threshold		% Areas differencing possible	% Differenced haloes sub-threshold	
				Persons	Hhlds		Persons	Hhlds
(1)	(2)	(3)	(4)	(5)	(6)	(7)	(8)	(9)
Postal sectors								
Using DBD	773	20.3	66.11	0.00	0.00	0.02	0.00	0.00
Using PC/ED	747	20.3	93.44	0.43	0.36	0.12	0.00	0.00
Regular grid								
1-km side	15986	1	5.40	0.29	0.00	37.77	12.67	4.19
5-km side	711	25	31.79	0.22	0.00	0.00	0.00	0.00
10-km side	197	100	32.99	0.00	0.00	0.00	0.00	0.00
Variable grid								
Min 32 hhlds	14992	1.7	0.99	0.36	0.00	3.02	0.00	0.00
Min 64 hhlds	6508	4.0	4.30	0.00	0.00	0.05	0.00	0.00
Min 128 hhlds	3172	8.0	15.01	0.00	0.00	0.00	0.00	0.00
Min 256 hhlds	1390	18.5	41.29	0.00	0.00	0.00	0.00	0.00

has a similar average area size to EDs, but with a significantly different distribution, so that there are more cells in the variable grid than there are EDs in the region. All the variable grids allow some cells to be differenced, although as the minimum number of households is reduced, and the average cell size becomes smaller, so a smaller proportion of the cells can be differenced. As with the regular grids, the reduction in cell size makes it more likely that there will be EDs that wholly contain grid cells, allowing the EDs to be differenced. The two smaller variable grids allow some EDs to be differenced (columns 7 and 9).

Table 24.4 shows, for each geography compared against EDs, the proportion of areas that could be differenced, together with the percentage of those areas that had below-threshold numbers of persons or of households. In the table, columns 4 and 7, labelled % Areas differencing possible', show the percentage of areas of the containing geography that were suitable for differencing. Thus, in the first row, which shows the results for the comparison of postal sectors with EDs, using digital boundary data (DBD), 66% of the postal sectors could be differenced by looking at EDs inside them, whereas only 0.02% of EDs could be differenced by looking at postal sectors inside them.

The columns marked 'persons' and 'hhlds' (5 and 6; 8 and 9), respectively, show the percentage of haloes created by differencing which contained fewer than 50 persons or 16 households. Thus in the second row, 0.4% of haloes generated for postal sectors (by fitting EDs inside them via lookup tables) contained a sub-threshold number of people, and 0.4% contained a sub-threshold number of households. In the same way, although haloes could be generated for 0.1% of EDs none of these haloes contained sub-threshold counts of persons or households.

The variable-sized grid cells offer very interesting results. The smallest grid, constructed with a minimum of 32 households, has an average cell population only slightly larger than for an ED, yet has rather more areas in total because of a different distribution of cell sizes. Few cells can be differenced and very few of those can produce haloes with sub-threshold populations. This is because of the nature of the grid. Having large grid cells in rural areas prevents the effective selection of a set of cells that will closely approximate the shape of the ED. This set of cells can be contrasted with the 1-km regular grid. Both have a similar number of cells and a (roughly) similar average size. However, the variable-sized grid would appear to be much 'safer' in that fewer sub-threshold haloes can be revealed. The variable-sized grids would therefore seem to offer a very suitable alternative geography, which could be supplied alongside the standard ED geography. However, it is recognized that they would probably prove difficult to work with for many users. A greater understanding of the uses to which grid-based geographies are put is required to assess whether such a grid would prove useful.

24.5 CONCLUSION AND FUTURE RESEARCH

The experiments described previously show how a software system can be used to compare multiple sets of geographies, to see whether it is 'safe' to release more than one geography. Although two sets of areas used—EDs and postal sectors—were extensively used to report the last census, we could in fact use any geography for which digital boundaries existed. In the case of the geographies tested earlier, it can be seen that the practical risks from differencing are few for most geographies apart from the smallest grids.

Such a system could therefore be used by National Statistical Offices to check 'user-requested geographies'. There are a number of issues that remain unresolved. These include the use of a more effective measurement of risk than the simple comparison of counts in halo areas with thresholds, and policy decisions about whether any user-requested geography should be compared against a single base set of areas, or all geographies which have previously been produced, including other user-requested geographies.

Acknowledgements

The research reported here was made possible by an award by the European Union under the ESPRIT programme administered by EUROSTAT. The award funded a network, coordinated by Statistics Netherlands, of projects on Statistical Disclosure Control, in which the University of Leeds project was concerned with the Flexible Geography Task.

Glossary

Address. Text-based description for a location. Usually, but not necessarily, defined in relation to the postal delivery system.

ADDRESS-POINT™. Ordnance Survey product, containing a 0.1-m resolution grid reference for every postal address in the UK.

AFPD. The All Fields Postcode Directory created by ONS since 1998, which contains a complete list of current and past postcodes in England and Wales and their links, on a best fit basis, to other official areas, including EDs in the 1991 Census.

Approximation. The use of aggregated counts, which relate to a grouping of geographical areas that are not precisely coincident with the area for which the data are reported. For example in Scotland in 1991 approximate ward data are based on aggregation of counts for the best-fitting group of postcodes.

Automated Zoning Procedure (AZP). An algorithm originally devised by Openshaw (1977) for the iterative recombination of geographical units into larger regions in such as way as to maximize the value of some objective function.

Basic spatial units (BSUs). Elemental geographical areas, which form the basis for higher level geographical divisions of the country. In reality, there are no generally accepted BSUs in the UK, and their definition has been the subject of extensive debate.

Boundary-Line. Ordnance Survey product, containing a digital representation of statutory and administrative geographies.

British Standard 7666 (BS7666). A multi-part standard for the consistent definition of geographical descriptions published by the British Standards Institute.

Cartogram. Cartographic representation in which some property other than the physical area of the units being mapped (e.g., their population) is used to determine their size in the image.

CAS. Census Area Statistics.

CDU. Census Dissemination Unit.

Census Access Project. Project for procurement, delivery, and dissemination of Standard Area Statistics from the 2001 Census to the public sector in England and Wales. Based on a cooperative agreement between ONS and public sector organizations (Department of Health, Department of Transport, Local Government and the Regions, the Local Government Association and the Economic and Social Research Council). Equivalent arrangements have been made by NISRA and by GROS (via the SCROL project in Scotland).

Census Area Statistics (CAS). Tables of univariate and crosstabulated counts produced from the 2001 Census for geographical level from Output Areas to national parts and the UK as a whole. Equivalent to the Small Area Statistics at the 1971, 1981, and 1991 Censuses.

Census Dissemination Unit (CDU). A unit located at MIMAS, Manchester Computing, University of Manchester, funded under the ESRC/JISC Census Programmes of 1991–1996, 1996–2001 and 2001–2006 to provide the online delivery of census data sets such area statistics and derived products to the UK academic community.

Census Offices. A collective term used to refer to the official bodies responsible for administration of the census in the UK. Currently the Census Offices are ONS, GROS and NISRA. Also known as the Census Organizations.

Census Organizations. A collective term used to refer to the official bodies responsible for administration of the census in the UK. Currently the Census Offices are ONS, GROS and NISRA. Also known as the Census Offices.

Chloropleth map. Shaded area map.

Constitutions. See Lookup table.

Correlation lists. See Lookup table.

Collection geography. The geographical division of the country into areas used for administration of data collection.

Confidentiality principle. The principle enshrined in census legislation, that the published results should not inadvertently release information relating to identifiable individuals.

Data modification. The deliberate statistical perturbation of published census counts in order to uphold the confidentiality principle.

Differencing. The derivation of census counts for geographical areas formed by the overlap of two other areas for which published counts are available.

Economic and Social Research Council. UK Government research council funding research in the social sciences and responsible for the purchase of census data for use by the academic community.

ED. Enumeration district.

Enumeration district. Geographical area, which is the responsibility of a single census enumerator at the time of data collection.

ESRC. Economic and Social Research Council.

Estimating with Confidence. A project spanning local authorities, academic and government staff which undertook the research and consultation leading to the EWCPOP data set, providing estimates and corrections for the undercount, which took place in the 1991 Census.

Exporting area. A restricted area in census output geography, whose counts are combined with another local area in order to bring them above the required threshold.

EWC. Estimating with Confidence.

Family Health Services Authority (FHSA). Organizational unit of the UK National Health Service responsible for NHS general medical practitioners, dentists and other primary care services. Functions are now carried out by Health Authorities.

Flexible outputs, geographies. The concept of census outputs and geographical areas being defined (in real time or near-real time) in relation to users' requirements, rather than being fixed in advance as a single published data set.

Geographic equivalency tables. See Lookup table.

General Register Office for Scotland (GROS). The Census Office responsible for Scotland.

Geographical Information System (GIS). A computer-based system for the capture, storage, retrieval, analysis and display of geographically referenced data.

Georeferencing. The inclusion of a geographical code within a data record, allowing it to be directly or indirectly related to a specific location.

GIS. Geographical Information System.

GROS. General Register Office for Scotland.

Household. A group of individuals, defined according to explicit rules, which may change over time (e.g. individuals living at the same address and usually eating together).

Importing area. A census output area into which the counts from another local area have been incorporated in order to bring them above the required threshold.

International Passenger Survey (IPS). Sample survey of passengers at ports of entry/departure across the UK.

IPS. International Passenger Survey.

JUVOS. System for reporting current unemployment statistics (claimant counts) for geographic areas. Access to JUVOS is available through NOMIS.

Key Statistics. The set of data products, tables of percentages and base counts, from the 2001 Census for geographical areas from OAs to national parts and the UK as a whole.

Labour Force Survey (LFS). A large national household survey administered by ONS on behalf of government departments. The LFS provides information on a rolling quarterly basis for geographical areas down to local authorities.

Land-Line™. Ordnance Survey product, containing a digital representation of the largest-scale topographic mapping.

LFS. Labour Force Survey.

Local Base Statistics. A data set introduced in 1991, referring to the set of census counts published for geographical units from wards upwards. The 2001 Census equivalent is the Standard Tables data set.

Lookup table. A file (generally, but not necessarily, computer-readable), which gives the precise or approximate coincidence of different geographical areas (e.g. the 1991 Census postcode/enumeration district directory). A link file, recording the allocation of census zones to target zones.

MAUP. Modifiable areal unit problem.

MIMAS. Manchester Information and Associated Services. A national data service for the academic community hosted by the University of Manchester, and the location for academic access to the 1981 and 1991 UK Census data sets.

Modifiable areal unit problem (MAUP). Commonly broken down into the scale and aggregation problems. Count data aggregated to imposed areal units such as those comprising a census geography will be heavily dependent on the size of those units (scale) and the placement of the boundaries (aggregation). Reconfiguration of the geography will produce different results, without any change in the underlying phenomena.

National Health Service Central Register (NHSCR). Database of all patients registered with the UK National Health Service, maintained by the Office for National Statistics.

NISRA. Northern Ireland Statistics and Research Agency.

NHSCR. National Health Service Central Register.

NOMIS. National Online Manpower Information System. A service for delivering government statistics to Local Government and other clients, located at the University of Durham.

Northern Ireland Statistics and Research Agency (NISRA). The Census Office responsible for Northern Ireland.

OAs. Output areas.

Office for National Statistics (ONS). The Census Office responsible for England and Wales, formed in 1997, merging the activities of OPCS and the Central Statistical Office.

Office of Population Censuses and Surveys (OPCS). The Census Office responsible for England and Wales until the creation of ONS in 1997.

ONS. Office for National Statistics.

OPCS. Office of Population Censuses and Surveys.

Ordnance Survey (GB). Britain's national mapping agency.

Ordnance Survey Northern Ireland (OSNI). Northern Ireland's national mapping agency.

Origin-Destination Statistics. A set of data products from the 2001 Census containing tables for flows of migrants between wards/postcode sectors and between local authorities and tables for flows of commuters between residence zones and workplace areas.

Output areas (OAs). Geographical areas designed specifically for the publication of census counts. Used only in Scotland in 1991, and throughout the UK in 2001.

Output geography. The geographical division of the country into areas used for the publication of census results.

Parish. Geographical area used in the organization of local government. Found only in England, and not covering the entire country. Many urban areas are not parished.

PC/ED Directory. Postcode to Enumeration District Directory. ONS product that identifies which ED(s) a unit postcode is associated with.

Point in polygon algorithm. An algorithm for calculating whether a single point (x,y) coordinate reference falls within the area of a polygon defined by a series of coordinate references.

Postcode. A 7-character hierarchical alphanumeric code devised for the automated sorting of mail, and used extensively as a georeferencing code.

Postcode sector. The second-smallest division of the postcode system, represented by the first 5 characters of the postcode. Around 9000 sectors cover the entire country.

Restricted area. A geographical area in the census output geography whose aggregated population falls below the required confidentiality threshold, and which therefore cannot be released.

SAS. Small Area Statistics.

SASPAC. Small Area Statistics Package. Widely used computer software for accessing the census small area statistics.

Safe data. Data that has passed statistical disclosure tests and has been released to users.

Safe setting. A secure environment for housing microdata from the census (usually a Census Office), at which safe data can be extracted for release to users.

SCROL. Scottish Census Results Output Library. Project administered by GROS to procure and disseminate Standard Area Statistics for Scotland to the public sector.

Small Area Statistics (SAS). The set of census counts published for the smallest census areas. In 1981 and 1991 this term was applied to the statistical data available in digital form for enumeration districts and for larger areas. The equivalent data set produced from the 2001 Census is the Census Area Statistics.

SMS. Special Migration Statistics.

Special Migration Statistics (SMS). A set of data products from the 1981 and 1991 Censuses containing information on flows of migrants between origins and destinations. Available at two geographical scales, electoral ward and local authority. The equivalent products from the 2001 Census are the set of migration tables, SMS, within the Origin-Destination Statistics.

Special Travel Statistics (STS). The set of data products for Scotland to be released as part of the Origin-Destination Statistics from the 2001 Census. The STS contain the SWS for Scotland and statistics on the journey to study (at school, college, institute or university).

Special Workplace Statistics (SWS). A set of data products from the 1981 and 1991 Censuses containing information on flows of workers from residence areas to workplace areas. Available at two geographical scales, electoral ward and local authority. The equivalent products from the 2001 Census are the set of journey to work tables, the SWS, within the Origin-Destination Statistics.

Standard Area Statistics. The set of data products from the 2001 Census for geographical areas throughout the UK. The Standard Area Statistics are made up of the Key Statistics, the Census Area Statistics and the Standard Tables.

Standard Tables. The set of data products, cross-tabulations, from the 2001 Census for geographical areas from wards to national parts. The data set consists of Theme Tables, harmonized UK Tables, specific tables for England and Wales, Wales, Scotland and Northern Ireland. The 1991 Census equivalent data products were the Local Base Statistics.

Statutory areas. Geographical areas representing the division of the country for administrative and political purposes, which have specific legal status.

STS. Special Travel Statistics (Scotland).

SWS. Special Workplace Statistics.

Theme Tables. One of the data products making up the Standard Tables data set. Theme tables are collections of socio-economic cross-tabulations for specific population groups (e.g. dependent children, the elderly, ethnic groups).

Threshold. A minimum count of persons or households, below which publication of detailed census outputs is not permitted in order to uphold the confidentiality principle.

Trade-off principle. The principle that it is necessary to accept some degree of trade-off between the detail possible in the geographical and statistical aggregation of individual results into census outputs.

Unit postcode. Smallest division of the postcode system, typically containing 14 addresses, represented by a full 7-character postcode (e.g. SO17 1BJ).

Ward. Statutory area used in the organization of local government for the election of council members. Local authorities and parliamentary constituencies are made up of electoral wards. Wards are subject to boundary change, splitting or amalgamation between censuses.

References

Aitkin, M., Anderson, T., Francis, B., and Hinde, J. (1989) *Statistical Modelling in GLIM.* Clarendon Press, Oxford.

Akkerman, A. (1995) The urban household pattern of daytime population change. *Annals of Regional Science* **29**, 1–16.

Alho, J. M. (1994) Analysis of sample-based capture-recapture experiments. *Journal of Official Statistics* **10**, 245–256.

Alvanides, S. (2000) *Zone Design Methods for Application in Human Geography*. Unpublished PhD Thesis, School of Geography, University of Leeds, Leeds (available from the author: *s.alvanides@newcastle.ac.uk*).

Alvanides, S. and Openshaw, S. (1999) Zone design for planning and policy analysis, In: Stillwell, J., Geertman, S., and Openshaw, S. (eds.) *Geographical Information and Planning*. Springer-Verlag, Berlin 299–315.

Alvanides, S., Openshaw, S., and Duke-Williams, O. (2000) Designing zoning systems for flow data, In Atkinson, P. and Martin, D. (eds.) *GIS and Geocomputation, Innovations in GIS 7*. Taylor and Francis, London 115–134.

Alvanides, S., Openshaw, S., and Macgill, J. (2001) Zone design as a spatial analysis tool, In Tate, N. and Atkinson, P. M. (eds.) *Modelling Scale in Geographical Information Science.* Taylor and Francis, London 141–157.

Alvanides, S., Openshaw, S., and Whalley, S. (2002) Further experiments and recommendations on the design of output areas for the 2001 UK Census, *Computers, Environment and Urban Systems* (forthcoming—draft available from the authors: *s.alvanides @newcastle.ac.uk*).

Andrienko, G. L. and Andrienko, N. V. (1999) Interactive maps for visual data exploration. *International Journal of Geographical Information Science* **13**, 355–374 *http:// allanon.gmd.de/and/icavis/*.

Armitage, B. and Bowman, J. (1995) *Accuracy of Rolled-Forward Population Estimates in England and Wales, 1981–91*, Occasional Paper 44, OPCS, London.

Atkins, D., Charlton, M., Dorling, D., and Wymer, C. (1993) *Connecting the 1981 and 1991 Censuses.* NE.RRL Research Report 93/9 CURDS, University of Newcastle upon Tyne.

Australian Bureau Of Statistics (1996) *1996 Census of Population and Housing: Digital Geography Technical Information Paper*, ABS, Canberra.

Australian Bureau of Statistics (1999) *How Australia Takes a Census.* Working Paper, ABS, Canberra. *http://www.abs.gov.au/websitedbs/D3110122.NSF/4a255eef008309e44 a255eef00061e57*.

Bader, M. and Weibel, R. (1997) Detecting and resolving size and proximity conflicts in the generalization of polygon maps. *Proceedings of the 18th ICA/ACI International Cartographic Conference*, ICA/ACI Stockholm, Sweden **3**, 1525–1532.

Bailey, S., Charlton, J., Dollamore, G., and Fitzpatrick, J. (1999a) Which authorities are alike? *Population Trends* **98**, 29–41.

Bailey, S., Charlton, J., Dollamore, G., and Fitzpatrick, J. (1999b) The ONS classification of local and health authorities of Great Britain: Revised for authorities in 1999. *Studies in Medical and Population Subjects 63.* ONS, London.

Bailey, T. C. and Gatrell, A. C. (1995) *Interactive Spatial Data Analysis.* Longman, Harlow.

Banister, D. and Banister, C. (1994) Energy consumption in transport in Great Britain: Macro level estimates. *Transportation Research A* **29A**, 21–32.

Barr, R. (1993a) Mapping and spatial analysis. In: Dale, A. and Marsh, C. (eds.) *The 1991 Census User's Guide.* HMSO, London 248–268.

Barr, R. (1993b) Census geography Part II: A review. In: Dale, A. and Marsh, C. (eds.) *The 1991 Census User's Guide.* HMSO, London 70–83.

Bartholomew, D. J. (1988) *Measuring Social Disadvantage and Additional Educational Needs: an Assessment of Methods carried out for the Department of the Environment.* Unpublished report, Department of Statistical and Mathematical Sciences, LSE, London.

Bates, J. and Bracken, I. (1987) Migration age profiles for local authority areas in England, 1971–1981. *Environment and Planning A* **19**, 521–535.

Batey, P. W. J., Brown, P. J. B., and Corver, M. (1999) Participation in higher education: a geodemographic perspective on the potential for further expansion in student numbers. *Geographical Systems: The International Journal of Geographical Information, Analysis, Theory and Decision* **1**, 277–303.

Batty, M. (1999) *Evaluation of the ESRC/JISC Census of Population Programme.* ESRC Policy and Evaluation Division, Swindon.

Beatty, R. and Rodgers, M. (1999) *Mid-year Population Estimates in Northern Ireland—Validation and Extension to Local Government Districts*, Occasional Paper 12, NISRA, Belfast.

Becker, R. A., Cleveland, W. S., and Wilks, A. R. (1987) Dynamic graphics for data analysis. *Statistical Science* **2**, 355–395.

Beckman, R. J., Baggerly, K. A., and McKay, M. D. (1996) Creating synthetic baseline populations. *Transportation Research-A* **30**, 415–429.

Belin, T. R. and Rolph, J. E. (1994) Can we reach consensus on census adjustment? *Statistical Science* **9**, 486–508.

Bell, M. and Rees, P. (2000) Lexis diagrams in the context of migration: a review and application to British and Australian data. Paper Presented at the Workshop on *Lexis in Context: German and Eastern & Northern European Contributions to Demography 1860–1910*, Max Planck Institute for Demographic Research, Rostock, Germany.

Bentham, G., Eimermann, J., Haynes, R., Lovett, A., and Brainard, J. (1995) Limiting long-term illness and its associations with mortality and indicators of social deprivation. *Journal of Epidemiology and Community Health* **49** (Supplement 2), S57–S64.

Birkin, M. and Clarke, M. (1988) SYNTHESIS—a synthetic spatial information system for urban and regional analysis: methods and examples. *Environment and Planning A* **20**, 1645–1671.

Blake, M. and Openshaw, S. (1995) *Selecting Census variables for Use in Classification Research*, Working Paper 95/5, School of Geography, University of Leeds, Leeds.

Blodgett, J. and Meij, H. (1997) MABLE / Geocorr: Tackling the Geographic Babel Problem. *Workshop on Research and Development Opportunities in Federal Information Services. http://www.census.gov/plue/geocorr/doc/article.html*.

Boden, P., Stillwell, J., and Rees, P. H. (1992) How good are the NHSCR data? In: Stillwell, J., Rees, P. H., and Boden, P. (eds.) *Migration Processes and Patterns*. Vol. 2. Population Redistribution in the United Kingdom. Belhaven, London 13–27.

Boots, B. N. (1986) *Voroni (Thiessen) Polygons. Concepts and Techniques in Modern Geography 54*, Geo Books, Norwich.

Boyle, P. J. (1998) Migration and housing tenure in South East England. *Environment and Planning A* **30**, 855–866.

Boyle, P. J., Flowerdew, R., and Shen, J. (1998) Modelling inter-ward migration in Hereford and Worcester: the importance of housing growth and tenure. *Regional Studies* **32**, 113–132.

Bracken, I. and Martin, D. (1989) The generation of spatial population distributions from census centroid data. *Environment and Planning A* **21**, 537–543.

Bracken, I. and Martin, D. (1995) Linkage of the 1981 and 1991 censuses using surface modelling concepts. *Environment and Planning A* **27**, 379–390.

Brackstone, G. J. (1987) Small area data: policy issues and technical challenges. In: Platek, R., Rao, J. N. K., Särndal, C. E., and Singh, M. P. (eds.) *Small Area Statistics: An International Symposium.* John Wiley & Sons, New York 3–20.

Bramley, G. and Lancaster, S. (1998) Modelling local and small-area income distributions in Scotland. *Environment and Planning C: Government and Policy* **16**, 681–706.

Bradford, M. G., Robson, B. T., and Tye, R. (1993) An urban deprivation index 1991. In: Simpson, S. (ed.) *Census Indicators of Local Poverty and Deprivation: Methodological Issues.* LARIA, Wokingham.

Bradford, M. G., Robson, B. T., and Tye, R. (1995) Constructing an urban deprivation index: a way of meeting the need for flexibility. *Environment and Planning A* **27**, 519–533.

Brainard, J., Lovett, A., and Parfitt, J. (1996) Assessing hazardous-waste transport risks using a GIS. *International Journal of Geographical Information Systems* **10**, 831–849.

Breheny, M., Foot, D., and Archer, S. (1998) Sustainable settlements and work trip closure. *Paper Presented to the Annual Conference of the Regional Science Association*, British and Irish Section, York.

Breiman, L. (1994) The 1991 Census adjustment: undercount or bad data? *Statistical Science* **9**, 458–475.

Brewer, C. A. (1994) Color use guidelines for mapping and visualization. In: MacEachren, A. M. and Taylor, D. R. F. (eds.) *Visualization in Modern Cartography.* Elsevier, New York 123–147. *http://www.personal.psu.edu/c/a/cab38/ColorSch/SchHome.html*.

British Standards Institute (1994) *BS7666: Part 3: 1994 Spatial Datasets for Geographical Referencing. Specification for Addresses*. IST/36, BSI, London.

Brown, A. (1986) Family circumstances of young children. *Population Trends* **43**, 18–23.

Brown, J., Chambers, R., Diamond, I., and Buckner, L. (undated) Modelling down to small areas. In: *2001: A One Number Census. Census Consultation Paper*, ONS, Titchfield.

Brown, P. J. B. (1990) *Geodemographics: A Review of Recent Developments and Emerging Issues—Towards an RRL Research Agenda*, ESRC Regional Research

Laboratory Initiative, Discussion Paper 5, Department of Town and Regional Planning, University of Sheffield, Sheffield.

Brown, P. J. B. (1991) Exploring geodemographics. In: Masser, I. and Blakemore, M. (eds.) *Handling Geographical Information.* Longman, London 221–258.

Brown, M. and Dale, A. (1997) A Survey of SAR Users, Their Requirements for 2001 SARs and Their Views on Dissemination and Support. *Working Paper No. 6*, Cathie Marsh Centre for Census and Survey Research. Available at *http://www.ccsr.ac.uk/publications/working/confpapers.htm*.

Brunsdon, C. A., Fotheringham, A. S., and Charlton, M. E. (1998) An investigation of methods for visualising highly multivariate datasets. In: Unwin, P. and Fisher, P. F. (eds.) *Case Studies of Visualization in the Social Sciences.* Technical Report Series 43, ESRC/JISC Advisory Group on Computer Graphics, Bristol, 55–80, Available at *http://www.aqcoq.ac.uk/reports/visual/casestud/contents.htm*.

Buck, N., Gershuny, J., Rose, D., and Scott, J. (eds.) (1994) *Changing Households: The British Household Panel Survey 1990–1992.* ESRC Research Centre on Micro-Social Change, University of Essex, Colchester.

Burnhill, P. and Morse, D. J. (1993) Census Geography. Paper given at *ESRC Workshop on the 1991 Census and Large Government Datasets*, September 1993, University of Manchester, Manchester.

Bush, J. A. (1996) *Screening and the Body: Surveillance, Regulation and the Cervical Screening Programme.* Unpublished PhD thesis. Department of Geography, University of Sheffield, Sheffield.

Bynner, J., Ferri, E., and Shepherd, P. (1997) *Twenty-something in the 90s: Getting on, Getting by; Getting Nowhere.* Dartmouth Press, Aldershot.

Campbell, M., Holdsworth, C., Payne, T., and Dale, A. (1996) *Sampling Variance and Design Factors in the Sample of Anonymised Records*, CCSR Occasional Paper 6, CCSR, University of Manchester, Manchester.

Campbell, D., Radford, J. M. C., and Burton, P. (1991) Unemployment rates: an alternative to the Jarman Index? *British Medical Journal* **303**, 750–755.

Carr-Hill, R. and Sheldon, T. (1991) Designing a deprivation payment for general practitioners: the UPA(8) wonderland. *British Medical Journal* **302**, 393–396.

Carr-Hill, R., Sheldon, T. A., Smith, P., Martin, S., Peacock, S., and Hardman, G. (1994) Allocating resources to health authorities: development of method for small area analysis of use of inpatient services. *British Medical Journal* **309**, 1046–1049.

Carstairs, V. and Morris, R. (1989) Deprivation: explaining differences in mortality between Scotland and England and Wales. *British Medical Journal* **299**, 886–889.

Carstairs, V. and Morris, R. (1991) *Deprivation and Health in Scotland.* Aberdeen University Press, Aberdeen.

Carter, J. (1998) *Cartographic Data Visualizer. http://www.kinds.ac.uk/kinds/janus.htm*.

Census Dissemination Unit (1996–1999) *Census Dissemination Unit Annual Reports 1996-1999* CDU, University of Manchester, Manchester. *http://census.ac.uk/cdu/about/about_cdu.htm#1*.

Census Microdata Unit (1994) *A User Guide to the SARS.* Second edition, CMU, University of Manchester, Manchester.

Central Statistical Office (1995) *Social Trends: 1995 Edition.* HMSO, London.

Chambers, R. L. (1996) Robust case-weighting for multipurpose establishment surveys. *Journal of Official Statistics* **12**, 3–32.

Champion, A. and Dorling, D. (1994) Population change for Britain's functional regions, 1951–1991. *Population Trends* **77**, 14–23.

Champion, A., Fotheringham, A. S., Rees, P. H., Boyle, P. J., and Stillwell, J. (1998) *The Determinants of Migration Flows in England: A Review of Existing Data and Evidence.* Department of Geography, University of Newcastle upon Tyne, Newcastle upon Tyne.

Champion, A., Green, A. E., Owen, D. W., Ellin, D., and Coombes, M. G. (1987) *Changing Places: Britain's Demographic, Economic and Social Complexion.* Edward Arnold, London.

Chappell, R., Vickers, L., and Evans, H. (2000) The use of patient registers to estimate migration. *Population Trends* **101**, 19–24.

Charlton, J. (1996) Which areas are healthiest? *Population Trends* **83**, 17–24.

Charlton, M., Rao, L., and Carver, S. (1995) GIS and the census. In: Openshaw, S. (ed.) *Census Users' Handbook.* GeoInformation International, Cambridge 133–166.

Chisholm, M. (1995) Some lessons from the review of local government in England. *Regional Studies* **29**, 563–580.

Clark, A. M. (1992) 1991 Census: data collection. *Population Trends* **70**, 22–27.

Clark, A. M. and Thomas, F. G. (1990) The geography of the 1991 Census. *Population Trends* **60**, 9–15.

Clarke, G. P. (1996) Microsimulation: an introduction. In: Clarke, G. P. (ed.) *Microsimulation for Urban and Regional Policy Analysis.* Pion, London.

Cohen, G. (1998) *Methodology for Producing 1997 Small Area Population Estimates (Lothian)*, Internal Paper CHI 98(5) presented to GRO(S) working party on use of patient data from the Central Health Index GRO(S), Edinburgh.

Cole, K. (1993) The 1991 local base and small area statistics. In: Dale, A. and Marsh, C. (eds.) *The 1991 Census User's Guide.* HMSO, London 201–247.

Cole, K. (1994) Data modification, data suppression, small populations and other features of the 1991 small area statistics. *Area* **26**, 69–78.

Cole, K. (1998) The need for a gold standard for centroids and lookup tables, In: Rees, P. H. (ed.) *The 2001 census: what Geography do we want?* Working Paper 98/8, School of Geography, University of Leeds, Leeds.

Cole, K. and Gatrell, A. C. (1986) Public libraries in Salford: a geographical analysis of provision and access. *Environment and Planning A* **18**, 253–268.

Commission of the European Communities (1992) *The Impact of Transport on the Environment.* Com (92) 46, European Commission, Brussels.

Compton, G. (2001) *1999 Small Area Population Estimates Scotland*, Occasional Paper 3, GROS, Edinburgh. *http://www.gro-scotland.gov.uk/grosweb/grosweb.nsf*.

CONI (1994) *1991 Census of Northern Ireland*. Migration Tables. HMSO, Belfast.

Connolly, C. and Chisholm, M. (1999) The use of indicators for targeting public expenditure: the index of local deprivation. *Environment and Planning C: Government and Policy* **17**, 463–482.

Coombes, M. G. (1995) Dealing with census geography: principles, practices and possibilities. In: Openshaw, S. (ed.) *Census Users' Handbook.* GeoInformation International, Cambridge 111–132.

Coombes, M. G. (1997) Monitoring equal employment opportunity. In: Kahn, V. (ed.) *Employment, Education and Housing Among the Ethnic Minority Populations of Britain.* HMSO, London 111–132.

Coombes, M. G., Dixon, J. S., Goddard, J. B., Openshaw, S., and Taylor, P. J. (1982) Functional regions for the population census of Britain. In: Herbert, D. T. and Johnston, R. J. (eds.) *Geography and the Urban Environment*. Vol. 5, John Wiley & Sons, London 63–112.

Coombes, M. G., Dorling, D., Atkins, D., and Raybould, S. (1995) Functional Region Coding for the 1991 Census. *http://census.ac.uk/cdu/Datasets/Lookup_tables/Morphological/*.

Coombes, M. G., Green, A. E., and Openshaw, S. (1986) An efficient algorithm to generate official statistical reporting areas: the case of the 1984 travel-to-work areas revision in Britain. *Journal of the Operational Research Society* **37**, 943–953.

Coombes, M. G. and Openshaw, S. (2001) Contrasting approaches to identifying 'localities' for research and public administration. In: Frank, A., Raper, J., and Cheylan, J. -P. (eds.) *Life and Motion of Socio-Economic Units.* Taylor and Francis, London 301–315.

Coombes, M. G., Raybould, S., Wong, C., and Openshaw, S. (1995) Towards an index of deprivation: a review of alternative approaches. In: *1991 Deprivation Index: A Review of Approaches and Matrix of Results.* Department of the Environment, HMSO, London.

Coombes, M. G., Wymer, C., Atkins, D., and Openshaw, S. (1996) Localities and City Regions. *http://census.ac.uk/cdu/Datasets/Lookup_tables/Morphological/*.

Craig, J. (1985) A 1981 Socio-economic classification of local and health authorities of Great Britain. *Studies in Medical and Population Subjects 48.* HMSO, London.

Creeser, R. and Gleave, S. (eds.) (2000) *Migration Within England and Wales Using the ONS Longitudinal Study.* Series LS 9, The Stationery Office, London.

CRU/OPCS/GROS (1980) *People in Britain: A Census Atlas.* HMSO, London.

Cruddas, M., Thomas, J., and Chambers, R. (1997) *Investigating neural networks as a possible means of imputation for the 2001 UK Census.* Paper Presented at the 1997 Statistics Canada Symposium, Copies available from Office for National Statistics, Titchfield.

Dahmann, D. C. and Fitzsimmons, J. D. (eds.) (1995) *Metropolitan and Nonmetropolitan Areas: New Approaches to Geographical Definition.* Population Division Working Paper 12, US Bureau of the Census, Washington DC.

Dale, A. (1994) Definitional issues in comparing household and family change between censuses. In: Howett, M. (ed.) *Research for Policy: Proceedings of the 1994 Annual Conference of LARIA.* LARIA, Wokingham 84–88.

Dale, A. and Marsh, C. (eds.) (1993) *The 1991 Census User's Guide.* HMSO, London.

Dale, A. and Elliot, M. (1998) *A Report on the Disclosure Risk of Proposals for SARs from the 2001 Census.* CCSR Working Paper 5, CCSR, University of Manchester, Manchester.

Dale, A. and Elliot, M. J. (2001) Proposals for 2001 SARs: an assessment of disclosure risk. *Journal of the Royal Statistical Society*, Series A **164**(3), 1–21.

Davies, H. (1995) Accessing the data via SASPAC91, In: Openshaw, S. (ed.) *Census Users' Handbook.* GeoInformation International, Cambridge 83–110.

Denham, C. (1980) The geography of the Census: 1971 and 1981. *Population Trends* **19**, 6–12.

Denham, C. (1993) Census geography: I an overview. In: Dale, A. and Marsh, C. (eds.) *The 1991 Census User's Guide.* HMSO, London 52–69.

Department of Energy (1990) *Energy Use and Energy Efficiency in UK Transport up to the Year 2010.* Energy Efficiency Series 10, HMSO, London.

DETR (1998) *Updating and Revising the Index of Local Deprivation.* DETR, London.

DETR (1999) *Projections of Households in England to 2021: 1996-Based Estimates of the Number of Households for Regions.* DETR, London.

DETR (2000a) Indices of deprivation 2000. *DETR Regeneration Research Summary 31.* DETR, London.

DETR (2000b) Response to the Formal Consultations on the Indices of Deprivation 2000 (ID2000). Prepared by the Index Team, Department of Social Policy and Social Work, University of Oxford, Oxford.

Dewdney, J. (1983) Censuses past and present. In: Rhind, D. (ed.) *A Census Users' Handbook.* Methuen, London 1–16.

DiBiase, D., Reeves, C., MacEachren, A. M., von Wyss, M., Krygier, J. B., Sloan, J., and Detweiler, M. C. (1994) Multivariate display of geographic data: applications in earth system science. In: MacEachren, A. M. and Taylor, D. R. F. (eds.) *Visualization in Modern Cartography.* Pergamon, Oxford, 287–312.

Dixie, J. (1999) *Planning for the 2001 Census coverage survey in England and Wales.* Paper Presented at the 1999 Research Conference of the Federal Committee on Statistical Methodology. *http://www.fcsm.gov/papers/dixie.pdf*.

DoE (1983) *Urban Deprivation.* Information Note 2, Inner Cities Directorate, DoE, London.

DoE (1987) *Handling Geographic Information: The Report of the Committee of Enquiry Chaired by Lord Chorley.* HMSO, London.

DoE (1995a) A 1991 Index of Local conditions. Department of the Environment, London.

DoE (1995b) *1991 Deprivation Index: A Review of Approaches and Matrix of Results.* HMSO, London.

Dorling, D. (1991) *The Visualisation of Spatial Social Structure.* Unpublished PhD Thesis, Department of Geography, University of Newcastle, Newcastle upon Tyne.

Dorling, D. (1993) Map design for census mapping. *Cartographic Journal* **30**, 167–183.

Dorling, D. (1995) *A New Social Atlas of Britain.* John Wiley & Sons, Chichester.

Dorling, D. (1996) Area cartograms: their use and creation. *Concepts and Techniques in Modern Geography 59*. Environmental Publications, Norwich.

Dorling, D. (1999) Commentary: Who's afraid of income inequality? *Environment and Planning A* **31**, 571–574.

Dorling, D. and Atkins, D. J. (1995) Population density, change and concentration in Great Britain 1971, 1981 and 1991. *Studies in Medical and Population Subjects 58*. HMSO, London.

Dorling, D. and Fairbairn, D. (1997) *Mapping, Ways of Representing the World.* Longman, Harlow.

Dorling, D. and Simpson, S. (eds.) (1999) *Statistics in Society: The Arithmetic of Politics.* Arnold, London.

Douglas, D. H. and Peucker, T. K. (1973) Algorithms for the reduction of the number of points required to represent a digitised line or its caricature. *Canadian Cartographer* **10**, 112–122.

Drewett, R., Goddard, J., and Spence, N. (1976) *British Cities: Urban Population and Employment Trends 1951–1971.* HMSO, London.

Drinkwater, S. and Leslie, D. (1996) Adding earnings variables to the SARs: accounting for the ethnicity effect. SARs Newsletter 8, September, available at *http://les1.man.ac.uk/ccsr/publications/newsletters/8/news8*.

Dugmore, K. (1996) What do Users want from the 2001 Census? In: *Looking Towards the 2001 Census*. Occasional Paper 46, OPCS, London 21–23.

Duguid, G. J. (1995) *Deprived Areas in Scotland: Results of an Analysis of the 1991 Census*. The Scottish Office Central Research Unit, Edinburgh.

Duke-Williams, O. and Rees, P. H. (1998) Can Census Offices publish statistics for more than one small area geography? An analysis of the differencing problem in statistical disclosure. *International Journal of Geographical Information Science* **12**, 579–605.

Duley, C. (1989) *A model for Updating Census-Based Household and Population Information for Inter-Censal Years*. Unpublished PhD thesis, School of Geography, University of Leeds, Leeds.

Dykes, J. A. (1996a) Dynamic maps for spatial science, a unified approach to cartographic visualization. In: Parker, D. (ed.) *Innovations in GIS 3*. Taylor and Francis, London, 177–187.

Dykes, J. A. (1996b) Exploring Spatial Data Representation with Dynamic Graphics. *http://www.mimas.ac.uk/argus/ICA/J.Dykes/*.

Dykes, J. A. (1998) *cdv Homepage*. *http://www.geog.le.ac.uk/jad7/cdv/*.

Dykes, J. A. and Unwin, D. J. (1998) *Maps of the Census: A Rough Guide*. *http://www.agocg.ac.uk/reports/visual/casestud/dykes/conten_1.htm*.

Edwardes, A., Mackaness, W. A., and Urwin, T. (1998) Self evaluating generalisation algorithms to automatically derive multi scale boundary sets. Paper Presented at 18th International Conference on Spatial Data Handling. International Geographical Union, Vancouver 361–373.

EEC (1999) *Providing Global Access to Distributed Data through Metadata Standardisation—The Parallel Stories of NESSTAR and the DDI*. Economic Commission for Europe Working Paper 10, *http://www.nesstar.org/papers/GlobalAccess.html*.

Elliot, D. (1991) *Weighting for Non-Response: A Survey Researcher's Guide*. OPCS, London.

Erikson, R. and Goldthorpe, J. (1993) *The Constant Flux: A Study of Class Mobility in Industrial Societies*. Clarendon Press, Oxford.

Evans, I. S. (1977) The selection of class intervals. *Transactions of the Institute of British Geographers* **2**, 98–124.

Experían (2000) *Postal Sector Data Sets for the Academic Community*. Experían, Nottingham.

Feng, Z. and Flowerdew, R. (1998) Fuzzy geodemographics: a contribution from fuzzy clustering methods. In: Carver, S. (ed.) *Innovations in GIS 5*. Taylor and Francis, London 119–127.

Ferreira, J., Jr. and Wiggins, L. L. (1990) The density dial: A visualization tool for thematic mapping. *Geo Info Systems* **10**, 69–71.

Ferri, E. (ed.) (1993) *Life at 33: The Fifth Follow-up of the National Child Development Study*. ESRC National Children's Bureau, City University, London.

Fieldhouse, E. (1996) Putting unemployment in its place: using the SARs to explore the risk of unemployment in GB in 1991. *Regional Studies* **30**, 119–133.

Fieldhouse, E. A. and Tye, R. (1996) Deprived people or deprived places? Exploring the ecological fallacy in studies of deprivation with the samples of anonymised records. *Environment and Planning A* **28**, 237–259.

Finkel, R. A. and Bentley, J. L. (1974) Quad trees: a data structure for retrieval on composite keys. *Acta Informatica* **4**, 1–9.

Fisher, P. F. (1998) Is GIS hidebound by the legacy of cartography? *The Cartographic Journal* **35**, 5–9.

Flowerdew, R. (ed.) (1994) Special issue on standard spending assessments. *Environment and Planning C: Government and Policy* **12**(1).

Flowerdew, R. and Green, A. E. (1993) Migration, transport and workplace statistics from the 1991 Census. In: Dale, A. and Marsh, C. (eds.) *The 1991 Census User's Guide.* HMSO, London 269–294.

Forrest, R. and Gordon, D. (1993) *People and Places: A 1991 Census Atlas of England.* School for Advanced Urban Studies, University of Bristol, Bristol.

Fotheringham, A. S. (1997) Trends in quantitative methods I: stressing the local. *Progress in Human Geography* **21**, 88–96.

Fotheringham, A. S., Charlton, M. E., and Brunsdon, C. (1998) Geographically-weighted regression: a natural evolution of the expansion method for spatial data analysis. *Environment and Planning A* **30**, 1905–1927.

Fox, A. J. (ed.) (1989) *Health Inequalities in European Countries.* Gower, Aldershot.

Fox, A. J. and Goldblatt, P. O. (1982) *1971–1975 Longitudinal Study: Socio-Demographic Mortality Differentials.* OPCS LS Series 1, HMSO, London.

Freedman, D. and Wachter, K. (1994) Heterogeneity and census adjustment for the intercensal base. *Statistical Science* **9**, 476–485.

Frost, M., Linneker, B., and Spence, N. (1997) The energy consumption implications of changing worktravel in London, Birmingham and Manchester, 1981 and 1991. *Transportation Research A* **31A**, 1–19.

Ganzeboom, H., De Graaf, P., and Treiman, D. (1992) A standard international socio–economic index of occupational status. *Social Science Research* **21**, 1–56.

Ganzeboom, H. and Treiman, D. (1996) Internationally comparable measures of occupational status for the 1988 international standard classification of occupations. *Social Science* **25**, 201–239.

Gatrell, A. C. (1994) Density estimation and the visualization of point patterns. In: Hearnshaw, H. M. and Unwin, D. J. (eds.) *Visualization in Geographical Information Systems.* John Wiley & Sons, Chichester 65–75.

Ghosh, M. and Rao, J. N. K. (1994) Small area estimation: an appraisal. *Statistical Science* **9**, 55–93.

Gibson, A. and Asthana, S. (1998) Schools, pupils and examination results: contextualising school performance. *British Educational Research Journal* **24**, 269–282.

Gomulka, J. (1992) Grossing-up revisited. In: Hancock, R. and Sutherland, H. (eds.) *Microsimulation Models for Public Policy Analysis: New Frontiers*, STICERD Occasional Paper, LSE, London 121–131.

Gonzales, M. E. (1973) Use and evaluation of synthetic estimators. In: *Proceedings of the Social Statistics Section of the American Statistical Association.* American Statistical Association, Washington DC, 33–36.

Goodchild, M. F., Anselin, L., and Deichmann, U. (1993) A framework for the areal interpolation of socioeconomic data. *Environment and Planning A* **25**, 383–397.

Gordon, D. (1995) Census based deprivation indices: their weighting and validation. *Journal of Epidemiology and Community Health* **49**, S39–S44.

Gordon, D. and Forrest, R. (1995) *People and Places 2: Social and Economic Distinctions in England.* School for Advanced Urban Studies, University of Bristol, Bristol.

Gould, M. and Jones, K. (1996) Analysing perceived limiting long-term illness using UK Census microdata. *Social Science and Medicine* **42**, 857–869.

Government Statistical Service (1996) Report of the Task Force on Imputation. GSS Methodology Series 3, ONS, London.

Gregory, I., Gilham, V., and Southall, H., (1998) Looking forward to the past. *Mapping Awareness* **12**(2), 26–29.

GROS/NISRA/ONS (2000) *Origin-Destination Statistics: A Discussion Paper*, GROS, Edinburgh.

Hakim, C. (1996) 1991 Occupational Segregation Variables. SARs Newsletter 8 *http://les1.man.ac.uk/ccsr/publications/newsletters/8/news8*.

Hall, P. (1971) Spatial structure of metropolitan England and Wales. In: Chisholm, M. and Manners, G. (eds.) *Spatial Policy Problems of the British Economy.* Cambridge University Press, Cambridge 96–125.

Harding, S., Bethune, B., Maxwell, R., and Brown, J. (1997) Mortality trends using the longitudinal study. In: Drever, F. and Whitehead, M. (eds.) *Health Inequalities: Decennial Supplement.* ONS Series DS 15, HMSO, London 143–155.

Haslett, J., Wills, G., and Unwin, A. (1990) SPIDER, an interactive statistical tool for the analysis of spatially distributed data. *International Journal of Geographical Information Systems* **4**, 285–296.

Hattersley, L. and Creeser, R. (1995) *The Longitudinal Study, 1971–1991: History, Organisation and Quality of Data.* OPCS LS Series 7, HMSO, London.

Haynes, R. M., Lovett, A. A., Bentham, G., Brainard, J. S., and Gale, S. H. (1995) Comparison of ward population estimates from FHSA patient registers with the 1991 Census. *Environment and Planning A* **27**, 1849–1858.

Heady, P., Smith, S., and Avery, V. (1994) *1991 Census Validation Survey: Coverage Report.* HMSO, London.

Heady, P., Smith, S., and Avery, V. (1996) *1991 Census Validation Survey: Quality Report.* HMSO, London.

Herzog, A. (1999) Dorling's Cartogram Algorithm. *http://www.statistik.zh.ch/map/dorling/dorling.html*.

Higgs, G., Bellin, W., Farrell, S., and White, S. (1997) Educational attainment and social disadvantage: contextualising school league tables. *Regional Studies* **31**, 775–789.

Higgs, G., Senior, M. L., and Williams, H. C. W. L. (1998) Spatial and temporal variation of mortality and deprivation 1: widening health inequalities. *Environment and Planning A* **30**, 1661–1682.

Higgs, G. and White, S. D. (2000) Alternatives to census-based indicators of social disadvantage in rural communities. *Progress in Planning* **53**, 1–81.

Hirschfield, A. (1994) Using the 1991 Population Census to study deprivation. *Planning Practice and Research* **9**, 43–54.

Hirschfield, A. and Bowers, K. J. (1997) The effect of social cohesion on levels of recorded crime in disadvantaged areas. *Urban Studies* **34**, 1275–1295.

Hogan, H. (1993) The 1990 post-enumeration survey: operations and results. *Journal of the American Statistical Association*, **88**, 1047–1060.

Holdsworth, C. (1995) Minimal Household Units SARs Newsletter 5 *http://les1.man.ac.uk/ccsr/publications/newsletters/5/news5*.

Hollis, J. (1998) International migration. In: Simpson, S. (ed.) *Making Local Population Statistics: A Guide for Practitioners.* LARIA, Wokingham 151–154.

Holm, E., Lindgren, U., Mäkilä, K., and Malmberg, G. (1996) Simulating an entire nation. In: Clarke, G. P. (ed.) *Microsimulation for Urban and Regional Policy Analysis.* Pion, London 164–186.

Holterman, S. (1975) Areas of urban deprivation in Great Britain: an analysis of census data. *Social Trends* **6**, 33–47.

HM Government (1999) *The 2001 Census of Population.* Cm 4253, The Stationery Office, London. *http://www.statistics.gov.uk/census2001/pdfs/whitepap.pdf*.

HMSO (2001) Web Site for Licensing and Copyright. *http://www.hmso.gov.uk/copyhome.htm*.

Index 99 Team, Firth, D., and Payne, C. (1999) *Report for Formal Consultation Stage 2: Methodology for an Index of Multiple Deprivation.* Department of Social Policy and Social Work, University of Oxford, Oxford.

Inselberg, A. (1985) The plane with parallel coordinates. *The Visual Computer* **1**, 69–91.

Isaki, C. T. (1990) Small-area estimation of economic statistics. *Journal of Business and Economic Statistics* **8**, 435–441.

Jackson, G. and Lewis, C. (1996) Local government reorganisation in Scotland and Wales. *Population Trends* **83**, 43–51.

Jarman, B. (1983) Identification of underprivileged areas. *British Medical Journal* **286**, 1705–1709.

Jarman, B. (1984) Underprivileged areas: validation and distribution of scores. *British Medical Journal* **289**, 1587–1592.

Jarman, B. (ed.) (1988) *Primary Care.* Heinemann Medical Books, Oxford.

Jenks, G. F. and Caspall, F. C. (1971) Error on choroplethic maps, definition, measurement, reduction. *Annals of the Association of American Geographers* **61**, 217–244.

Johnston, R. J. and Pattie, C. J. (1996a) Local government in local governance: the 1994–1995 restructuring of local government in England. *International Journal of Urban and Regional Research* **20**, 671–696.

Johnston, R. J. and Pattie, C. J. (1996b) Intra-local conflict, public opinion and local government restructuring in England, 1993–1995. *Geoforum* **27**, 97–114.

Kalton, G. (1983) *Introduction to survey sampling.* Sage University Paper series on Quantitative Applications in the Social Sciences. 07–035 Sage, London.

Kilbey, T. and Scott, A. (1997) New ways to track migration flows. In: *Proceedings of the 1997 annual conference of the Local Authorities Research and Intelligence Association.* LARIA, Wokingham 35–44.

Kohonen, T. (1984) *Self-organization and Associative Memory.* Springer-Verlag, Berlin.

Langford, M. and Unwin, D. J. (1994) Generating and mapping population density surfaces within a geographical information system. *The Cartographic Journal* **31**, 21–26.

Leach, R. (1994) Restructuring local government. *Local Government Studies* **20**, 345–360.

Lee, P., Murie, A. and Gordon, D. (1995) Area measures of deprivation: a study of current methods and best practices in the identification of poor areas in Great Britain. *Centre for Urban and Regional Studies.* University of Birmingham, Birmingham.

London Research Centre (1992) *SASPAC User Manual Part 1.* Manchester Computing, University of Manchester, Manchester.

Longley, P. and Clarke, G. (eds.) (1995) *GIS for Business and Service Planning.* GeoInformation International, Cambridge.

Lunn, D. J., Simpson, S. N., Diamond, I., and Middleton, E. (1998) The accuracy of age-specific population estimates for small areas in Britain. *Population Studies* **52**, 327–344.

MacDougal, E. B. (1992) Exploratory analysis, dynamic statistical visualization, and geographical information systems. *Cartography and Geographic Information Systems* **19**, 237–246.

MacEachren, A. M. (1995) *How Maps Work: Representation, Visualization and Design.* Guilford Press, New York.

Macgill, J., Bailey, A., Clarke, G., and Alvanides, S. (2001) A zone design tool for use with the 2001 Census data: progress report. Paper Presented at the ESRC 2001 Census Development Programme, Fourth Workshop, 15–16th May 2001, Fairbairn House, University of Leeds, Leeds.

Mackaness, W. A. (1994) An algorithm for conflict identification and feature displacement in automated map generalization. *Cartography and Geographic Information Systems* **21**, 219–232.

Mackaness, W. A. and Beard, M. K. (1993) Use of graph theory to support map generalization. *Cartography and Geographic Information Systems* **20**, 210–221.

Madden, M., Stevenson, M. A., Brown, P. J. B., and Batey, P. W. J. (1996) Developing a small-area electricity demand forecasting system. *The Journal of Energy and Development* **20**, 1–23.

Marsh, C. (1993a) Privacy, confidentiality and anonymity in the 1991 Census. In: Dale, A. and Marsh, C. (eds.) *The 1991 Census User's Guide.* HMSO, London 111–128.

Marsh, C. (1993b) The sample of anonymised records. In: Dale, A. and Marsh, C. (eds.) *The 1991 Census User's Guide.* HMSO, London 295–311.

Marsh, C. (1993c) An overview. In: Dale, A. and Marsh, C. (eds.) *The 1991 Census User's Guide.* HMSO, London 1–15.

Marsh, C., Dale, A., and Skinner, C. (1994) Safe data versus safe settings: access to microdata from the British census. *International Statistical Review* **62**, 35–53.

Marsh, C., Skinner, C., Arber, S., Penhale, B., Openshaw, S., Hobcraft, J., Lievesley, D., and Walford, N. (1991) The case for samples of anonymized records from the 1991 Census. *Journal of the Royal Statistical Society Series A* **154**, 305–340.

Martin, D. (1989) Mapping population data from zone centroid locations. *Transactions of the Institute of British Geographers NS* **14**, 90–97.

Martin, D. (1996a) An assessment of surface and zonal models of population. *International Journal of Geographical Information Systems* **10**, 973–989.

Martin, D. (1996b) Depicting changing distributions through surface estimation. In: Longley, P. and Batty, J. M. (eds.) *Spatial Analysis: Modelling in a GIS Environment.* GeoInformation International, Cambridge 105–122.

Martin, D. (1997) From enumeration districts to output areas: experiments in the automated design of a census output geography. *Population Trends* **88**, 36–42.

Martin, D. (1998a) Optimizing census geography: the separation of collection and output geographies. *International Journal of Geographical Information Science* **12**, 673–685.

Martin, D. (1998b) Automatic neighbourhood identification from population surfaces. *Computers, Environment and Urban Systems* **22**, 107–120.

Martin, D. (2000) Towards the geographies of the 2001 Census. *Transactions of the Institute of British Geographers NS* **25**, 321–332.

Martin, D. and Higgs, G. (1997) Population georeferencing in England and Wales. *Environment and Planning A* **29**, 333–347.

Martin, D., Harris, J., Sadler, J., and Tate, N. (1998) Putting the census on the web: lessons from two case studies. *Area* **30**, 311–320.

Martin, D., Roderick, P., Brigham, P., Barnett, S., and Diamond, D. (2000) The (mis)representation of rural deprivation. *Environment and Planning A* **32**, 735–751.

Martin, D., Senior, M. L., and Williams, H. C. W. L. (1994) On measures of deprivation and the spatial allocation of resources for primary health care. *Environment and Planning A* **26**, 1911–1929.

Martin, D., Tate, N. J., and Langford, M. (2000) Refining population surface models: experiments with Northern Ireland census data. *Transactions in GIS* **4**, 343–360.

MCC (1994) *Guide to Accessing the Census 1991 (SAS/LBS) & 1981 (SAS) Data Using SASPAC.* CSS 604, Manchester Computing Centre, University of Manchester, Manchester.

McCormick, B. H., DeFanti, T. A., and Brown, M. D. (eds.) (1987) Visualization in scientific computing. *Computer Graphics* **1**, 6.

McKee, C. (1989) *The 1971/1981 Census Change Files: A User's View and Projection to 1991.* SERRL Working Report 13, Birkbeck College, University of London, London.

McLoone, P. and Boddy, F. A. (1994) Deprivation and mortality in Scotland, 1981 and 1991. *British Medical Journal* **309**, 1465–1474.

Mersey, J. A. (1990) Colour and thematic map design: the role of colour scheme and map complexity in choropleth map communication. *Monograph 41 Cartographica* **27**(3).

Merz, J. (1986) Structural adjustment in static and dynamic microsimulation models. In: Orcutt, G. H., Merz, J., and Quinke, H. (eds.) *Microanalytic Simulation Models to Support Social and Financial Policy.* North-Holland, Amsterdam 423–446.

Merz, J. (1994) *Microdata Adjustment by the Minimum Information Loss Principle*, Discussion Paper 10, Forschungsinstitut Freie Befufe, Universtät Lüneburg, Lüneburg, Germany.

Mesev, V., Longley, P., Batty, M., and Xie, Y. (1995) Morphology from imagery—detecting and measuring the density of urban land-use. *Environment and Planning A* **27**, 759–780.

MIDAS (1997) *Accessing the 1991 Census Digitised Boundary Data.* CSS 624, Manchester Computing, University of Manchester, Manchester.

Mills, I. (1987) Developments in census-taking since 1841. *Population Trends* **48**, 37–44.

Mills, I. and Teague, A. (1991) Editing and imputing data for the 1991 Census. *Population Trends* **64**, 30–37.

Mitchell, R., Martin, D., and Foody, G. M. (1998) Unmixing aggregate data: estimating the social composition of enumeration districts. *Environment and Planning A* **30**, 1929–1941.

Monmonier, M. S. (1989) Geographic brushing, enhancing exploratory analysis of the scatterplot matrix. *Geographical Analysis* **21**, 81–84.

Monmonier, M. S. (1991) *How to Lie With Maps.* University of Chicago Press, Chicago.

Monmonier, M. S. (1995) *Drawing the Line: Tales of Maps and Cartocontroversy.* Henry Holt, New York.

Morphet, C. (1992) The interpretation of small area census data. *Area* **24**, 63–72.

Morphet, C. (1993) The mapping of small-area census data—a consideration of the role of enumeration district boundaries. *Environment and Planning A* **25**, 267–278.

Morris, R. and Carstairs, V. (1991) Which deprivation? A comparison of selected deprivation indices. *Journal of Public Health Medicine* **13**, 318–326.

Morse, D. J. M., Towers, A. L., and Bedworth, P. (1998) *Interfaces to digital boundary data*. Paper Presented at the ESRC/JISC 4th Workshop: Planning for the 2001 Census, School of Geography, University of Leeds, Leeds.

Moser, C. A. and Scott, W. (1961) *British Towns: A Statistical Study of Their Social and Economic Differences.* Oliver and Boyd, Edinburgh.

Moss, C. (1999) Selection of topics and questions for the 2001 Census. *Population Trends* **97**, 28–36.

Mugglin, A. S., Carlin, B. P., Zhu, L., and Conlon, E. (1999) Bayesian areal interpolation, estimation, and smoothing: an inferential approach for geographic information systems. *Environment and Planning A* **31**, 1337–1352.

Muller, J. C. (1991) Generalisation of spatial databases. In: Maguire, D. J., Goodchild, M., and Rhind, D. (eds.) *Geographical Information Systems: Principles and Application.* Longman, Harlow 457–475.

Muller, J. C. and Honsaker, J. L. (1978) Choropleth map production by facsimile. *The Cartographic Journal* **15**, 14–19.

MVA Systematica with London Research Centre (1996) *England Sub-National Projections: Model Dynamics*. Report prepared for ONS. MVA/LRC, London.

NAfW (National Assembly for Wales) (2000) *Welsh Index of Multiple Deprivation.* National Statistics, London.

ONS (1998) *Mid-1997 Population Estimates for England and Wales.* ONS Monitor PP1 98/1, ONS, Titchfield.

ONS (1999a) *Gazetteer of the Old and New Geographies of the United Kingdom.* ONS, London.

ONS (1999b) *Making a Population Estimate in England and Wales.* Update to Occasional Paper 37, Version 1, Population Estimate Unit, ONS, Titchfield.

ONS (1999c) *Mid-1998 Population Estimates: England and Wales*. SERIES PE 1, ONS, Titchfield.

ONS (1999d) Tables. *Population Trends* **98**, 53–86.

ONS (1999e) *Revised Output Proposals.* Advisory Group Paper (99) 16, ONS, Titchfield.

ONS (1999f) *Data Quality Information.* Advisory Group Paper (99) 06, ONS, Titchfield.

ONS (2000) *Consultations Census 2001: Standard Tables*, Third Revision. ONS, Titchfield.

ONS (2001) Web Site for Census News. *http://www.statistics.gov.uk/census2001/cennews.asp*.

ONS and Coombes, M. G. (1998) *1991-Based Travel to Work Areas.* ONS (mimeo) London.

ONS/GROS/NISRA (1999a) *2001 Census: Output. A Discussion Paper. Supplement C. Origin-Destination Output*. Census Consultation Paper 2, ONS, Titchfield.

ONS/GROS/NISRA (1999b) *2001 Census Output: A Discussion Paper*. Census Consultation Paper 2, ONS, Titchfield.

ONS/GROS/NISRA (1999c) *2001 Census: Output. A Discussion Paper. Supplement G. Proposed Data Classifications*. Census Consultation Paper 2, ONS, Titchfield.

ONS/GROS/NISRA (2000a) *2001 Census: Key Statistics and Standard Tables Including Theme Tables, Ethnic Group and Armed Forces Tables. A Discussion Paper*. ONS, Titchfield. *http://www.statistics.gov.uk/*.

ONS/GROS/NISRA (2000b) *Census 2001 Consultations: Standard Tables*, Third Revision, August 2000. ONS, Titchfield. *http://www.statistics.gov.uk/*.

ONS/GROS/NISRA (2000c) *2001 Census News 44.* ONS, Titchfield.

ONS/GROS/NISRA (2000d) *2001 Census: Census Area Statistics—Discussion Paper.* ONS, Titchfield. *http://www.statistics.gov.uk/*.

ONS/GROS/NISRA (2000e) *Census 2001 Consultations: Census Area Statistics*. First Draft. ONS, Titchfield. *http://www.statistics.gov.uk/*.

ONS/GROS/NISRA (2001a) *2001 Census. Standard Area Statistics: Key Statistics, Standard Tables, Census Area Statistics (CAS). A Discussion Paper.* Census Consultation Paper. Office for National Statistics, Titchfield.

ONS/GROS/NISRA (2001b) *2001 Census. Key Statistics. Geographic Levels: Output Area to National.* Census Consultation Paper. Office for National Statistics, Titchfield.

ONS/GROS/NISRA (2001c) *2001 Census. Census Area Statistics*. Geographic Levels: Output Area to National. Office for National Statistics, Titchfield.

ONS/GROS/NISRA (2001d) *2001 Census Standard Tables*. Geographic Levels: Ward to National. Office for National Statistics, Titchfield.

ONS/GROS/NISRA (2001e) 2001 Census. Standard Area Statistics - Master Index. Census Consultation Paper. Office for National Statistics, Titchfield.

ONS/GROS/NISRA (2001f) *Census 2001. Origin-Destination Statistics (Final Specifications).* Office for National Statistics, Titchfield.

ONS/GROS/NISRA (2001g) *2001 Census Standard Tables*. Theme Tables. Geographic Levels: Ward to National. Office for National Statistics, Titchfield.

OPCS (1993a) Area constitutions: Districts within counties in England and Wales. *1991 Census User Guide 40.* OPCS, Titchfield.

OPCS (1993b) How complete was the 1991 Census? *Population Trends* **71**, 22–25.

OPCS (1993c) Rebasing the annual population estimates. *Population Trends* **73**, 27–31.

OPCS/GROS (1982) Special workplace statistics, cell numbering layouts. *1981 Census User Guide 156.* OPCS, Titchfield.

OPCS/GROS (1991) Topic statistics. Regional migration prospectus. *1991 Census User Guide 22.* OPCS, Titchfield.

OPCS/GROS (1992a) *1991 Census Definitions Great Britain.* CEN 91 DEF, HMSO, London.

OPCS/GROS (1992b) ED/Postcode directory: prospectus. *1991 Census User Guide 26.* OPCS, Titchfield.

OPCS/GROS (1992c) Special workplace statistics, cell numbering layouts. *1991 Census User Guide 52.* OPCS, Titchfield.

OPCS/GROS (1992d) Special workplace statistics, prospectus. *1991 Census User Guide 36.* OPCS, Titchfield.

OPCS/GROS (1992e) Local base statistics. *1991 Census User Guide 24.* OPCS, Titchfield.

OPCS/GROS (1992f) Local base statistics, small area statistics: explanatory notes. *1991 Census User Guide 38.* OPCS, Titchfield.

OPCS/GROS (1993a) *1991 Census. Ethnic Group and Country of Birth Great Britain.* Vol. 2, CEN 91 EGCB, HMSO, London.

OPCS/GROS (1993b) Special migration statistics prospectus. *1991 Census user guide 35.* OPCS, Titchfield.

OPCS/GROS (1993c) Special migration statistics. Cell numbering. *1991 Census User Guide 51.* OPCS, Titchfield.

OPCS/GROS (1993d) File specification and instructions for submitting orders on floppy disk, special migration statistics. *1991 Census User Guide 53*, OPCS, Titchfield.

OPCS/GROS (1993e) *1991 Census Historical Tables: Great Britain.* CEN 91 HT, HMSO, London.

OPCS/GROS (1994a) *1991 Census Migration: Great Britain Part 1 (100% tables).* Volume 1 of 2, CEN 91 MIG, HMSO, London.

OPCS/GROS (1994b) *1991 Census Migration: Great Britain Part 1 (100% Tables).* Volume 2 of 2, CEN 91 MIG, HMSO, London.

OPCS/GROS (1994c) *1991 Census Migration: Great Britain Part 2 (10% Tables).* CEN 91 MIG, HMSO, London.

OPCS/GROS (1994d) *1991 Census: Workplace and Transport to Work, Great Britain.* CEN 91 WT, HMSO, London.

OPCS/GROS (1995a) *1991 Census Regional Migration: Great Britain Part 1 (100% Tables).* Computer readable files. OPCS, Titchfield.

OPCS/GROS (1995b) *1991 Census Regional Migration: Great Britain Part 2 (10% Tables).* Computer readable files. OPCS, Titchfield.

OPCS/GROS (1995c) *1991 Census General Report: Great Britain.* CEN 91, HMSO, London.

Openshaw, S. (1977a) A geographical solution to scale and aggregation problems in region-building, partitioning and spatial modelling. New Series, *Transactions of the Institute of British Geographers* **2**, 459–472.

Openshaw, S. (1977b) Algorithm 3: a procedure to generate pseudo-random aggregations of N zones into M zones, where M is less than N. *Environment and Planning A* **9**, 1423–1428.

Openshaw, S. (1978) An optimal zoning approach to the study of spatially aggregated data, In: Masser, I. and Brown, P. J. B. (eds.) *Spatial Representation and Spatial Interaction.* Martinus Nijhoff, Leiden 95–113.

Openshaw, S. (1983) Multivariate analysis of census data: the classification of areas. In: Rhind, D. (ed.) *A Census User's Handbook.* Methuen, London 243–264.

Openshaw, S. (1984) The modifiable areal unit problem. *Concepts and Techniques in Modern Geography 38.* Geo Books, Norwich.

Openshaw, S. (1990) Spatial referencing for the user in the 1990s. *Mapping Awareness* **4**(2), 24–29.

Openshaw, S. (ed.) (1995a) *Census Users' Handbook.* GeoInformation International, Cambridge.

Openshaw, S. (1995b) The future of the census. In: Openshaw, S. (ed.) *Census Users' Handbook.* GeoInformation International, Cambridge 389–411.

Openshaw, S. (1996) Developing GIS-relevant zone-based spatial analysis methods, In: Longley, P. and Batty, M. (eds.) *Spatial Analysis: Modelling in a GIS Environment.* GeoInformation International, Cambridge 55–73.

Openshaw, S. and Alvanides, S. (1999a) Applying geocomputation to the analysis of spatial distributions. In: Longley, P. A., Goodchild, M. F., Maguire, D. J., and Rhind, D. W. (eds.) *Geographical Information Systems: Principles, Techniques, Applications and Management.* 2nd Edition, Wiley, Chichester 267–282.

Openshaw, S. and Alvanides, S. (1999b), *A Zone Design Tool for use with 2001 Census Data.* ESRC Project research proposal (H507 25 5153), School of Geography, University of Leeds, Leeds. (available from the authors: *s.alvanides@newcastle.ac.uk*).

Openshaw, S. and Alvanides, S. (2001) Designing zoning systems for the representation of socio-economic data, In: Frank, A., Raper, J., and Cheylan, J-P. (eds.) *Life and Motion of Socio-Economic Units.* ESF-GISDATA 8, Taylor and Francis, London 281–300.

Openshaw, S., Blake, M., and Wymer, C. (1995) Using neurocomputing methods to classify Britain's residential areas. In: Fisher, P. (ed.) *Innovations in GIS 2.* Taylor and Francis, London 97–111.

Openshaw, S. and Openshaw, C. (1997) *Artificial Intelligence in Geography.* John Wiley, Chichester.

Openshaw, S. and Rao, L. (1995) Algorithms for reengineering 1991 Census geography. *Environment and Planning A* **27**, 425–446.

Openshaw, S. and Taylor, P. (1979) A million or so correlation coefficients: three experiments on the modifiable areal unit problem. In: Wrigley, N. (ed.) *Statistical Applications in the Spatial Sciences.* Pion, London 127–144.

Openshaw, S. and Turton, I. (1996) New opportunities for geographical census analysis using individual level data. *Area* **28**, 167–176.

Openshaw, S. and Wymer, C. (1995) Classifying and regionalizing census data. In: Openshaw, S. (ed.) *Census Users' Handbook.* GeoInformation International, Cambridge 239–270.

Orford, S., Dorling, D., and Harris, R. (1998) Review of Visualization in the Social Sciences: A State of the Art Survey and Report. AGOCG Technical Report Series 41, AGOCG ESRC/JISC Advisory Group on Computer Graphics, Bristol. *http://www.agocg.ac.uk/train/review/cover.htm*.

Overton, E. and Ermisch, J. (1984) Minimal household units. *Population Trends* **35**, 18–22.

Owen, D. W., Green, A. E., and Coombes, M. G. (1986) Using the social and economic data on the BBC Domesday interactive videodisc. *Transactions of the Institute of British Geographers NS* **11**, 305–314.

Owen, D. W. and Johnson, M. (1996) Ethnic minorities in the Midlands. In: Ratcliffe, P. (ed.) *Ethnicity in the 1991 Census*. Vol. 3. Social Geography and Ethnicity in Britain: Geographical Spread, Spatial Concentration and Internal Migration. HMSO, London 227–270.

Owen, D. W. and Ratcliffe, P. (1996) *Estimating Local Change in the Population of Minority Ethnic Groups, 1981–1991.* Working Paper 1, Changing Spatial Location Patterns of ethnic minorities in Great Britain 1981–1991 project, CRER, University of Warwick, Warwick.

Pearman, H. (1993) Designing a land and property gazetteer. *Mapping Awareness* **7**(5), 19–21.

Peck, J. (1997) *Workplace.* Guilford Press, New York.

Peloe, A. and Rees, P. H. H. (1999) Estimating ethnic change in London 1981–1991 using a variety of census data. *International Journal of Population Geography* **5**, 179–194.

Penhale, B., Noble, M., Smith, G., and Wright, G. (1999a) *Ward Level Population Estimates for the 1999 Index of Local Deprivation.* Department of Applied Social Studies and Social Research, University of Oxford, Oxford.

Penhale, B., Noble, M., Smith, G., and Wright, G. (1999b) *Final Consultation on Ward Level Population Estimates.* Department of Applied Social Studies and Social Research, University of Oxford, Oxford.

Perkins, C. R. and Parry, R. B. (1996) *Mapping the UK.* Bowker Saur, London.

Phillimore, P. and Reading, R. (1992) A rural advantage? Urban-rural health differences in Northern England. *Journal of Public Health* **14**, 290–299.

Phillimore, P., Beatty, A., and Townsend, P. (1994a) Widening inequality of health in Northern England, 1981–1991. *British Medical Journal* **308**, 1125–1128.

Phillimore, P., Beatty, A., and Townsend, P. (1994b) *Health and Inequality: The Northern Region 1981–1991: A Report.* Department of Social Policy, University of Newcastle-upon-Tyne, Newcastle-upon-Tyne.

Prandy, K. (1990) Revised Cambridge scale of occupation. *Sociology* **24**, 629–655.

Prandy, K. (1992) *Cambridge Scale Scores for CASOC Groupings.* Working paper 11, Social and Political Sciences, Cambridge.

Raper, J. F., Rhind, D. W., and Shepherd, J. W. (1992) *Postcodes: The New Geography.* Longman, Harlow.

Redmond, G., Sutherland, H., and Wilson, M. (1998) *The Arithmetic of Tax and Social Security Reform: A User's Guide to Microsimulation Methods and Analysis.* Cambridge University Press, Cambridge.

Rees, P. H. (1972) Problems of classifying subareas within cities. In: Berry, B. (ed.) *City Classification Handbook: Methods and Applications.* Wiley, New York 265–330.

Rees, P. H. (1985) Does it really matter which migration data you use in a population model? In: White, P. and van der Knaap, G. (ed.) *Contemporary Studies of Migration.* Geo Books, Norwich 55–77.

Rees, P. H. (1989) Research policy and review 30. How to add value migration data from the 1991 Census. *Environment and Planning A* **21**, 1363–1379.

Rees, P. H. (1994) Estimating and projecting the populations of urban communities. *Environment and Planning A* **26**, 1671–1697.

Rees, P. H. (1995a) Putting the census on the researcher's desk. In: Openshaw, S. (ed.) *Census Users' Handbook.* GeoInformation International, Cambridge 27–82.

Rees, P. H. (1995b) Research using the 1991 Census: findings on deprivation, unemployment, ethnicity and religion. *Environment and Planning A* **27**, 515–518.

Rees, P. H. (ed.) (1997a) *The Debate About the Geography of the 2001 Census: Collected Papers from 1995–1996.* Working Paper 97/1, School of Geography, University of Leeds, Leeds.

Rees, P. H. (1997b) *Migration Statistics from the 2001 Census: what do we want?* Working Paper 97/6, School of Geography, University of Leeds, Leeds.

Rees, P. H. (1997c) Migration statistics from the 2001 Census: what do we want? In Rees, P. H. (ed.) *Third Workshop on the 2001 Census—Special Datasets: what do we want?* Working Paper 97/9, School of Geography, University of Leeds, Leeds.

Rees, P. H. (ed.) (1997d) *Third Workshop the 2001 Census—Special Datasets: what do we want?* Working Paper 97/9, School of Geography, University of Leeds, Leeds.

Rees, P. H. (1998a) What do you want from the 2001 Census? Results of an ESRC/JISC survey of user views. *Environment and Planning A* **30**, 1775–1796.

Rees, P. H. (ed.) (1998b) *Fourth Workshop: The 2001 Census—What do we really, really want?* Working Paper 98/7, School of Geography, University of Leeds, Leeds.

Rees, P. H. (1998c) *Final Report to ESRC on Workshops Planning for the 2001 Census* ESRC Award H507 265031. Available from ESRC, Swindon or from the author at the School of Geography, University of Leeds, Leeds.

Rees, P. H. (ed.) (1998d) *First Workshop: The 2001 Census—what Geography do we want?* Working Paper 98/8, School of Geography, University of Leeds, Leeds.

Rees, P. H. (ed.) (1998e) *Second Workshop: The 2001 Census—What interfaces do we want?* Working Paper 98/9, School of Geography, University of Leeds, Leeds.

Rees, P. H. (1999) The 2001 Census of population: what does the White Paper propose? *Environment and Planning A* **31**, 1141–1148.
Rees, P. H., Bell, M., Duke-Williams, O. W., and Blake, M. (2000) Problems and solutions in the measurement of migration intensities: Australia and Britain compared. *Population Studies*, **54**(2), 207–222.
Rees, P. H. and Duke-Williams, O. (1994) *The Special Migration Statistics: A Vital Resource for Research into British Migration.* Working Paper 94/20, School of Geography, University of Leeds, Leeds.
Rees, P. H. and Duke-Williams, O. (1995a) The Story of the British Special Migration Statistics. *Scottish Geographical Magazine* **111**, 13–26.
Rees, P. H. and Duke-Williams, O. (1995b) *Inter-district migration by ethnic groups.* Paper Presented at the International Conference on Population Geography, University of Dundee, Dundee.
Rees, P. H. and Duke-Williams, O. (1997) Methods for estimating missing data on migrants in the 1991 British Census. *International Journal of Population Geography* **3**, 323–368.
Rees, P. H., Kupiszewski, M., and Durham, H. (1996) *Internal Migration and Regional Population Dynamics in Europe: United Kingdom Case Study.* Working Paper 96/20, School of Geography, University of Leeds, Leeds. *http://www.geog.leeds.ac.uk/wpapers/96-20.pdf*.
Rees, P. and Kupiszewski, M. (1999) Internal Migration and Regional Population Dynamics in Europe: a Synthesis. *Population Studies 32.* Council of Europe Publishing, Strasbourg.
Rees, P. H. and Martin, D. (1997) Flexible Geographies and Area Aggregation: Designing Small Areas for Outputs from the 2001 Census. In: Rees, P. H. (ed.) *The Debate About the Geography of the 2001 Census: Collected Papers from 1995–1996.* Working Paper 97/1, School of Geography, University of Leeds, Leeds.
Research International Labour Office (1990) *International Standard Classification of Occupations 1988.* International Labour Office Publications, Geneva.
Rhind, D. (ed.) (1983) *A Census User's Handbook.* Methuen, London.
Rhind, D. (1997) *Framework for the World.* GeoInformation International, Cambridge.
Rhind, D. W., Cole, K., Armstrong, M., Chow, L., and Openshaw, S. (1990) *An Online, Secure and Infinitely Flexible Database System for the National Census of Population.* SERRL Working Report 14, Birkbeck College, University of London, London.
Robinson, A. H., Morrison, J. L., Muehrcke, P. C., Kimerling, A. J., and Guptill, S. C. (1995) *Elements of Cartography.* Sixth Edition, John Wiley & Sons, Chichester.
Robinson, G. M. (1998) *Methods and Techniques in Human Geography.* John Wiley & Sons, Chichester.
Robinson, J. M. and Zubrow, E. (1997) Restoring continuity: exploration of techniques for reconstructing the spatial distribution underlying polygonized data. *International Journal of Geographical Information Science* **11**, 633–648.
Royall, R. M. (1970) On finite population sampling under certain linear regression models. *Biometrika* **57**, 377–387.
Royall, R. M. and Cumberland, W. G. (1978) Variance estimation in finite population sampling. *Journal of the American Statistical Association* **73**, 351–361.
Royston, G. H. D., Hurst, J. W., Lister, E. G., and Stewart, P. A. (1992) Modelling the use of health services by populations of small areas to inform the allocation of central resources to larger regions. *Socio-Economic Planning Sciences* **26**, 169–180.

Sandhu, A. (1993) *Problems of Imputation in the 1991 Census*. Occasional Paper 1, CMU, University of Manchester, Manchester.

Saunders, J. (1998) Weighted census-based deprivation indices: their use in small areas. *Journal of Public Health Medicine* **20**, 253–260.

Schafer, J. L. (1997) *Analysis of Incomplete Multivariate Data.* Chapman and Hall, London.

Schuman, J. (1999) The ethnic minority populations of Great Britain—latest estimates. *Population Trends* **96**, 33–43.

Scott, A. and Kilbey, T. (1999) Can patient registers give an improved measure of internal migration in England and Wales? *Population Trends* **96**, 44–55.

Scott, A. J. and Holt, D. (1982) The effect of two-stage sampling on ordinary least squares methods. *Journal of the American Statistical Association* **77**, 848–854.

Senior, M. L. (1991) Deprivation payments to GPs: not what the doctor ordered. *Environment and Planning C: Government and Policy* **9**, 79–94.

Senior, M. L. (1998) Area variations in self-perceived limiting long-term illness in Britain, 1991: is the Welsh experience exceptional? *Regional Studies* **32**, 265–280.

Senior, M. L., Williams, H. C. W. L., and Higgs, G. (1998) Spatial and temporal variation of mortality and deprivation 2: statistical modelling. *Environment and Planning A* **30**, 1815–1834.

Shepherd, I. D. H. (1995) Putting time on the map, dynamic displays in data visualization and GIS. In: Fisher, P. F. (ed.) *Innovations in GIS 2.* Taylor & Francis, London, 169–187.

Shevky, E. and Bell, M. (1955) *Social Area Analysis: Theory, Illustrative Application and Computational Procedure.* Stanford University Press, Stanford.

Shevky, E. and Williams, M. (1949) *The Social Areas of Los Angeles: Analysis and Typology.* University of California Press, Berkeley.

Shouls, S., Congdon, P., and Curtis, S. (1996) Modelling inequality in reported long term illness in the UK: combining individual and area characteristics. *Journal of Epidemiology and Community Health* **50**, 366–376.

Simpson, S. (1996) Resource allocation by measures of relative social need in geographical areas: the relevance of the signed chi-square, the percentage and the raw count. *Environment and Planning A* **28**, 537–554.

Simpson, S. (1997) Demography and ethnicity: case studies from Bradford. *New Community* **23**, 89–107.

Simpson, S. (ed.) (1998a) *Making Local Population Statistics: A Guide for Practitioners.* LARIA, Wokingham.

Simpson, S. (1998b) Case study: apportionment using counts of patients and electors. In: Simpson, S. (ed.) *Making Local Population Statistics: A Guide for Practitioners.* LARIA, Wokingham 61–65.

Simpson, S. (1999) Statistical exclusion and social exclusion. *Radical Statistics* **71**, 45–60.

Simpson, S. (2001) *Producing and using geography conversion tables.* Paper Presented at the ESRC 2001 Census Development Programme, Fourth Workshop, University of Leeds, Leeds. *http://convert.mimas.ac.uk/afpd/main.cfm*.

Simpson, S., Cossey, R., and Diamond, I. (1997a) 1991 population estimates for areas smaller than districts. *Population Trends* **90**, 31–39.

Simpson, S., Diamond, I., Middleton, L., Lunn, D., and Cossey, R. (2000) How can a national set of small area population estimates gain acceptance? Lessons from the estimating with confidence project. In: Arnold, R., Elliott, P., Wakefield, J., and Quinn, M. (eds.) *Population Counts in Small Areas: Implications for Studies of Environment and Health.* Studies in Medical and Population Subjects 62, The Stationery Office, London.

Simpson, S., Diamond, I., Tonkin, P., and Tye, R. (1996) Updating small area population estimates in England and Wales. *Journal of the Royal Statistical Society A* **159**, 235–247.

Simpson, S. and Middleton, E. (1997) *Who is Missed by a National Census? A Review of Empirical Results from Australia, Britain, Canada and the USA*. CCSR Working Paper 2, CCSR, University of Manchester, Manchester. *http://les.man.ac.uk/ccsr/conference/workingpaper2.htm*.

Simpson, S. and Middleton, E. (1998) Characteristics of census undercount; what do we know already? Adjusting census output for undercount. In: Simpson, S. (ed.) *A One Number Census: Proceedings of a research workshop.* Occasional Paper 15, CCSR, University of Manchester, Manchester 1–14.

Simpson, S. and Middleton, E. (1999) Undercount of migration in the UK 1991 Census and its impact on counterurbanisation and population projections. *International Journal of Population Geography* **5**, 387–405.

Simpson, S., Middleton, E., Diamond, I., and Lunn, D. (1997b) Small-area population estimates: a review of methods used in Britain in the 1990s. *International Journal of Population Geography* **3**, 265–280.

Simpson, S., Tye, R., and Diamond, I. (1995) *What was the Real Population of Local Areas in Mid-1991?* Estimating with Confidence Project. Working paper 10, Department of Social Statistics, University of Southampton, Southampton.

Simpson, L. and Yu, A. (2000) Data Conversion and Look up Tables. Paper Presented at the ESRC 2001 Census Development Programme, Third Workshop, Royal Statistical Society, London. *http://convert.mimas.ac.uk/afpd/main.cfm*.

Sleight, P. (1993) *Targeting Customers.* NTC Publications, Henley-on-Thames.

Smith, D. (1985) *Victorian Maps of the British Isles.* Batsford, London.

Smith, P., Sheldon, T. A., Carr-Hill, R. A., Martin, S., Peacock, S., and Hardman, G. (1994) Allocating resources to health authorities: results and implications of small area analysis of use of inpatient services. *British Medical Journal* **309**, 1050–1054.

Stillwell, J., Duke-Williams, O., and Rees, P. H. (1995) Time series migration in Britain: the context for 1991 Census Analysis. *Papers in Regional Science* **74**, 341–359.

Storkey, M. (1998) *Population Projections of Ethnic Minority Groups in London.* London Research Centre, London.

Thomas, F. (2001) *Origin-Destination Statistics: Summary Report of Second Analysis of Response to the Discussion Paper*, May 2000. GROS, Edinburgh.

Thomas, R. and Warren, A. (1997) Evaluating the role of regression methods in the determination of standard spending assessments. *Environment and Planning C: Government and Policy* **15**, 53–72.

Thunhurst, C. (1985) The analysis of small area statistics and planning for health. *Statistician* **34**, 93–106.

Tobler, W. R. (1973) Choropleth maps without class intervals. *Geographical Analysis* **5**, 26–28.

Tobler, W. R. (1979) Smooth pycnophylactic interpolation for geographical regions. *Journal of the American Statistical Association* **74**, 519–530.

Townsend, P. (1987) Deprivation. *Journal of Social Policy* **16**, 125–146.

Townsend, P., Davidson, N., and Whitehead, M. (eds.) (1988) *Inequalities in Health: The Black Report and the Health Divide.* Penguin Books, Harmondsworth.

Townsend, P., Phillimore, P., and Beatty, A. (1986) *Inequalities in Health in the Northern Region: An Interim Report.* Northern Region Health Authority and Bristol University, Newcastle upon Tyne.

Townsend, P., Phillimore, P., and Beatty, A. (1988) *Health and Deprivation: Inequality and the North.* Croom Helm, London.

Tranmer, M., Pickles, A., Fieldhouse, E., Elliot, M., Dale, A., Brown, M., Martin, D., Steel, D., and Gardner, C. (2000) *Microdata for Small Areas.* Report to ESRC under the 2001 Census Development Programme Award no. H507255161. Available as Occasional Paper No. 20, CCSR, University of Manchester, Manchester at *http://les.man.ac.uk/ccsr/cmu/2001/Occ20.pdf*.

Tranmer, M. and Steel, D. G. (1998) Using census data to investigate the causes of the ecological fallacy. *Environment and Planning A* **30**, 817–831.

Tryon, R. (1955) *The Identification of Social Areas by Cluster Analysis.* University of California Publications in Psychology VIII University of California Press, Berkeley, CA (1).

Tufte, E. R. (1990) *Envisioning Information.* Graphics Press, Cheshire CT.

Turton, I. (1999) A Pilot Flexible Census Output System. ESRC Award H507 25 5155 Proposal. Available from ESRC, Swindon or from the author.

Turton, I. and Openshaw, S. (1994) *A Step by Step Guide to Accessing the 1991 SAR Using USAR.* Working Paper 94/6, School of Geography, University of Leeds, Leeds. *http://www.geog.leeds.ac.uk/ wpapers/working.html*.

Turton, I. and Openshaw, S. (1995) Putting the 1991 Census sample of anonymised records on your UNIX workstation. *Environment and Planning A* **27**, 391–411.

Unwin, D. and Dykes, J. (1996) Using a cartographic data visualiser (cdv) with the population census. *MIDAS Newsletter* **9**, 22–24.

Vickers, P. and Yar, M. (1998) *The development and evaluation of the donor imputation system (DIS) for the 2001 UK Census of population and housing.* Paper Presented at the Joint IASS/IAOS Conference Statistics for Economic and Social Development, September 1998. Copies available from the authors at ONS.

Voas, D. and Williamson, P. (1998) *Evaluating goodness-of-fit measures for synthetic microdata.* Working Paper 98/1, Population Microdata Unit, Department of Geography, University of Liverpool, Liverpool.

Voas, D. and Williamson, P. (2001) The diversity of diversity: a critique of geodemographic classification. *Area* **33**, 63–76.

Wallace, M., Charlton, J., and Denham, C. (1995) The new OPCS area classification. *Population Trends* **79**, 15–30.

Wallace, M. and Denham, C. (1996) The ONS classification of local and health authorities of Great Britain. *ONS Studies on Medical and Population Subjects 59.* HMSO, London.

Wanders, A-C. (1998) *Integrating aggregate and disaggregate migration data from the 1991 population census.* Paper Presented at the Annual Conference of the RGS-IBG, University of Surrey, Surrey.

Ware, C. and Beaty, J. C. (1988) Using color dimensions to display data dimensions. *Human Factors* **30**, 127–142.

Webber, R. (1975) *Liverpool Social Area Study 1971 Data: Final Report*. PRAG Technical Papers TP14 PRAG, Centre for Environmental Studies, London.

Webber, R. (1989) Using multiple data sources to build an area classification system, operational problems encountered by MOSAIC. *Journal of the Market Research Society* **31**, 107–109.

Webber, R. and Craig, J. (1976) Which local authorities are alike. *Population Trends* **5**, 13–19.

Webber, R. and Craig, J. (1978) A socioeconomic classification of local authorities in Great Britain. *Studies in Medical and Population Subjects 35.* HMSO, London.

Wegman, E. J. (1990) Hyperdimensional data analysis using parallel coordinates. *Journal of the American Statistical Association* **85**, 664–675.

Williamson, P. (1992) *Community Health Care Policies for the Elderly: A Microsimulation Approach*. Unpublished Ph.D. thesis, School of Geography, University of Leeds, Leeds.

Williamson, P. (1996) Community care policies for the elderly, 1981 and 1991: a microsimulation approach. In: Clarke, G. P. (ed.) *Microsimulation for Urban and Regional Policy Analysis.* Pion, London 64–87.

Williamson, P., Birkin, M., and Rees, P. H. (1998) The estimation of population microdata by using data from small area statistics and samples of anonymised records. *Environment and Planning A* **30**, 785–816.

Williamson, P., Rees, P. H., and Birkin, M. (1995) Indexing the census: a by-product of the simulation of whole populations by means of SAS and SAR data. *Environment and Planning A* **27**, 413–424.

Wilson, T. and Rees, P. H. (1998) *Lookup Tables to Link 1991 Population Statistics to the 1998 Local Government Areas*. Working paper 98/5, School of Geography, University of Leeds, Leeds. *http://www.geog.leeds.ac.uk/wpapers/98-5.pdf*.

Wilson, T. and Rees, P. H. (1999) Linking 1991 population statistics to the 1998 local government geography of Great Britain. *Population Trends* **97**, 37–45.

Wise, S., Haining, R., and Ma, J. (1997) Regionalisation tools for the exploratory spatial analysis of health data, In: Fischer, M. M. and Getis, A. (eds.) *Recent Developments in Spatial Analysis.* Springer, Berlin 83–100.

Wood, J. D., Fisher, P. F., Dykes, J. A., Unwin, D. J., and Stynes, K. (1999) The use of the Landscape Metaphor in understanding population data. *Environment and Planning B: Planning and Design* **26**, 281–295.

Wright, J. K. (1933) A method of mapping densities of population with Cape Cod as an example. *Geographical Review* **26**, 103–110.

Wrigley, N. (1990) ESRC and the 1991 Census. *Environment and Planning A* **22**, 573–582.

Yu, A. and Simpson, L. (2000) *The All-Fields Postcode Directory (AFPD): validation for use as a look-up tables*. Paper Presented at the ESRC 2001 Census Development Programme, Second Workshop, University of Leeds, Leeds. *http://convert.mimas.ac.uk/afpd/main.cfm*.

Index